ILLUSTRATED ANATOMY *of the* HEAD *and* NECK

ILLUSTRATED ANATOMY *of the* HEAD *and* NECK

Margaret J. Fehrenbach, RDH, MS
Dental Hygienist and Oral Biologist
Private Practice in Seattle, Washington
Instructor of Dental Hygiene Program
Pierce College
Tacoma, Washington
Former Instructor of Dental Hygiene Program
Shoreline Community College
Seattle, Washington

Susan W. Herring, PhD
Anatomist and Researcher
Professor of Orthodontics, School of Dentistry
Adjunct Professor of Biological Structure, School of Medicine
University of Washington
Seattle, Washington

Illustrated by
Pat Thomas, CMI
Certified Medical Illustrator, AMI
Oak Park, Illinois

W.B. SAUNDERS COMPANY
A Division of Harcourt Brace & Company
Philadelphia London Toronto Montreal Sydney Tokyo

W.B. SAUNDERS COMPANY
A Division of Harcourt Brace & Company

The Curtis Center
Independence Square West
Philadelphia, Pennsylvania 19106

Library of Congress Cataloging-in-Publication Data

Fehrenbach, Margaret J.
Illustrated anatomy of the head and neck/Margaret J. Fehrenbach, Susan W. Herring; illustrated by Pat Thomas.

p. cm.
Includes bibliographical references and index.

ISBN 0–7216–4082–6

1. Head—Anatomy. 2. Neck—Anatomy. I. Herring, Susan W. II. Title.
[DNLM: 1. Head—anatomy & histology. 2. Neck—anatomy & histology. WE 700 F296i 1996]

QM535.F44 1996 611′91—dc20

DNLM/DLC 95–11491

ILLUSTRATED ANATOMY OF THE HEAD AND NECK ISBN 0–7216–4082–6

Printed in the United States of America

Last digit is the print number: 9 8 7 6 5 4

Preface

Overview

To meet the needs of today's dental professional, this textbook on head and neck anatomy has more than the basic information on the topic. Special emphasis is placed on the specific anatomy of the temporomandibular joint in order to allow the dental professional to understand the disorders associated with this joint. The textbook also includes a chapter on the anatomical basis of local anesthesia for pain control. Another chapter explains the routes of the spread of dental infection.

The authors and illustrator of this textbook have tried to make it easy to understand, as well as interesting to read. We hope that it challenges the reader to incorporate the information presented into clinical situations.

Features

Each chapter topic has been chosen to be relevant to the present needs of the dental professional and to build on former topics. Each chapter begins with an outline of subjects covered, chapter objectives, and a list of key words with a pronunciation guide. This pronunciation guide is based on *Dorland's Medical Dictionary,* 28th edition, W.B. Saunders Company, Philadelphia. The chapters continue with an organized text and pronunciation guide for each anatomical structure, original illustrations where possible, and clinical photographs where applicable. The anatomical terms follow those outlined in the internationally approved official body of anatomical nomenclature, *Nomina Anatomica,* 6th edition, Churchill and Livingstone, New York.

Further study of the subject has been made easier with tables that summarize information, identification exercises and review questions for each chapter. A bibliography, answers to each chapter's review questions, a glossary of key words and anatomical terms, and an index are at the end of the textbook. Within each chapter, there may be references to other chapters so that the reader can review or investigate interesting interrelated subjects.

Acknowledgments

The authors would like to thank photographer Jim Clark, University of Washington, Seattle, who provided many of the skull photographs. Kathy Basset, RDH, BS, Pierce College, Tacoma, provided much-needed skills of reviewing and consulting during the project.

We would also like to thank Doreen Naughton, RDH, BS, Dental Hygiene Health Services, Seattle, and Dental Public Health Sciences Department, University of Washington Dental School, Seattle, for her clinical expertise on the local anesthesia chapter. Bruce Rothwell, DMD, MSD, Hospital Dentistry, University of Washington, Seattle, provided valuable input for many chapters. Information concerning temporomandibular joint disorders from Ed Truelove, DDS, MSD, Professor and Chairman of the Department of Oral Medicine, University of Washington Dental School, Seattle, was used in compiling this text. We are also indebted to the original author, the late Peggy Thomas, who was instrumental in getting the project started. Finally, we would like to thank editors Selma Ozmat and Scott Weaver, and the staff from W.B. Saunders Company for making the textbook possible.

Contents

ILLUSTRATED ANATOMY *of the* HEAD *and* NECK

▼ *This chapter discusses the following topics:*

I. Clinical Applications
II. Anatomical Nomenclature
III. Normal Anatomical Variation

▼ *Key words*

Anterior (an-**tere**-ee-or) Front of an area of the body.

Apex (**ay**-peks) Pointed end of a conical structure.

Contralateral (kon-trah-**lat**-er-il) Structures on the opposite side of the body.

Deep (deep) Structures located inward, away from the body surface.

Distal (**dis**-tl) Area that is farther away from the median plane of the body.

Dorsal (**dor**-sal) Back of an area of the body.

External (eks-**tern**-il) Outer side of the wall of a hollow structure.

Frontal plane (**frunt**-il) Plane created by an imaginary line that divides the body at any level into anterior and posterior parts.

Frontal section (**frunt**-il) Section of the body through any frontal plane.

Horizontal plane (hor-i-**zon**-tal) Plane created by an imaginary line that divides the body at any level into superior and inferior parts.

Inferior (in-**fere**-ee-or) Area that faces away from the head and toward the feet of the body.

Internal (in-**tern**-il) Inner side of the wall of a hollow structure.

Ipsilateral (ip-see-**lat**-er-il) Structures on the same side of the body.

Lateral (**lat**-er-il) Area that is farther away from the median plane of the body or structure.

Medial (**me**-dee-il) Area that is closer to the median plane of the body or structure.

Median (**me**-dee-an) Structure at the median plane.

Median plane (**me**-dee-an) Plane created by an imaginary line dividing the body into right and left halves.

Midsagittal section (mid-**saj**-i-tl) Section of the body through the median plane.

Posterior (pos-**tere**-ee-or) Back of an area of the body.

Proximal (**prok**-si-mil) Area closer to the median plane of the body.

Sagittal plane (**saj**-i-tl) Any plane of the body created by an imaginary plane parallel to the median plane.

Superficial (soo-per-**fish**-al) Structures located toward the surface of the body.

Superior (soo-**pere**-ee-or) Area that faces toward the head of the body, away from the feet.

Transverse section (**trans**-vers) Section of the body through any horizontal plane.

Ventral (**ven**-tral) Front of an area of the body.

Chapter 1

Introduction

▼ ***After studying this chapter, the reader should be able to:***

1. Define and pronounce all the key words and anatomical terms in this chapter.
2. Discuss the clinical applications of the study of head and neck anatomy by dental professionals.
3. Apply the correct anatomical nomenclature during the study of head and neck anatomy.
4. Discuss normal anatomical variation and how it applies to different structures of the head and neck.
5. Correctly complete the review questions for this chapter.

Clinical Applications

The dental professional must have a thorough knowledge of head and neck anatomy when performing patient examination procedures, both extraoral and intraoral (Figure 1–1). The knowledge of normal anatomy will help determine if any abnormalities or lesions exist and possibly indicate the etiology and amount of involvement. This knowledge will also provide a basis for the description of the lesion for record-keeping purposes.

Head and neck anatomy is also useful when performing dental radiology procedures. Landmarks are used by the dental professional in the placement of the films, and knowledge of anatomy is important in the mounting and analysis of the films.

A patient may also present features of a temporomandibular joint disorder. A dental professional must have knowledge of the normal anatomy of the joint in order to understand the various disorders associated with it.

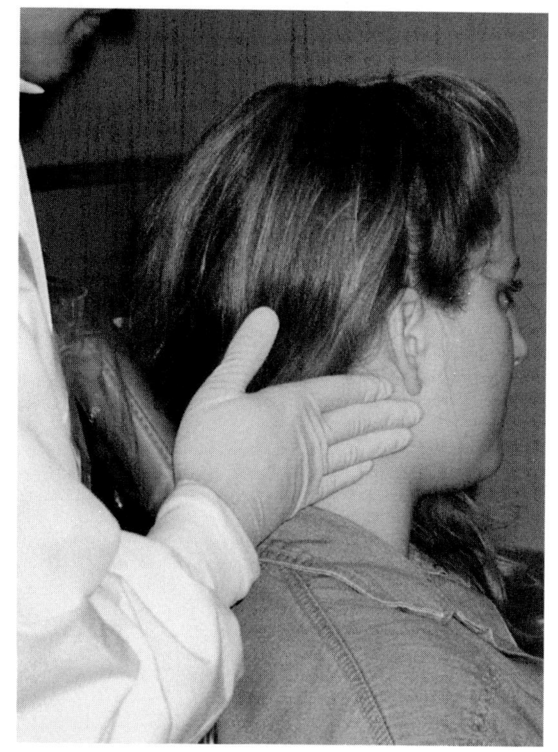

Figure 1–1
Extraoral, as well as intraoral, examination of the patient is based on knowledge of head and neck anatomy.

More specifically, the administration of local anesthesia is based on landmarks of the head and neck. The knowledge of anatomy will facilitate treatment planning of local anesthesia by the dental professional for the reduction of pain during various dental procedures. This knowledge will also allow for the correct placement of the syringe, avoiding certain complications.

During the examination of the patient, the dental professional may also note the presence of a dental infection. It is important to know the source of the infection as well as the areas it could spread to by way of certain anatomical features of the head and neck. Knowledge of the anatomy will supply the background for understanding the spread of dental infection.

Anatomical Nomenclature

Before beginning the study of head and neck anatomy, the dental professional may need to review the basic ***anatomical nomenclature*** (an-ah-**tom**-ik-al **no**-men-kla-cher), which is the system of names of anatomical structures. This review will allow for the easy application of these terms to the head and neck area.

The nomenclature of anatomy is based on the body's being in ***anatomical position*** (Figure 1–2). In anatomical position, the body is standing erect. The arms are at the sides with the palms and toes directed forward and the eyes looking forward. This position is assumed even when the body may be supine (on the back) or prone (on the front).

When studying the body in anatomical position, certain terms are used to refer to areas in relationship to other areas (Figure 1–3). The front of an area in relationship to the entire body is its ***anterior*** (an-**tere**-ee-or) portion. The back of an area is its ***posterior*** (pos-**tere**-ee-or) portion. The ***ventral*** (**ven**-tral) portion is directed toward the anterior and is the opposite of the ***dorsal*** (**dor**-sal) portion when considering the entire body.

Other terms can be used to refer to areas in relationship to other areas of the body (Figure 1–3). An area that faces toward the head and away from the feet is its ***superior*** (soo-**pere**-ee-or) portion. An area that faces away from the head and toward the feet is its ***inferior*** (in-**fere**-ee-or) portion. As an example, the face is on the anterior side of the head, and the hair is superior and posterior to the face. The ***apex*** (**ay**-peks) or tip is the pointed end of a conical structure, such as the tongue's apex or tip.

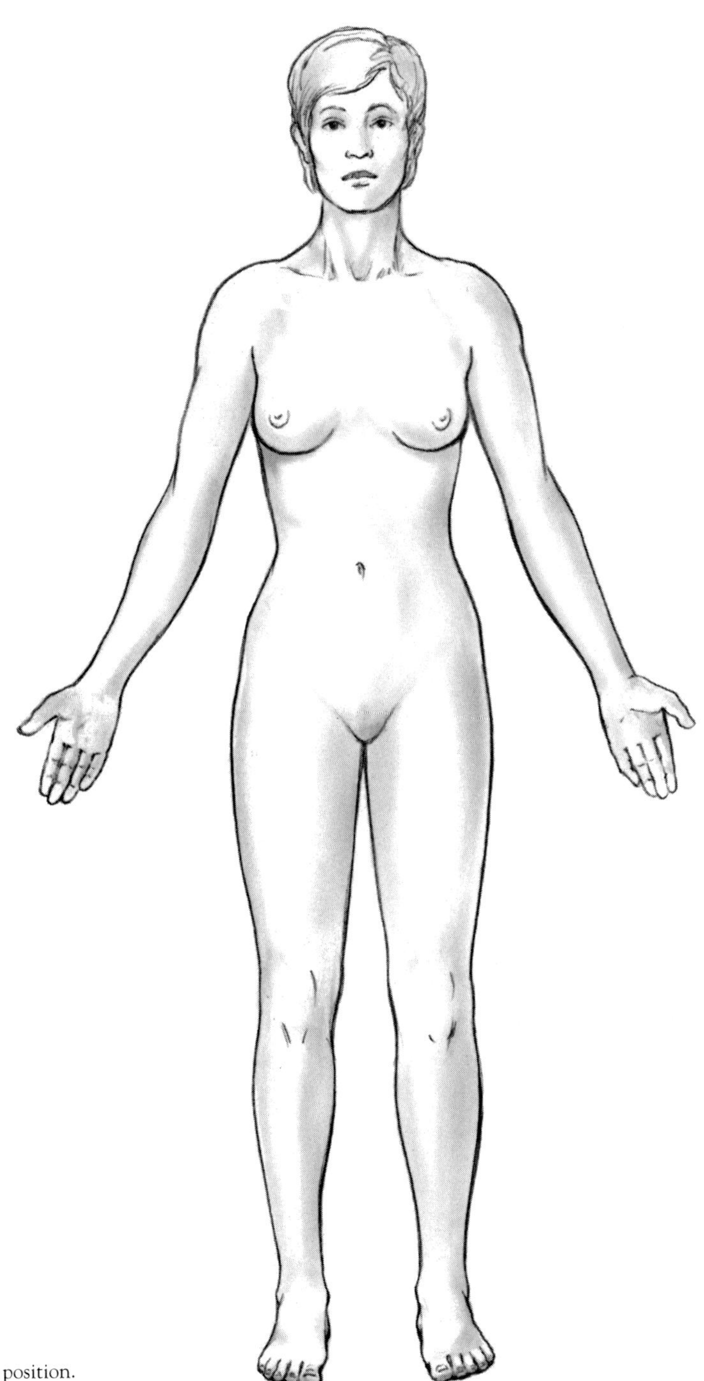

Figure 1–2
Body in anatomical position.

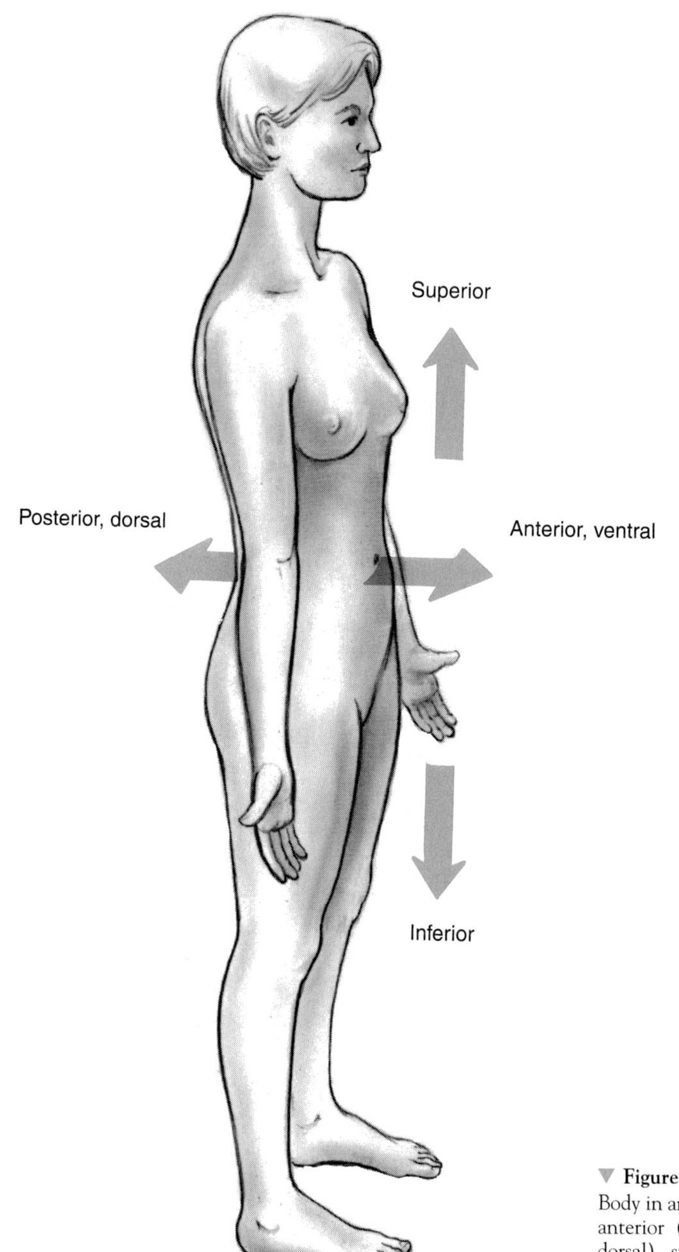

Figure 1–3
Body in anatomical position with the anterior (or ventral), posterior (or dorsal), superior, and inferior areas noted.

The body in anatomical position can be divided by planes or flat surfaces (Figure 1–4). The ***median plane*** (**me**-dee-an) or midsagittal plane is created by an imaginary line dividing the body into right and left halves. On the surface of the body, these halves are generally symmetrical in structure, yet the same symmetry does not apply to all internal structures.

Other planes can be created by different imaginary lines (Figure 1–4). A ***sagittal plane*** (**saj**-i-tl) is any plane created by an imaginary plane parallel to the

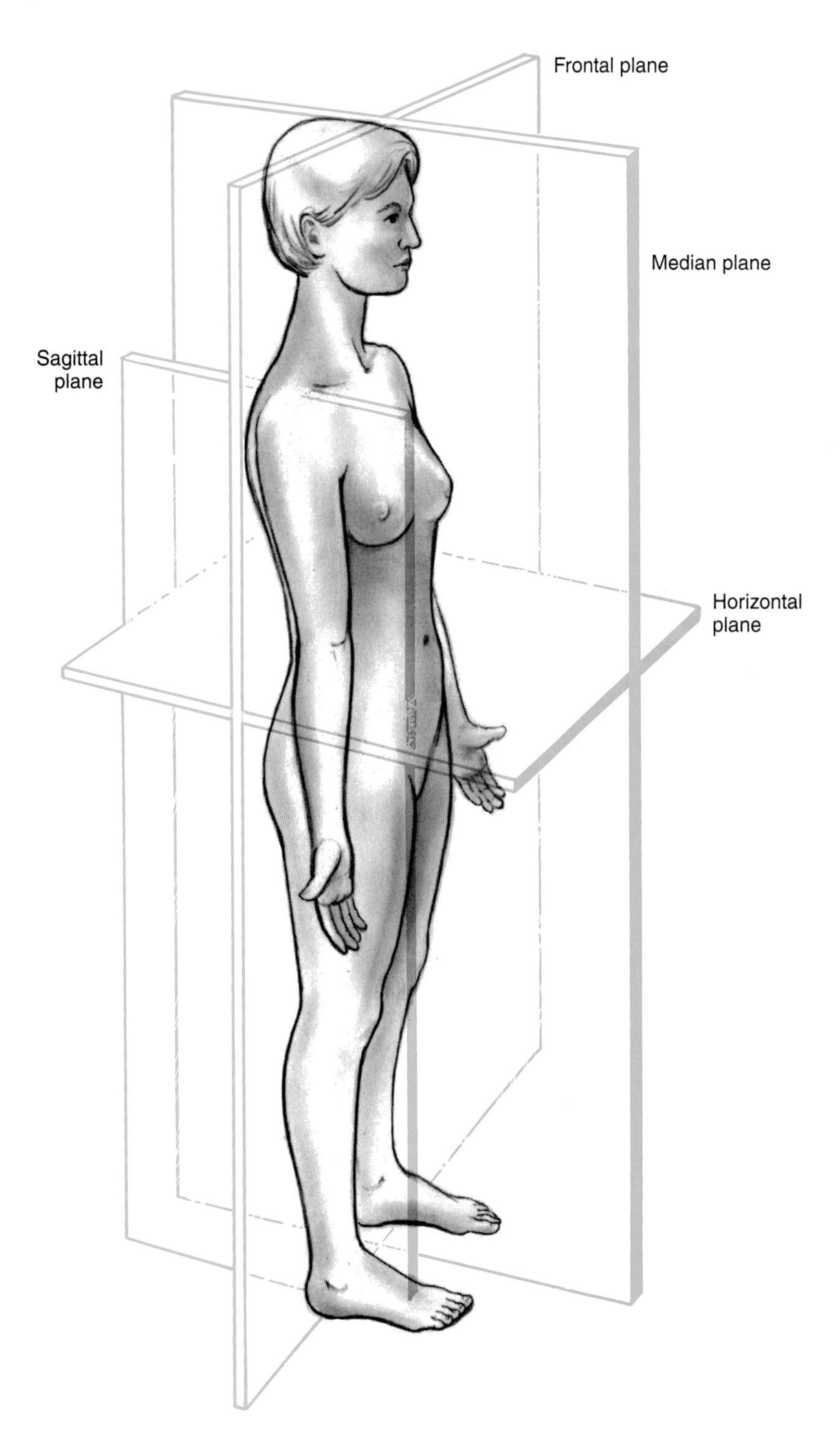

Figure 1–4
Body in anatomical position with the median, sagittal, horizontal, and frontal planes noted.

median plane. A ***horizontal plane*** (hor-i-**zon**-tal) is created by an imaginary line dividing the body at any level into superior and inferior parts. A ***frontal plane*** (**frunt**-il) or coronal plane is created by an imaginary line dividing the body at any level into anterior and posterior parts.

Portions of the body in anatomical position can also be described in relationship to these planes (Figure 1–5). A structure located at the median plane, for example the nose, is considered ***median*** (**me**-dee-an). An area closer to the median plane of the body or structure is considered ***medial*** (**me**-dee-il). An area

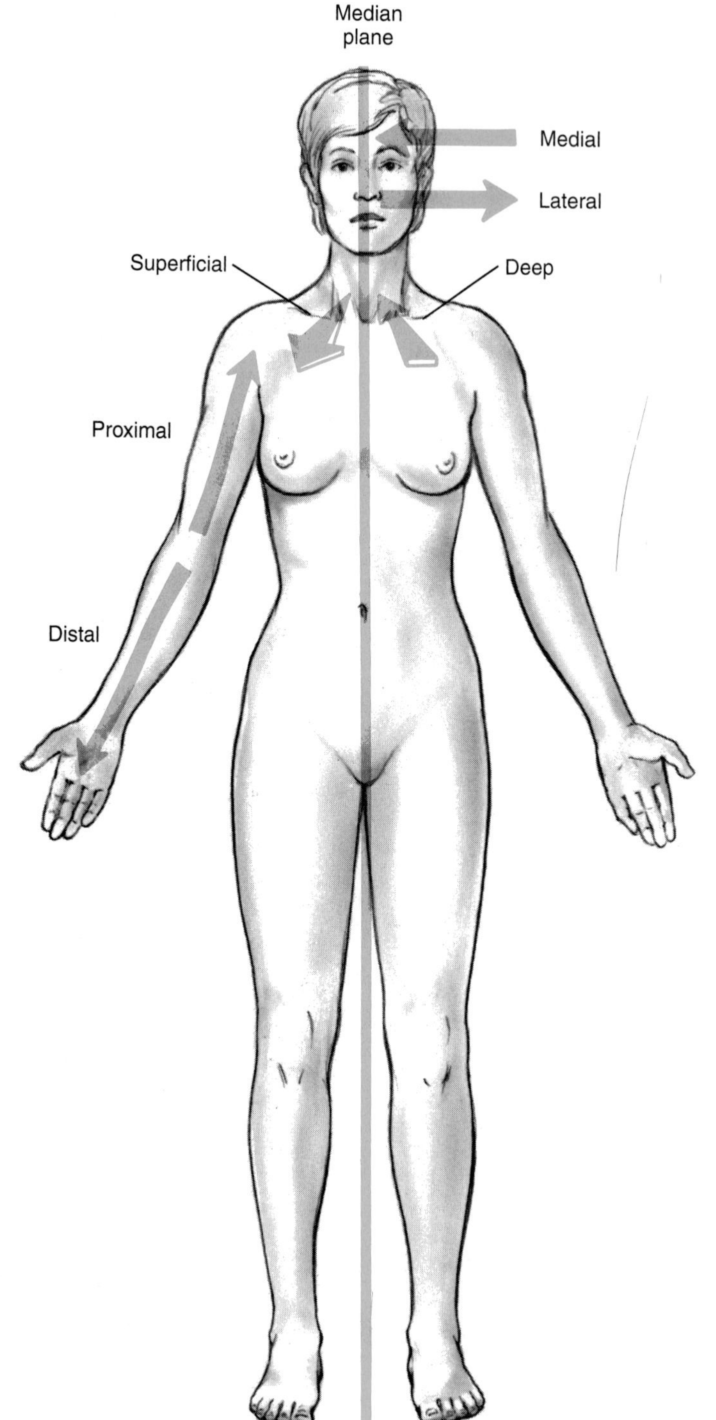

▼ **Figure 1–5**
Body in anatomical position with the medial (or proximal), lateral (or distal), and superficial (or deep) areas noted.

farther from the median plane of the body or structure is considered ***lateral*** (**lat**-er-il). For example, the eyes are medial to the ears, and the ears are lateral to the eyes.

Terms can be used to describe the relationship of portions of the body in anatomical position (Figure 1–5). An area closer to the median plane is considered ***proximal*** (**prok**-si-mal), and an area farther from the median plane is ***distal*** (**dis**-tl). For example, in the upper limb, the shoulder is proximal, and the fingers are distal.

Additional terms can be used to describe relationships between structures. Structures on the same side of the body are considered ***ipsilateral*** (ip-see-**lat**-er-il). Structures on the opposite side of the body are considered ***contralateral*** (kon-trah-**lat**-er-il). For example, the right leg is ipsilateral to the right arm but contralateral to the left arm.

Certain terms can be used to give information about the depth of a structure in relationship to the surface of the body (see Figure 1–5). The structures located toward the surface of the body are ***superficial*** (soo-per-**fish**-al). The structures located inward, away from the body surface, are ***deep*** (deep). For example, the skin is superficial, and the bones are deep.

Terms also can be used to give information about location in hollow structures such as the braincase of the skull. The inner side of the wall of a hollow structure is referred to as ***internal*** (in-**tern**-il). The outer side of the wall of a hollow structure is ***external*** (eks-**tern**-il).

The body or portions of it in anatomical position can also be cut or divided into sections along various planes in order to study the specific anatomy of a region (Figure 1–6). The ***midsagittal section*** (mid-**saj**-i-tl) or median section is a cut through the median plane. The ***transverse section*** (**trans**-vers) or horizontal section is a cut through a horizontal plane. The ***frontal section*** (**frunt**-il) or coronal section is a cut through any frontal plane.

Normal Anatomical Variation

When studying anatomy, the dental professional must understand that there can be anatomical variation of head and neck structures that is still within normal limits. The number of bones and muscles in the head and neck is usually constant, but specific details of these structures can vary from patient to patient. Bones may have different sizes of processes. Muscles may differ in size and details of their attachments. Joints, vessels, nerves, glands, lymph nodes, and fascial planes and spaces of an individual can vary in size, location, and even presence. The most common variations of the head and neck that affect dental treatment are discussed in this text.

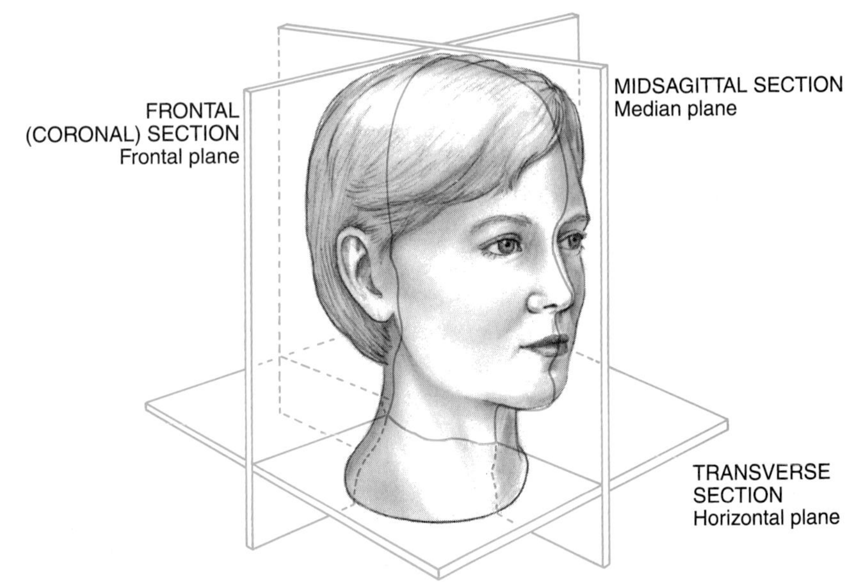

▼ **Figure 1–6**
Head and neck in anatomical position showing the midsagittal, transverse, and frontal sections.

REVIEW QUESTIONS

1. Which of the following planes *divides* the body in anatomical position into right and left halves?
   A. Horizontal plane
   B. Median plane
   C. Coronal plane
   D. Frontal plane

2. Which of the following is used to describe an area of the body that is farther from the median plane?
   A. Proximal
   B. Lateral
   C. Medial
   D. Ipsilateral
   E. Contralateral

3. Structures on the *same* side of the body are considered:
   A. Proximal
   B. Lateral
   C. Medial
   D. Ipsilateral
   E. Contralateral

4. An area of the body in anatomical position that faces toward the head is considered:
   A. Inferior
   B. Superior
   C. Proximal
   D. Distal
   E. Dorsal

5. Through which plane of the body in anatomical position is a midsagittal section taken?
   A. Horizontal plane
   B. Median plane
   C. Coronal plane
   D. Frontal plane

Chapter 2

Surface Anatomy

▼ ***This chapter discusses the following topics:***

I. Surface Anatomy

II. Regions of the Head
  A. Frontal Region
  B. Parietal and Occipital Regions
  C. Temporal Region
  D. Orbital Region
  E. Nasal Region
  F. Infraorbital, Zygomatic, and Buccal Regions
  G. Oral Region
  H. Mental Region

III. Regions of the Neck

▼ ***Key words***

Buccal (**buk**-al) Structures closest to the inner cheek.

Facial (**fay**-shal) Structures closest to the facial surface.

Labial (**lay**-be-al) Structures closest to the lips.

Lingual (**ling**-gwal) Structures closest to the tongue.

Palatal (**pal**-ah-tal) Structures closest to the palate.

Chapter 2

Surface Anatomy

▼ *After studying this chapter, the reader should be able to:*

1. Define and pronounce all the key words and anatomical terms in this chapter.
2. Discuss the anatomical considerations for patient examination and dental radiology of the head and neck region.
3. Locate and identify the regions and associated surface landmarks of the head and neck on a diagram and a patient.
4. Correctly complete the review questions for this chapter.
5. Integrate the knowledge of surface anatomy into the clinical practice of patient examination and dental radiology of the head and neck region.

Surface Anatomy

The dental professional must be thoroughly familiar with the surface anatomy of the head and neck in order to examine patients. The features of the surface provide essential landmarks for many of the deeper anatomical structures. Thus the examination of these accessible surface features by visualization and palpation can give information about the health of deeper tissues. Any changes in these surface features need to be recorded by the dental professional.

Many of the extraoral radiographs taken by the dental professional also use surface landmarks. Portions of the eye, nose, and ear are used for this purpose. This use of landmarks allows for easy film placement and consistency in taking the films.

A certain amount of variation in surface features is within a normal range. However, a change in a surface feature in a given person may signal a condition of clinical significance. Thus it is not the variations among individuals that should be noted, but the changes in a particular individual.

The study of anatomy of the head and neck begins with the division of the surface into regions. Within each region are certain surface landmarks. Practice finding these landmarks in each region on yourself to improve the skills of examination.

Regions of the Head

The **regions of the head** include the frontal, parietal, occipital, temporal, orbital, nasal, infraorbital, zygomatic, buccal, oral, and mental regions (Figure 2–1). The superficial to deep relationships of the head are relatively simple over most of its posterior and superior surfaces but are more difficult in the region of the face. The underlying bony structure of the head is

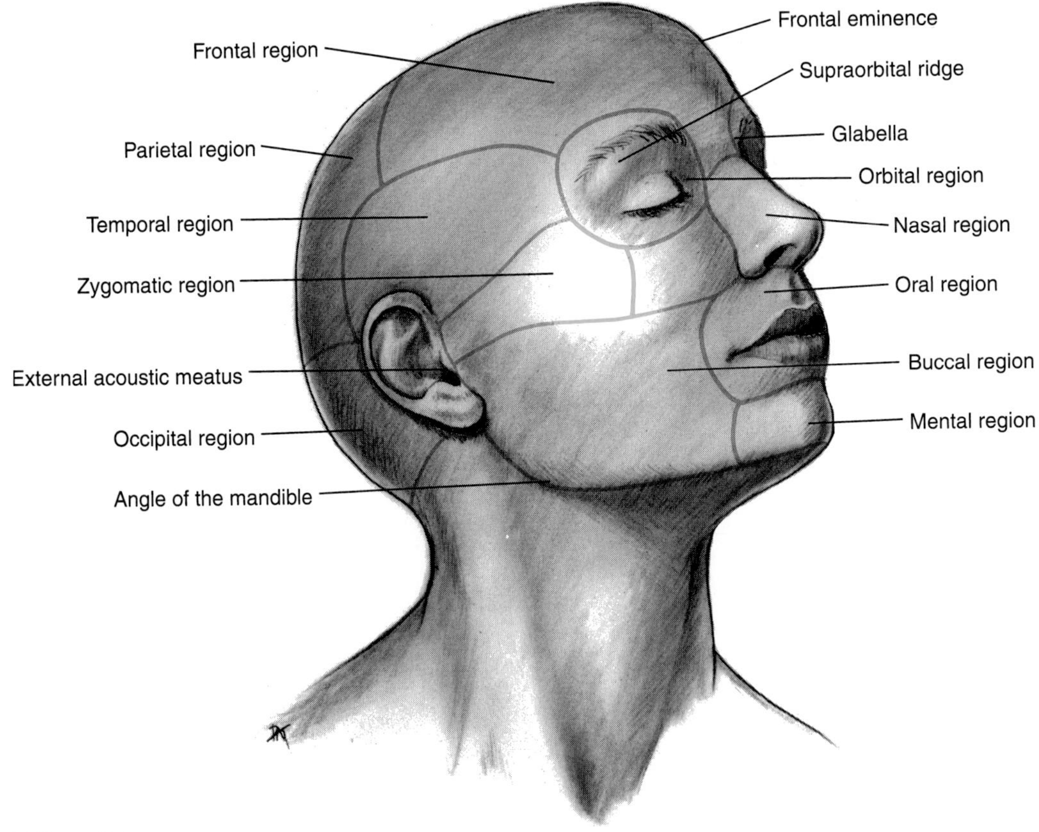

Figure 2–1
Regions of the head: frontal, parietal, occipital, temporal, orbital, nasal, infraorbital, zygomatic, buccal, oral, and mental.

covered in Chapter 3. The underlying glandular tissue, such as the salivary and lacrimal glands and thyroid gland, is covered in Chapter 7. Lymph nodes that are located throughout the tissues of the head are covered in Chapter 10.

▼ FRONTAL REGION

The **frontal region** (**frunt**-il) of the head includes the forehead and the area above the eyes (Figure 2–2). Just under each eyebrow is the **supraorbital ridge** (soo-prah-**or**-bit-al) or superciliary ridge. The smooth elevated area between the eyebrows is the **glabella** (glah-**bell**-ah), which tends to be flat in children and adult females and to form a rounded prominence in adult males. The prominence of the forehead, the **frontal eminence** (**frunt**-il **em**-i-nins), is also evident. The frontal eminence is typically more pronounced in children and adult females, and the supraorbital ridge is more prominent in adult males.

▼ PARIETAL AND OCCIPITAL REGIONS

The **parietal region** (pah-**ri**-it-al) and **occipital region** (ok-**sip**-it-al) of the head are covered by the scalp. The **scalp** consists of layers of soft tissue overlying the bones of the braincase. Large areas of the scalp may be covered by hair. It is still important to try to survey these areas during an extraoral examination since many lesions may be hidden visually from the clinician, as well as the patient.

▼ TEMPORAL REGION

Within the **temporal region** (**tem**-poh-ral) of the head, the external ear is a prominent feature (Figure 2–3). The external ear is composed of an **auricle** (**aw**-ri-kl) or oval flap of the ear and the **external acoustic meatus** (ah-**koos**-tik me-**ate**-us). The auricle collects sound waves. The external acoustic meatus is a tube

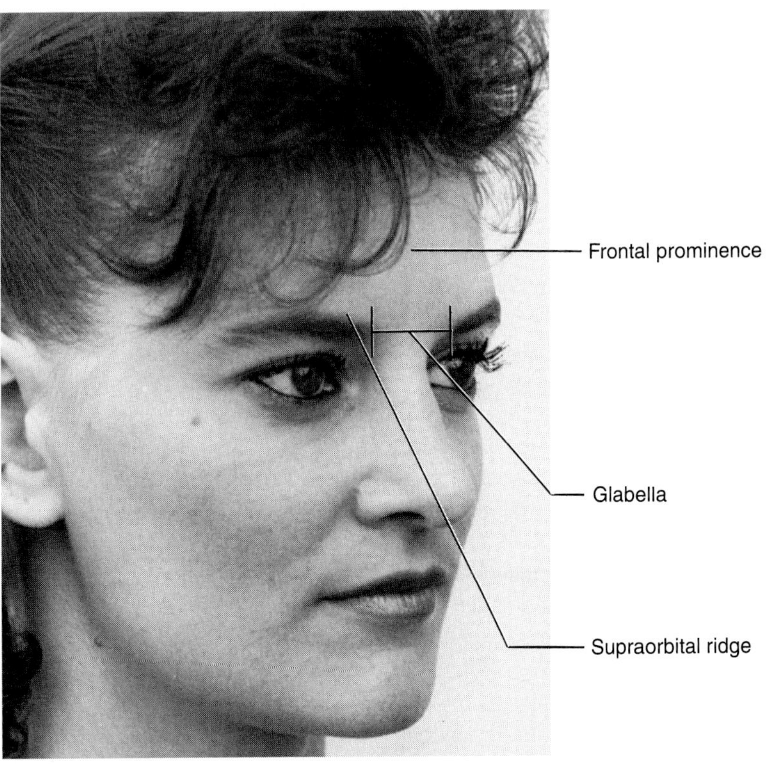

▼ **Figure 2–2**
Frontal view of the head with the landmarks of the frontal region noted.

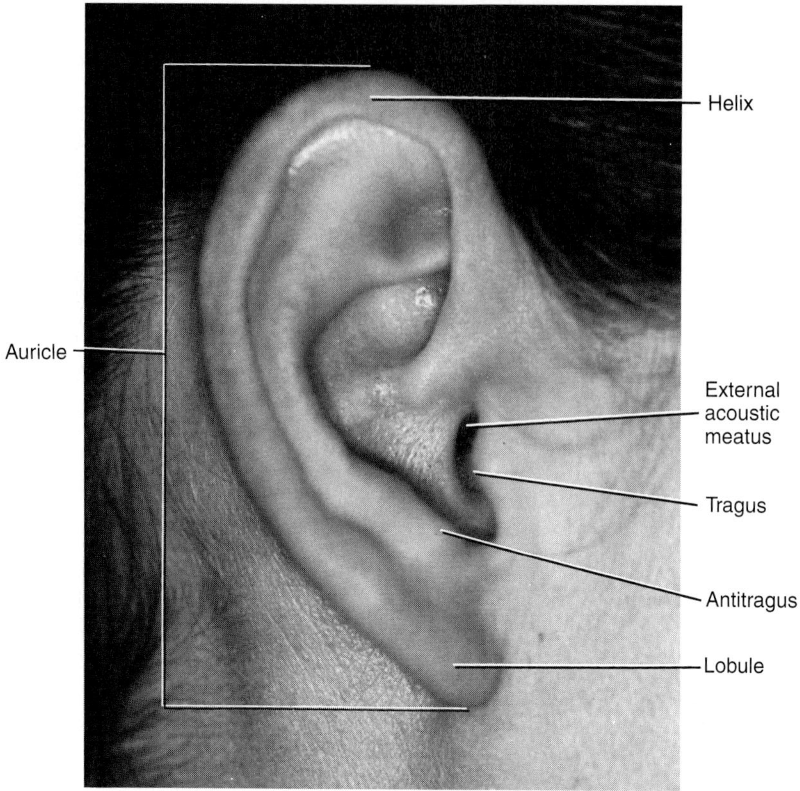

Figure 2–3
Lateral view of the external ear and its landmarks within the temporal region noted.

through which sound waves are transmitted to the middle ear within the skull.

The superior and posterior free margin of the auricle is the **helix** (**heel**-iks), which ends inferiorly at the **lobule** (**lob**-yule), the fleshy protuberance of the earlobe (Figure 2–3). The upper apex of the helix is typically level with the eyebrows and the glabella, and the lobule is approximately at the level of the apex of the nose.

The part of the auricle anterior to the external acoustic meatus is a smaller flap of tissue called the **tragus** (**tra**-gus) (Figure 2–3). The tragus, as well as the rest of the auricle, is flexible when palpated due to its underlying cartilage. The other flap of tissue opposite the tragus is the **antitragus** (an-tie-**tra**-gus). The external acoustic meatus and tragus are important landmarks when taking extraoral radiographs.

ORBITAL REGION

In the **orbital region** (**or**-bit-al) of the head, the eyeball and all its supporting structures are contained in the bony socket called the **orbit** (**or**-bit) (Figure 2–4). The eyes are usually near the midpoint of the vertical height of the head. The width of each eye is typically the same as the distance between the eyes. On the eyeball is the white area or **sclera** (**skler**-ah) with its central area of coloration, the circular **iris** (**eye**-ris). The opening in the center of the iris is the **pupil** (**pew**-pil), which changes size as the iris responds to changing light conditions.

Two movable **eyelids,** upper and lower, cover and protect each eyeball (Figure 2–4). Behind each upper eyelid and within the orbit is the **lacrimal gland** (**lak**-ri-mal), which produces lacrimal fluid or tears.

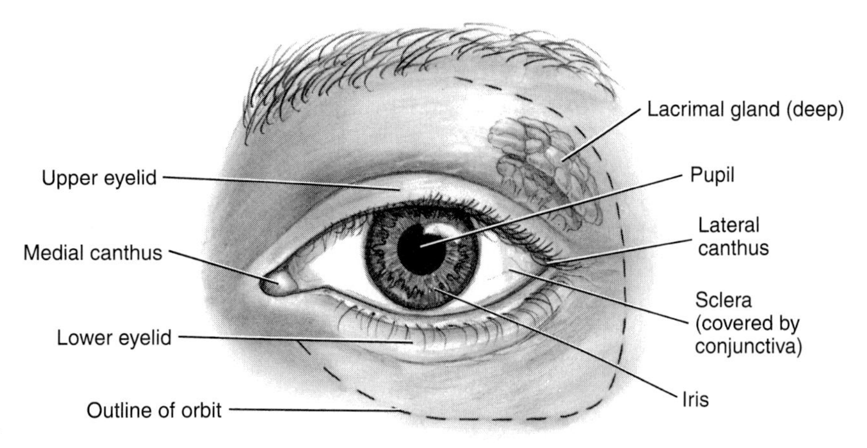

Figure 2–4
Frontal view of the left eye with the landmarks of the orbital region noted.

The **conjunctiva** (kon-junk-**ti**-vah) is the delicate and thin membrane lining the inside of the eyelids and the front of the eyeball. The outer corner where the upper and lower eyelids meet is called the **lateral canthus** (plural, **canthi**) (**kan**-this, **kan**-thy) or outer canthus. The inner angle of the eye is called the **medial canthus** or inner canthus. These canthi are important landmarks when taking extraoral radiographs.

NASAL REGION

The main feature of the **nasal region** (**nay**-zil) of the head is the external nose (Figure 2–5). The **root of the nose** is located between the eyes. Inferior to the glabella is a midpoint landmark of the nasal region that corresponds to the junction between the underlying bones, the **nasion** (**nay**-ze-on). Inferior to the nasion

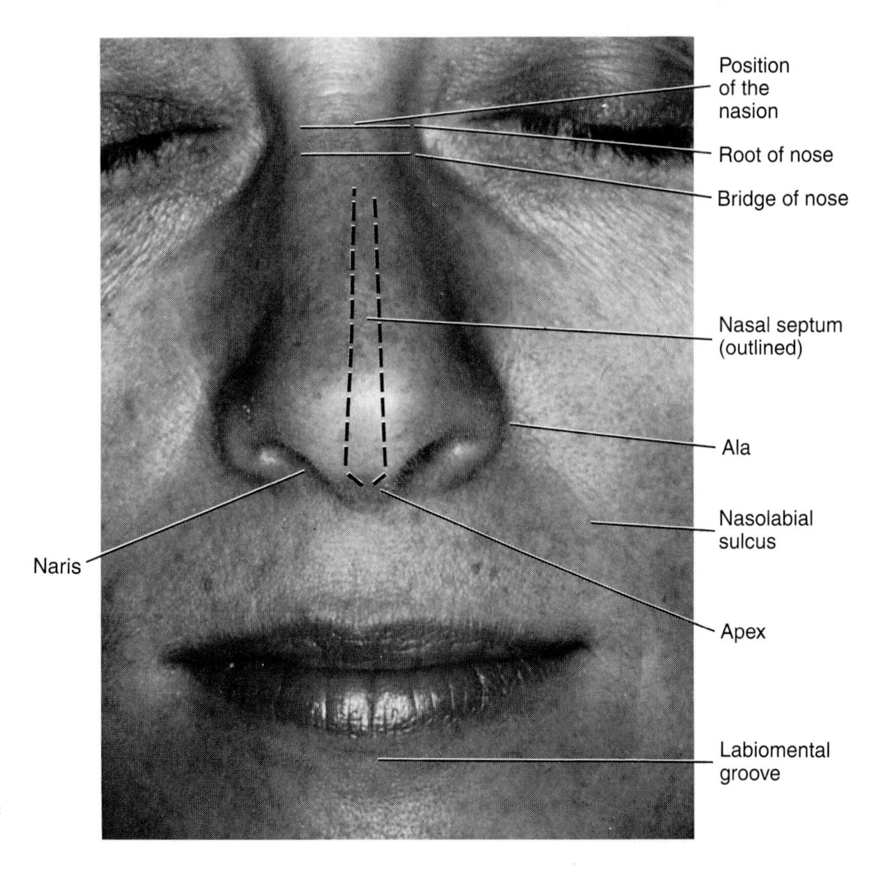

Figure 2–5
Frontal view of the face with the landmarks of the nasal region noted.

is the bony structure that forms the **bridge of the nose.** The tip or **apex of the nose** is flexible when palpated since it is formed from cartilage.

Inferior to the apex on each side of the nose is a nostril or **naris** (plural, **nares**) (**nay**-ris, **nay**-rees) (Figure 2–5). The nares are separated by the midline **nasal septum** (**nay**-zil **sep**-tum). The nares are bounded laterally by winglike cartilaginous structures, the **ala** (plural, **alae**) (**a**-lah, **a**-lay) of the nose. The width between the alae should be about the same width as one eye or the space between the eyes. The nasion and the alae of the nose are important landmarks when taking extraoral radiographs.

INFRAORBITAL, ZYGOMATIC, AND BUCCAL REGIONS

The infraorbital, zygomatic, and buccal regions of the head are all located on the facial aspect (Figure 2–6). The **infraorbital region** (in-frah-**or**-bit-al) of the head is located inferior to the orbital region and lateral to the nasal region. Farther laterally is the **zygomatic region** (zy-go-**mat**-ik), which overlies the cheek bone, the **zygomatic arch.** The zygomatic arch extends from just below the lateral margin of the eye toward the upper part of the ear.

Inferior to the zygomatic arch, and just anterior to the ear, is the **temporomandibular joint** (tem-poh-ro-man-**dib**-you-lar) (Figure 2–6). This is where the upper skull forms a joint with the lower jaw. The movements of the joint can be felt when one opens and closes the mouth or moves the lower jaw to the right or left. One way to feel the lower jaw moving at the temporomandibular joint is to place a finger into the outer portion of the external acoustic meatus.

The **buccal region** (**buk**-al) of the head is composed of the soft tissues of the cheek (Figure 2–6). The **cheek** forms the side of the face and is a broad area of the face between the nose, mouth, and ear. Most of the upper cheek is fleshy, mainly formed by a mass of fat and muscles. One of these is the strong **masseter muscle** (**mass**-et-er), which is felt when a patient clenches the teeth together. The sharp angle of the lower jaw inferior to the ear's lobule is termed the **angle of the mandible.**

ORAL REGION

The **oral region** of the head has many structures within it, such as the lips, oral cavity, palate, tongue, floor of the mouth, and portions of the throat. The lips are the gateway of the oral region, and each lip's **vermilion**

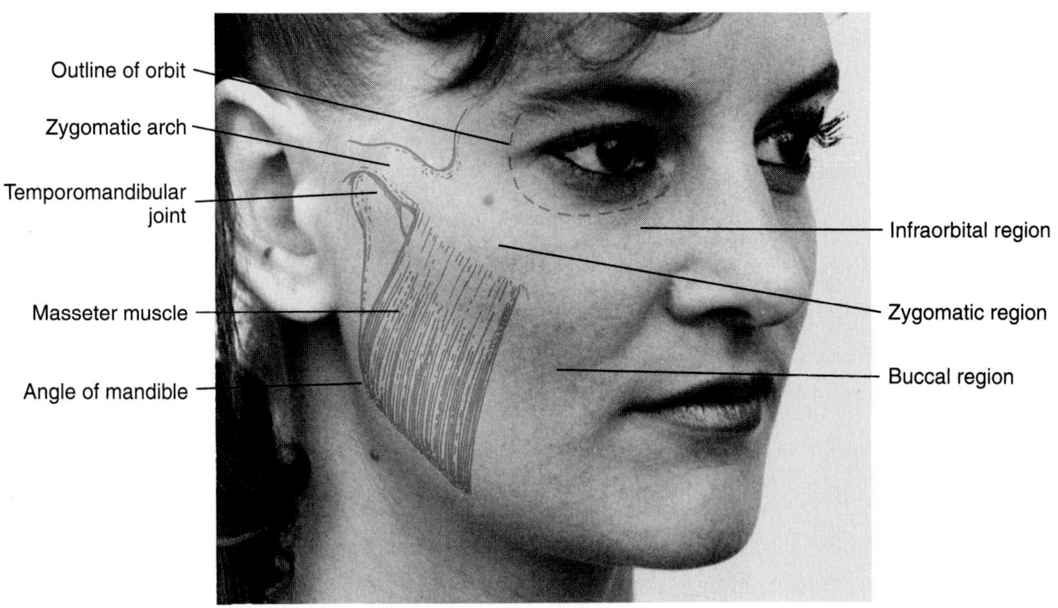

Figure 2–6
Landmarks of the zygomatic and buccal regions noted.

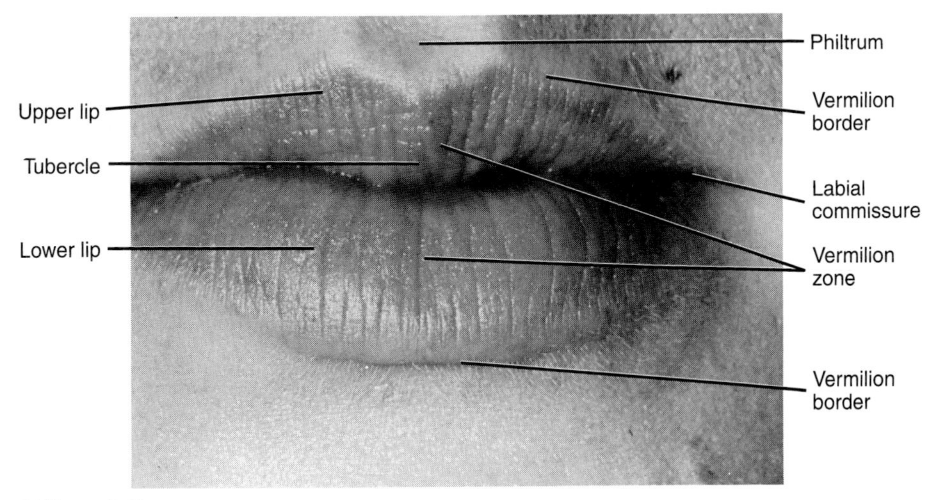

▼ **Figure 2–7**
Frontal view of the lips within the oral region.

zone (ver-**mil**-yon) has a darker appearance than the surrounding skin (Figure 2–7). The lips are outlined from the surrounding skin by a transition zone, the **vermilion border.** The width of the lips at rest should be about the same distance as between the irises of the eyes.

On the midline of the upper lip, extending downward from the nasal septum, is a vertical groove called the **philtrum** (**fil**-trum) (Figure 2–7). The philtrum terminates in a thicker area or **tubercle of the upper lip** (**too**-ber-kl). The upper and lower lips meet at each corner of the mouth or **labial commissure** (**kom**-i-shoor). The groove running upward between the labial commissure and the ala of the nose is called the **nasolabial sulcus** (nay-zo-**lay**-be-al **sul**-kus) (see Figure 2–5). The lower lip extends to the horizontal **labiomental groove** (lay-bee-o-**ment**-il), which separates the lower lip from the chin in the mental region.

▼ Oral Cavity

The inside of the mouth is known as the **oral cavity.** The jaws are within the oral cavity and deep to the lips (Figure 2–8). Underlying the upper lip is the upper jaw or **maxilla** (mak-**sil**-ah). The bone underlying the lower lip is the lower jaw or **mandible** (**man**-di-bl).

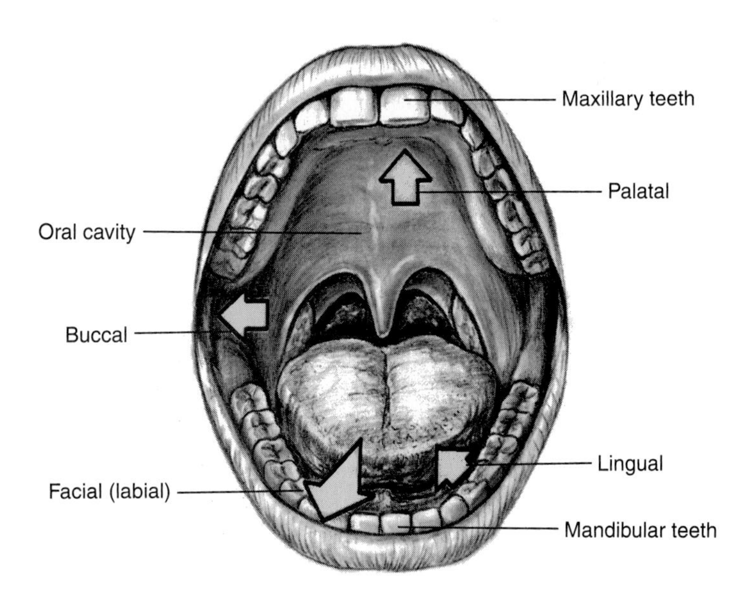

▼ **Figure 2–8**
The oral cavity and jaws with the designation of the terms lingual, palatal, buccal, facial, and labial within the oral cavity.

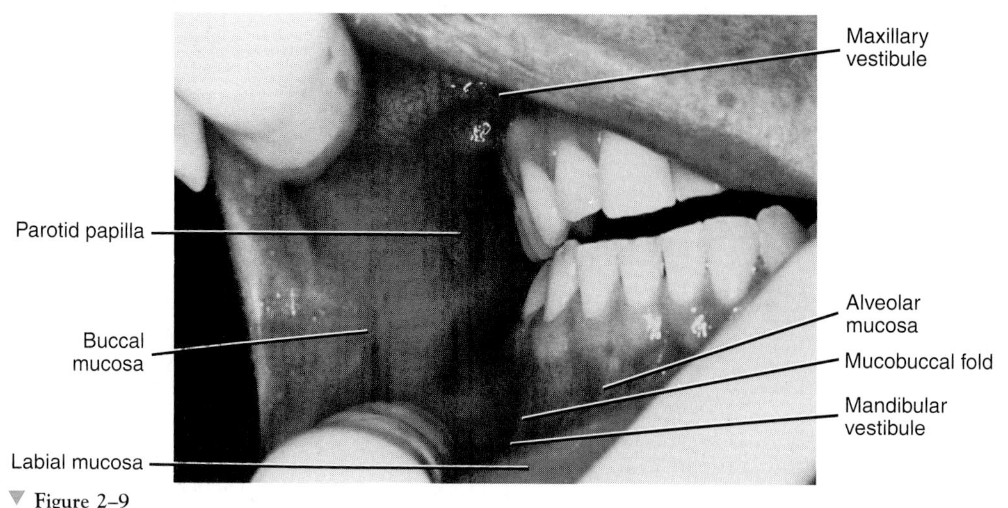

Figure 2–9
View of the buccal and labial mucosa of the oral cavity with landmarks noted.

Many areas in the oral cavity are termed or identified by their relationship to the tongue, palate, cheek, facial surface, or lips (Figure 2–8). Those structures closest to the tongue are termed ***lingual*** (**ling**-gwal). Those structures closest to the palate are termed ***palatal*** (**pal**-ah-tal). Those structures closest to the inner cheek are considered ***buccal*** (**buk**-al). Those structures closest to the facial surface or lips are termed ***facial*** (**fay**-shal) or ***labial*** (**lay**-be-al).

The oral cavity is lined by a mucous membrane or **mucosa** (mu-**ko**-sah) (Figure 2–9). The inner portions of the lips are lined by a pink and thick **labial mucosa.** The labial mucosa is continuous with the equally pink and thick **buccal mucosa** that lines the inner cheek. The buccal mucosa covers a dense pad of inner tissue, the **buccal fat pad.**

Further landmarks can be noted in the oral cavity (Figure 2–9). On the inner portion of the buccal mucosa, just opposite the maxillary second molar, is a small elevation of tissue called the **parotid papilla** (pah-**rot**-id pah-**pil**-ah), which contains the duct opening from the parotid salivary gland.

The upper and lower spaces between the cheeks, lips, and gums are the maxillary and mandibular **vestibules** (**ves**-ti-bules) (Figure 2–9). Deep within each vestibule, the pink and thick labial or buccal mucosa meets the redder and thinner **alveolar mucosa** (al-**ve**-o-lar) at the **mucobuccal fold** (mu-**ko**-buk-al).

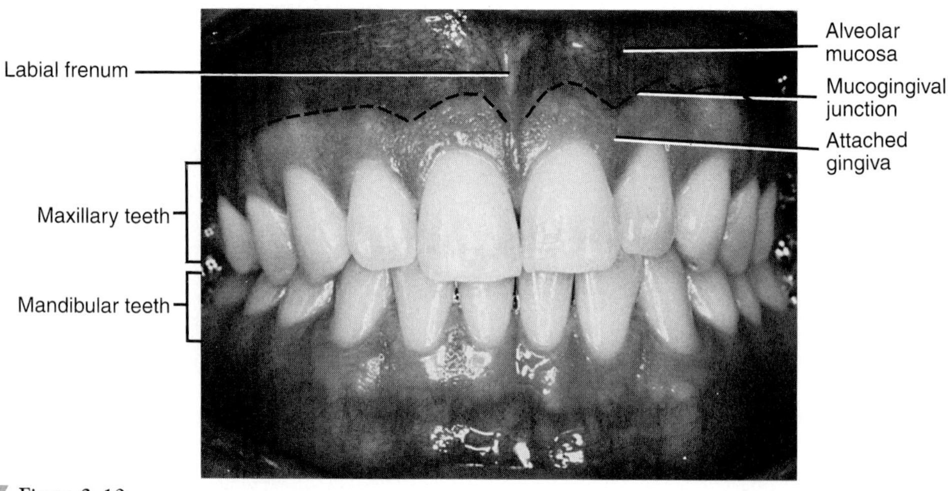

Figure 2–10
Frontal view of the oral cavity with its landmarks noted.

The **labial frenum** (**free**-num) or frenulum is a fold of tissue located at the midline between the labial mucosa and the alveolar mucosa of the maxilla and mandible (Figure 2–10).

Teeth of the oral cavity are located within the upper and lower jaws of the oral cavity (Figure 2–10). The teeth of the maxilla are the **maxillary teeth** (**mak**-sil-lare-ee), and the teeth of the mandible are the **mandibular teeth** (man-**dib**-you-lar). The maxillary anterior teeth should overlap the mandibular anterior teeth, and posteriorly, the maxillary buccal cusps should overlap the mandibular buccal cusps. Both dental arches in the adult have permanent teeth that include the **incisors** (in-**sigh**-zers), canines (**kay**-nines), premolars (pre-**mol**-ers), and **molars** (**mol**-ers).

Surrounding the maxillary and mandibular teeth are the gums or **gingiva** (jin-**ji**-vah), composed of a firm pink mucosa (Figures 2–10 and 2–11). The gingiva that tightly adheres to the bone around the roots of the teeth is the **attached gingiva.** The attached gingiva may have areas of pigmentation. The line of demarcation between the firmer and pinker attached gingiva and the movable and redder alveolar mucosa is the scallop-shaped **mucogingival junction** (mu-ko-**jin**-ji-val).

At the gingival margin of each tooth is the nonattached or **marginal gingiva** (**mar**-ji-nal) (Figure 2–11). The inner surface of the marginal gingiva faces a space or **sulcus** (plural, **sulci**) (**sul**-kus, **sul**-ky). The gingiva between the teeth is an extension of attached gingiva and is called the **interdental gingiva** (in-ter-**den**-tal) or interdental papilla.

Palate

The roof of the mouth or **palate** (**pal**-it) has two parts—an anterior portion and a posterior portion (Figure 2–12). The firmer anterior portion is called the **hard palate.** A midline ridge of tissue on the hard palate is the **median palatine raphe** (**pal**-ah-tine **ra**-fe). A small bulge of tissue at the most anterior portion of the hard palate, lingual to the anterior teeth, is the **incisive papilla** (in-**sy**-ziv pah-**pil**-ah). Directly posterior to this papilla are **palatine rugae** (**pal**-ah-tine **ru**-ge), which are firm, irregular ridges of tissue.

The looser posterior portion of the palate is called the **soft palate** (Figures 2–12 and 2–18). A midline muscular structure, the **uvula of the palate** (**u**-vu-lah), hangs from the posterior margin of the soft palate. The **pterygomandibular fold** (teh-ri-go-man-**dib**-yule-lar) is a fold of tissue that extends from the junction of hard and soft palates down to the mandible, just behind the most distal mandibular tooth, and stretches when the patient opens the mouth wider. This fold covers a deeper fibrous structure and separates the cheek from the throat. Just distal to the last tooth of the mandible is a dense pad of tissue, the **retromolar pad** (re-tro-**moh**-ler).

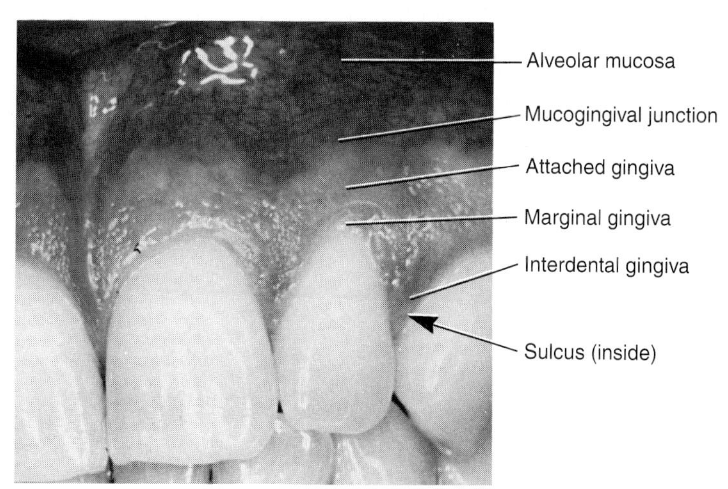

Figure 2–11
Close-up view of the gingiva and its associated landmarks.

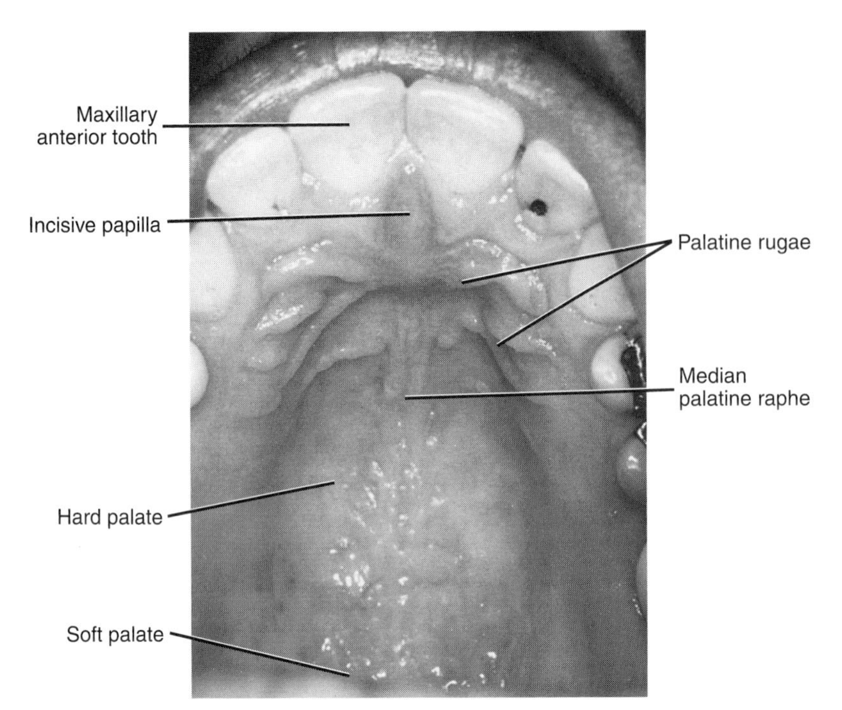

▼ **Figure 2–12**
View of the palate with its landmarks noted.

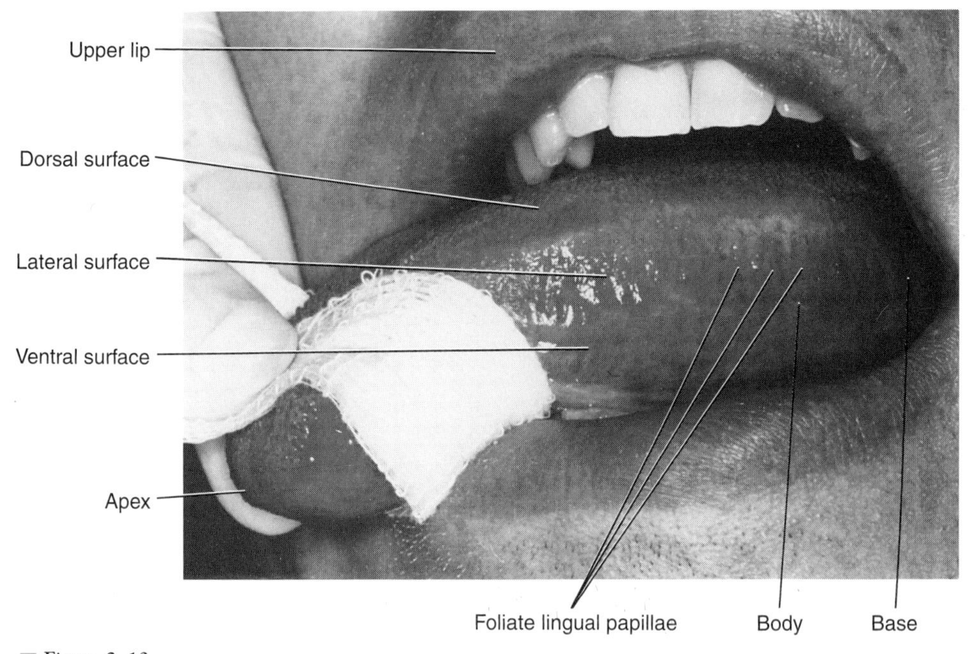

▼ **Figure 2–13**
Lateral view of the tongue with its portions and landmarks noted.

Tongue

The **tongue** is a prominent feature of the oral region (Figure 2–13). The posterior one third is the **base of the tongue** or pharyngeal portion. The base of the tongue attaches to the floor of the mouth. The base of the tongue does not lie within the oral cavity but within the oral part of the throat. The anterior two thirds of the tongue is termed the **body of the tongue** or oral portion and lies within the oral cavity. The tip of the tongue is the **apex of the tongue.** Certain surfaces of the tongue have small elevated structures of specialized mucosa called **lingual papillae** (pah-**pil**-ay), some of which are associated with taste buds.

The side or **lateral surface of the tongue** is noted for its vertical ridges of lingual papillae called **foliate lingual papillae** (**fo**-le-ate), some of which contain taste buds. These lingual papillae are more prominent in children.

The top surface or **dorsal surface of the tongue** (Figure 2–14) has a midline depression, the **median lingual sulcus,** corresponding to the position of a midline fibrous structure deeper in the tongue. In the tongue, dorsal and posterior are not equivalent terms, nor are ventral and anterior the same. Instead, they are four different locations. This is because the tongue of humans still has the same orientation as the tongue of four-footed animals, in which anterior and posterior mean toward the nose and tail, respectively, and dorsal and ventral refer to the back and belly, respectively. Upright posture is the reason that dorsal and posterior have become synonyms in the rest of the body.

The dorsal surface of the tongue also has many lingual papillae (Figure 2–14). The slender, threadlike lingual papillae are the **filiform lingual papillae** (**fil-i**-form), which give the dorsal surface its velvety texture. The red mushroom-shaped dots are called **fungiform lingual papillae** (**fun**-ji-form). These lingual papillae are more numerous on the apex and contain taste buds.

Farther posteriorly on the dorsal surface of the tongue and more difficult to see clinically is a V-shaped groove, the **sulcus terminalis** (**ter**-mi-nal-is) (Figure 2–14). The sulcus terminalis separates the base from the body of the tongue. Where the sulcus terminalis points backward toward the throat is a small pitlike depression called the **foramen cecum** (for-**ay**-men **se**-kum). The **circumvallate lingual papillae** (serk-um-**val**-ate), which are 10-14 in number, line up along the anterior side of the sulcus terminalis on the body. These large mushroom-shaped lingual papillae have taste buds. Even farther posteriorly on the dorsal surface of the tongue base is an irregular mass of tonsillar tissue, the **lingual tonsil** (**ton**-sil).

The underside or ventral surface of the tongue is noted for its visible large blood vessels, the deep

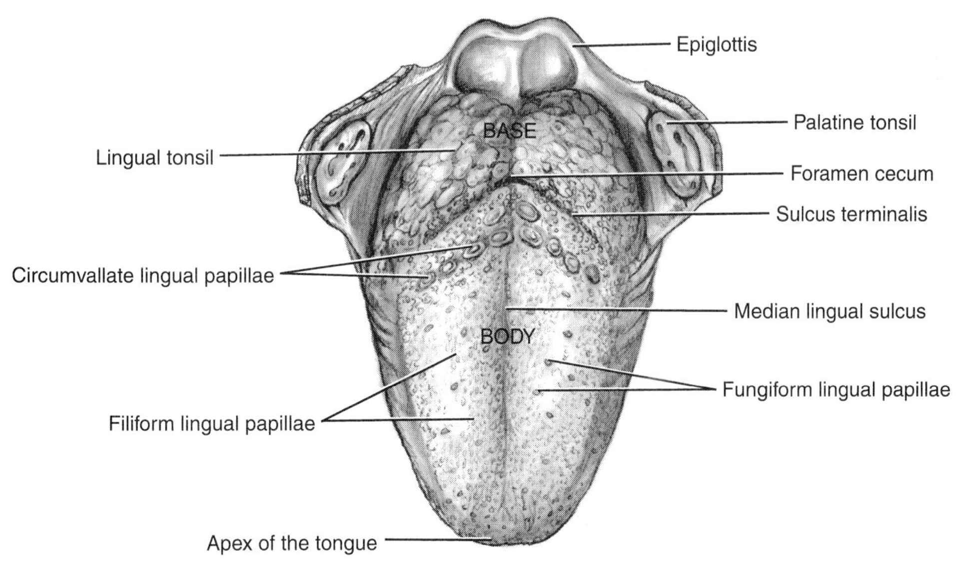

Figure 2–14
Dorsal view of the tongue with its landmarks noted.

lingual veins (vanes), that run close to the surface (Figure 2–15). Lateral to each deep lingual vein is the **plica fimbriata** (plural, **plicae fimbriatae**) (**pli**-kah fim-bree-**ay**-tah, **pli**-kay fim-bree-**ay**-tay), a fold with fringelike projections. Again, the term used for the underside of the tongue, ventral, was in regard to four-footed animals in anatomical position.

Floor of the Mouth

The floor of the mouth is located inferior to the ventral surface of the tongue (Figure 2–16). The **lingual frenum** (**free**-num) or frenulum is a midline fold of tissue between the ventral surface of the tongue and the floor of the mouth.

There is also a ridge of tissue on each side of the floor of the mouth, the **sublingual fold** (sub-**ling**-gwal) or plica sublingualis. Together these folds are arranged in a V-shaped configuration from the lingual frenum to the base of the tongue (Figure 2–16). The sublingual folds contain duct openings from the sublingual salivary gland. The small papilla or **sublingual caruncle** (sub-**ling**-gwal **kar**-unk-el) at the anterior end of each sublingual fold contains the duct openings from both the submandibular and sublingual salivary glands.

Pharynx

The oral cavity also provides the entrance into the throat or **pharynx** (**far**-inks). The pharynx is a muscular tube that serves both the respiratory and digestive systems. There are three portions to the pharynx: the nasopharynx, the oropharynx, and the laryngopharynx (Figure 2–17). Portions of the nasopharynx and oropharynx are visible on an intraoral examination. The **laryngopharynx** (lah-ring-gah-**far**-inks) is more inferior, close to the laryngeal opening, and thus is not visible on an intraoral examination.

The portion of the pharynx that is superior to the level of the soft palate is the **nasopharynx** (nay-zo-**far**-inks) (Figure 2–18). The nasopharynx is continuous with the nasal cavity. The portion of the pharynx that is between the soft palate and the opening of the larynx is the **oropharynx** (or-o-**far**-inks).

The opening from the oral region into the oropharynx is the **fauces** (**faw**-seez) or the faucial isthmus (Figure 2–18). The fauces are formed laterally by the

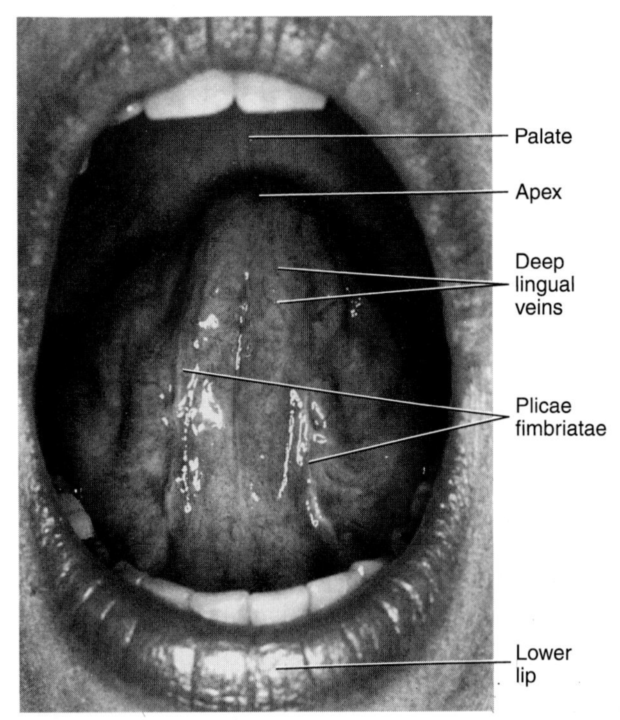

Figure 2–15
Ventral surface of the tongue with its landmarks noted.

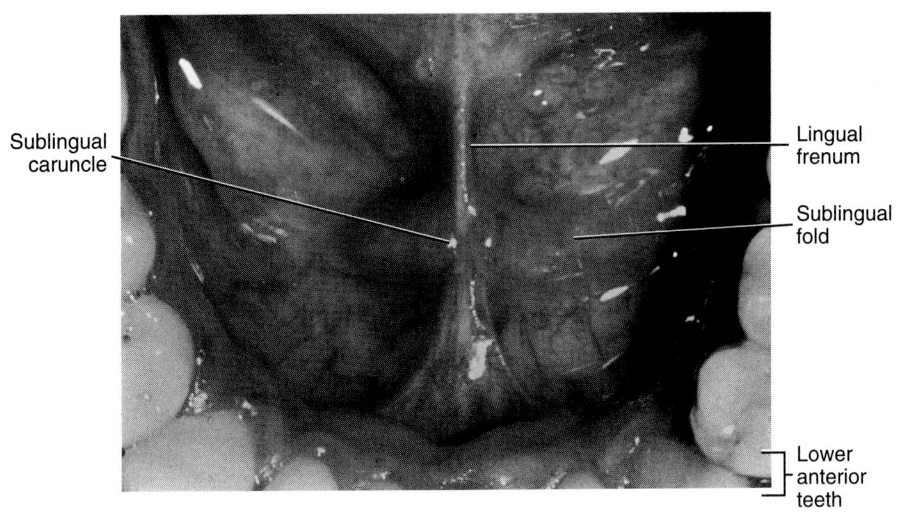

Figure 2–16
View of the floor of the mouth with its landmarks noted.

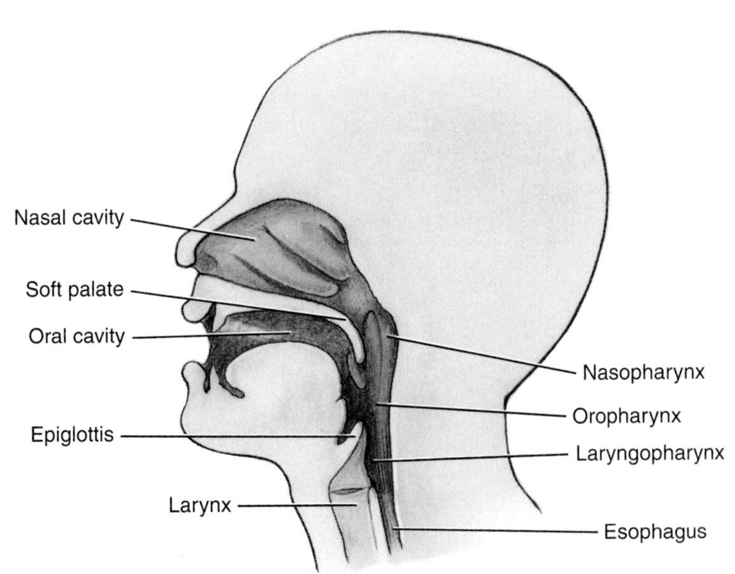

Figure 2–17
Midsagittal section of the head and neck illustrating portions of the pharynx.

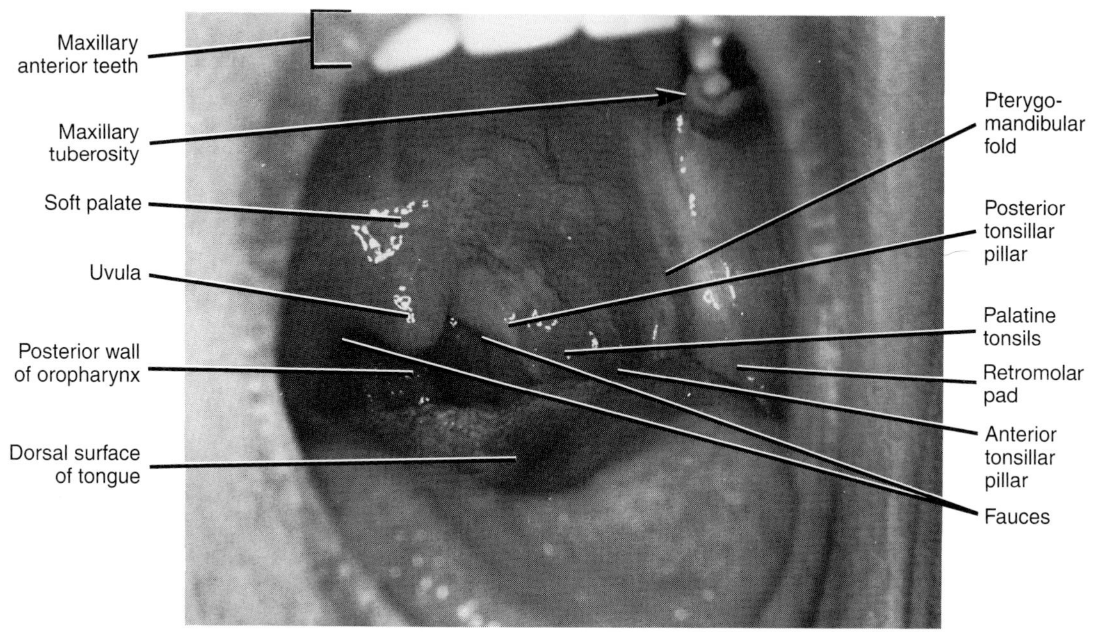

▼ Figure 2–18
Oral view of the nasopharynx and oropharynx.

anterior tonsillar pillar (**ton**-sil-ar **pil**-er) and the **posterior tonsillar pillar.** Tonsillar tissue, composing the **palatine tonsils** (**pal**-ah-tine), is located between these pillars or folds of tissue created by underlying muscles. The palatine tonsils are the tonsillar tissue that patients call their "tonsils."

▼ MENTAL REGION

The chin is the major feature of the **mental region** (**ment**-il) of the head (Figure 2–19). The prominence of the chin is called the **mental protuberance** (pro-**too**-ber-ins). The mental protuberance is often more pronounced in adult males but can also be visualized and palpated in adult females. The **labiomental groove** (lay-bee-o-**ment**-il), a horizontal groove between the lower lip and the chin mentioned in the description of the oral region, should be approximately midway between the apex of the nose and the chin and level with the angle of the mandible. Also present on the chin in some individuals is a midline depression or dimple that marks the underlying bony fusion of the lower jaw.

Regions of the Neck

The neck extends from the skull and mandible down to the clavicles and sternum. The **regions of the neck**

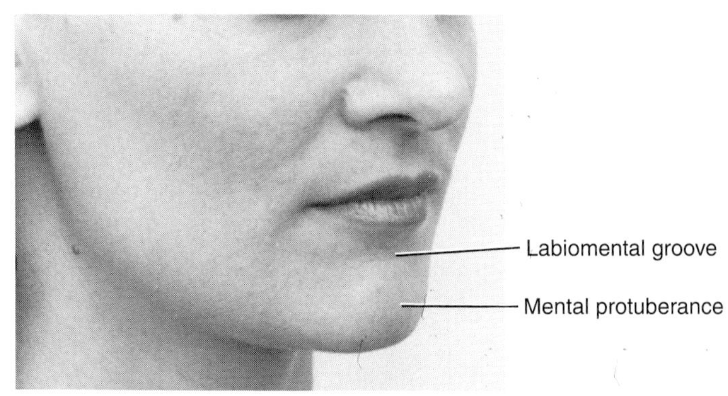

▼ Figure 2–19
Frontal view of the head showing the mental region with its landmarks noted.

can be divided into different cervical triangles based on the large bones and muscles in the area (Figure 2–20). Chapters 3 and 4 further describe these bones and muscles, respectively. Structures deep to the surface of these cervical triangles are also discussed in other chapters.

The large strap muscle, the **sternocleidomastoid muscle** (stir-no-kli-do-**mass**-toid), divides each side of the neck diagonally into an **anterior cervical triangle** (**ser**-vi-kal) and **posterior cervical triangle** (Figure 2–20). The anterior region of the neck corresponds to the two anterior cervical triangles, which are separated by a midline. The lateral region of the neck, posterior to the sternocleidomastoid muscle, is considered the posterior cervical triangle on each side.

At the anterior midline, the prominence of the larynx, the **thyroid cartilage** (**thy**-roid), is visible as the "Adam's apple," especially in adult males (Figure 2–21). The vocal cords or ligaments are attached to the posterior surface of the thyroid cartilage.

The **hyoid bone** (**hi**-oid) is also found in the anterior midline, superior to the thyroid cartilage (Figure 2–21). Many muscles attach to the hyoid bone, which controls the position of the base of the tongue. The hyoid bone can be effectively palpated by feeling below and medial to the angles of the mandible. Do not confuse the hyoid bone with the inferiorly placed thyroid cartilage when palpating the neck.

The anterior cervical triangle can be further subdivided into smaller triangular regions by muscles in the area that are not as prominent as the sternocleidomastoid muscle (Figure 2–22). For example, the superior portion of each anterior cervical triangle is demarcated by portions of the digastric muscle (both bellies) and the mandible, forming a **submandibular triangle** (sub-mand-**dib**-you-lar). The inferior portion of each

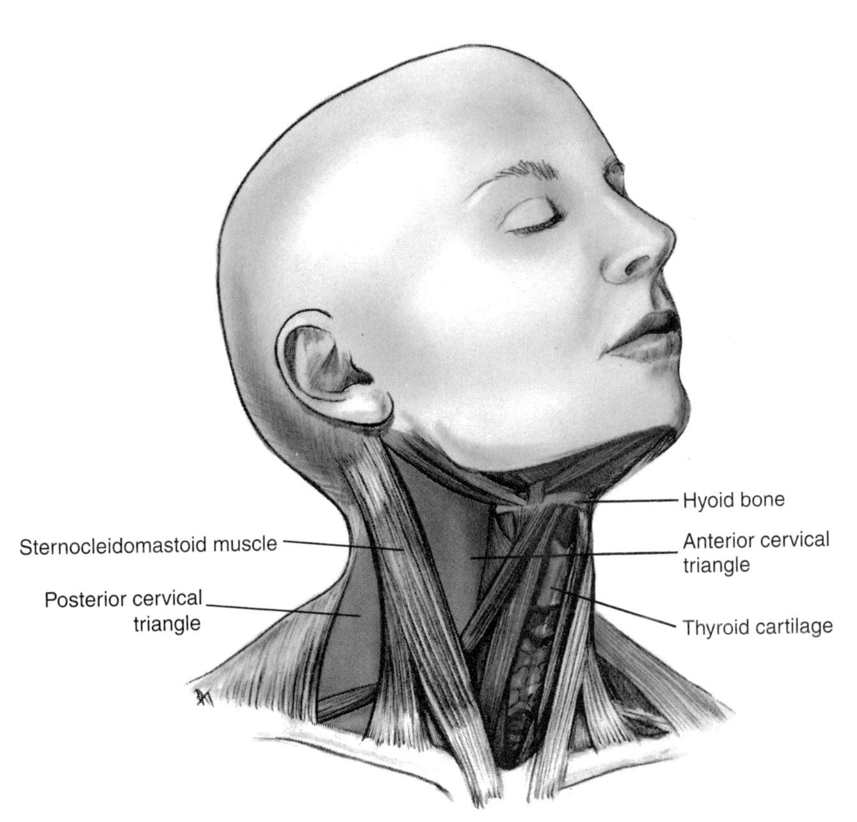

Figure 2–20
Regions of the neck: the anterior cervical triangle and posterior cervical triangle.

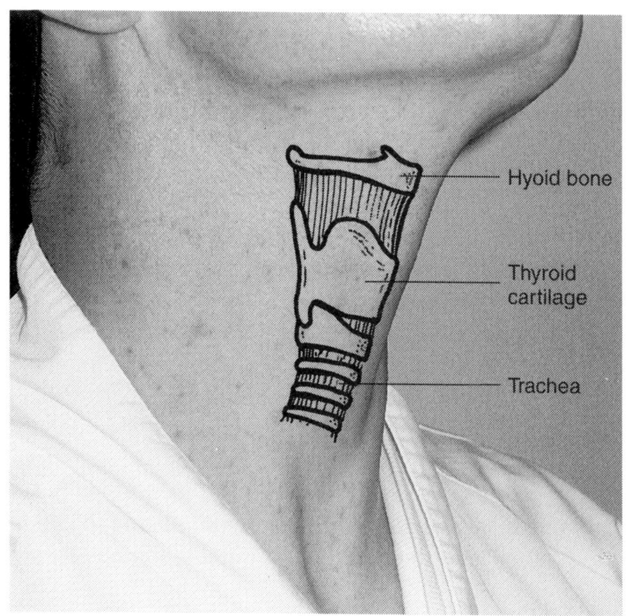

▼ Figure 2–21
Lateral view of the anterior cervical triangle of the neck highlighting its skeletal landmarks.

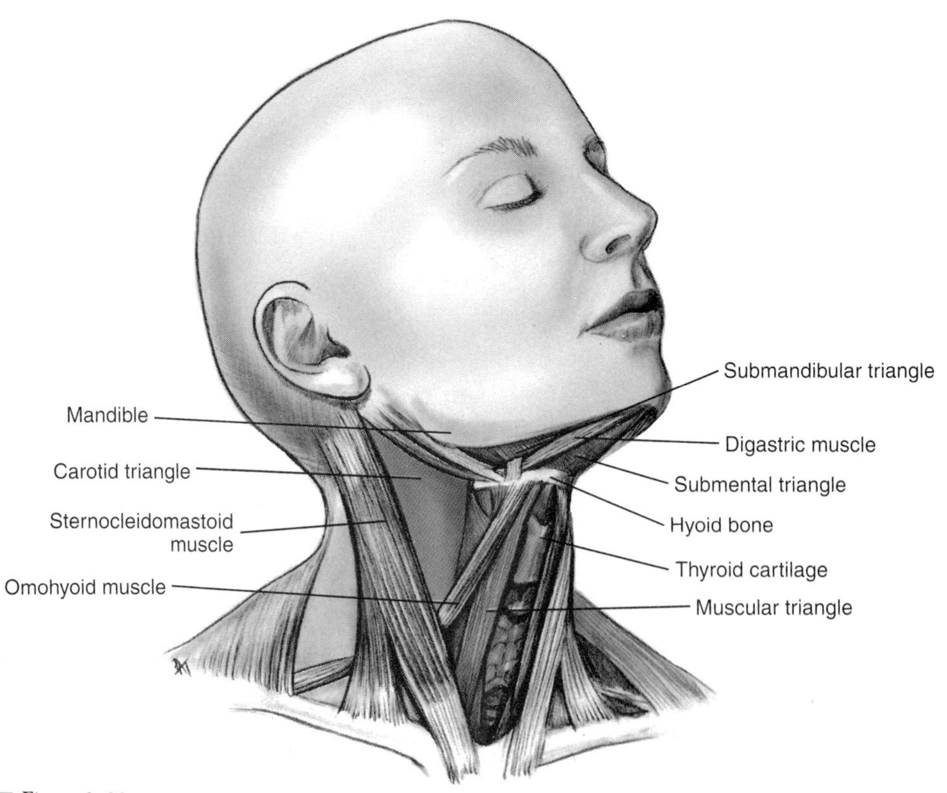

▼ Figure 2–22
Divisions of the anterior cervical triangle: submandibular, submental, carotid, and muscular.

anterior cervical triangle is further subdivided by the omohyoid muscle into a carotid triangle (kah-**rot**-id) above and a muscular triangle below. A midline **submental triangle** (sub-**men**-tal) is also formed by portions of the digastric muscle (right and left anterior belly) and the hyoid bone. These triangles are discussed further in Chapter 4.

Each posterior cervical triangle can also be further subdivided into smaller triangular regions by muscles in the area (Figure 2–23). The omohyoid muscle divides the posterior cervical triangle into the occipital triangle (ok-**sip**-it-al) above and the subclavian triangle (sub-**klay**-vee-an) below on each side. These triangles are discussed further in Chapter 4.

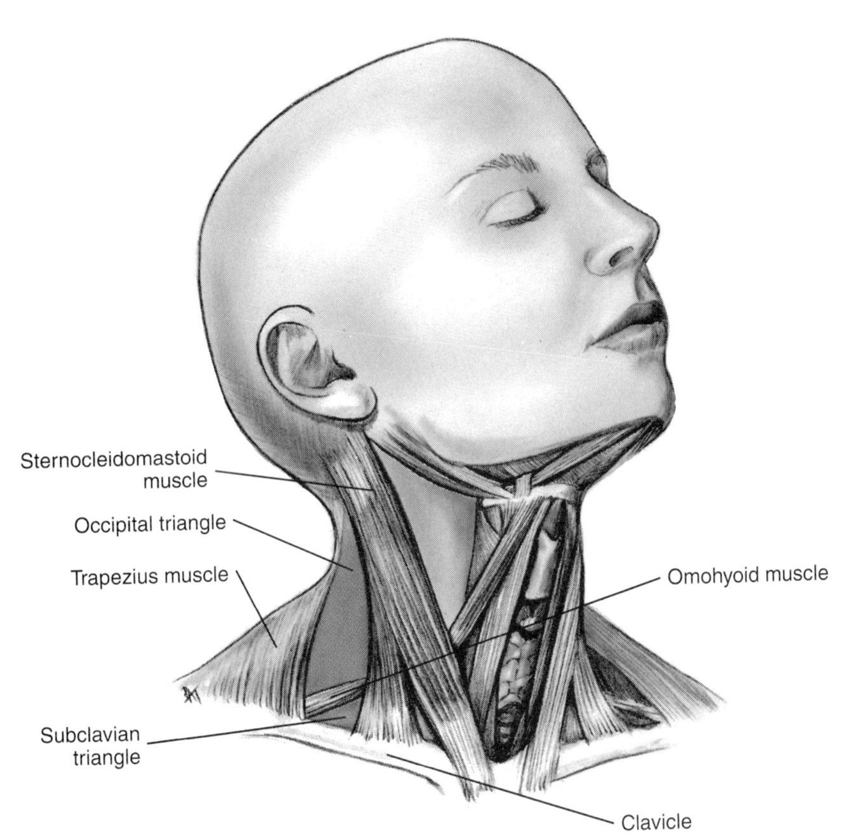

▼ **Figure 2–23**
Divisions of the posterior cervical triangle: occipital and subclavian.

REVIEW QUESTIONS

1. Those structures in the oral region that are closest to the tongue are called:
   A. Buccal
   B. Facial
   C. Lingual
   D. Pharyngeal
   E. Palatal

2. Which of the following terms is used to describe the smooth elevated area between the eyebrows in the frontal region?
   A. Supraorbital ridge
   B. Medial canthus
   C. Glabella
   D. Alae
   E. Auricle

3. In which region of the head and neck is the tragus located?
   A. Frontal region
   B. Nasal region
   C. Temporal region
   D. Anterior cervical triangle
   E. Submandibular triangle

4. The tissue at the junction between the labial or buccal mucosa and the alveolar mucosa is the:
   A. Mucogingival junction
   B. Mucobuccal fold
   C. Vermilion border
   D. Labial frenum
   E. Labiomental groove

5. Into which cervical triangles does the sternocleidomastoid muscle divide the neck?
   A. Superior and inferior
   B. Medial and lateral
   C. Anterior and posterior
   D. Proximal and distal

Chapter 3

Bones

▼ *This chapter discusses the following topics:*

I. Skeletal System
   - A. Bony Prominences
   - B. Bony Depressions
   - C. Bony Openings
   - D. Skeletal Articulations

II. Bones of the Head and Neck
   - A. Skull
   - B. Cranial Bones
   - C. Facial Bones
   - D. Paranasal Sinuses
   - E. Fossae of the Skull
   - F. Bones of the Neck

III. Abnormalities of Bone

▼ *Key words*

Arch (arch) Prominent bridgelike bony structure.

Articulation (ar-tik-you-**lay**-shin) Area where the bones are joined to each other.

Aperture (**ap**-er-cher) Opening or orifice in bone.

Bones (bones) Mineralized structures of the body that protect internal soft tissues and serve as the biomechanical basis for movement.

Canal (kah-**nal**) Opening in bone that is long, narrow, and tubelike.

Condyle (**kon**-dyl) Oval bony prominence typically found at articulations.

Cornu (**kor**-nu) Small hornlike prominence.

Crest (krest) Roughened border or ridge on the bone surface.

Eminence (**em**-i-nins) Tubercle or rounded elevation on a bony surface.

Epicondyle (ep-ee-**kon**-dyl) Small prominence that is located above or upon a condyle.

Fissure (**fish**-er) Opening in bone that is narrow and cleftlike.

Foramen/foramina (for-**ay**-men, for-**am**-i-nah) Short windowlike opening in bone.

Fossa/fossae (**fos**-ah, **fos**-ay) Depression on a bony surface.

Head (hed) Rounded surface projecting from a bone by a neck.

Incisura (in-si-**su**-rah) Indentation or notch at the edge of the bone.

Line (line) Straight small ridge of bone.

Meatus (me-**ate**-us) Opening or canal in the bone.

Notch (noch) Indentation at the edge of the bone.

Ostium/ostia (**os**-tee-um, **os**-tee-ah) Small opening in bone.

Perforation (per-fo-**ray**-shun) Abnormal hole in a hollow organ, such as in the wall of a sinus.

Plate (plate) Flat structure of bone.

Process (**pros**-es) General term for any prominence on a bony surface.

Spine (spine) Abrupt small prominence of bone.

Sulcus/sulci (**sul**-kus, **sul**-ky) Shallow depression or groove such as that on the bony surface or that between the tooth and the inner surface of the marginal gingiva.

Suture (**su**-cher) Generally immovable articulation in which bones are joined by fibrous tissue.

Tubercle (**too**-ber-kl) Eminence or small rounded elevation on the bony surface.

Tuberosity (too-beh-**ros**-i-tee) Large, often rough, prominence on the surface of bone.

Chapter 3

Bones

▼ ***After studying this chapter, the reader should be able to:***

1. Define and pronounce all the key words and anatomical terms in this chapter.
2. Locate and identify the bones of the head and neck and their landmarks on a diagram, a skull, and a patient.
3. Describe in detail the various parts and landmarks of the maxilla and mandible.
4. Discuss certain abnormalities of bone.
5. Correctly complete the identification exercise and review questions for this chapter.
6. Integrate the knowledge about the bones into the overall study of the head and neck anatomy.

Skeletal System

The ***bones*** (bones) of the skeletal system are mineralized structures in the body. Bones protect the internal soft tissues. Bones also serve as the biomechanical basis for movement of the body along with muscles, tendons, and ligaments. Bones are also a consideration in the spread of dental infection (see Chapter 12 for more information).

The prominences and depressions on the bony surface are landmarks for the attachments of associated muscles, tendons, and ligaments. The openings in the bone are also landmarks where various nerves and blood vessels enter or exit. Areas of the bones can also be demarcated that are not prominences or depressions such as a ***plate*** (plate), which is a flat bony structure.

▼ BONY PROMINENCES

A general term for any prominence on the bony surface is a ***process*** (**pros**-es). One specific type of prominence located on the bony surface is a ***condyle*** (**kon**-dyl), a relatively large, convex prominence usually involved in joints. A rounded surface projecting from a bone by a neck is a ***head*** (hed). Another large, often rough prominence is a ***tuberosity*** (too-beh-**ros**-i-tee). Tuberosities are typically attachment areas for muscles or tendons. An ***arch*** (arch) is shaped like a bridge with a bowlike outline. A ***cornu*** (**kor**-nu) is a hornlike prominence.

Other prominences of the bone include epicondyles, tubercles, crests, lines, and spines. These primarily serve as muscle and ligament attachments. An ***epicondyle*** (ep-ee-**kon**-dyl) is a prominence above or upon a condyle. A ***tubercle*** (**too**-ber-kl) or ***eminence*** (**em**-i-nins) is a rounded elevation on the bony surface. A ***crest*** (krest) is a prominent, often roughened border or ridge. A ***line*** (line) is a straight small ridge. An abrupt prominence of the bone that may be a blunt or sharply pointed projection is a ***spine*** (spine).

▼ BONY DEPRESSIONS

One type of depression on the bony surface is an ***incisura*** (in-si-**su**-rah) or ***notch*** (noch), an indentation at the edge of the bone. Another depression on the bony surface is a ***sulcus*** (plural, ***sulci***) (**sul**-kus, **sul**-ky), which is a shallow depression or groove that usually marks the course of an artery or nerve. A generally deeper depression on a bony surface is a ***fossa*** (plural, ***fossae***) (**fos**-ah, **fos**-ay). Fossae can be parts of joints, be attachment areas for muscles, or have other functions.

▼ BONY OPENINGS

The bone can have openings such as a foramen or canal. A ***foramen*** (plural, ***foramina***) (for-**ay**-men, for-**am**-i-nah) is a short windowlike opening in the bone. A ***canal*** (kah-**nal**) is a longer, narrow tubelike opening in the bone. A ***meatus*** (me-**ate**-us) is a type of canal. Another opening in a bone is a ***fissure*** (**fish**-er), which is a narrow cleftlike opening. A small opening, especially as an entrance into a hollow organ or canal, is an ***ostium*** (plural, ***ostia***) (**os**-tee-um, **os**-tee-ah). Another opening or orifice is an ***aperture*** (**ap**-er-cher).

▼ SKELETAL ARTICULATIONS

An ***articulation*** (ar-tik-you-**lay**-shin) is an area of the skeleton where the bones are joined to each other. An articulation of the bones can be either a movable or an immovable type of joint. A ***suture*** (**su**-cher) is the union of bones joined by fibrous tissue. Sutures appear on the dry skull as jagged lines. Sutures are considered to be generally immovable but may provide mechanical protection from the force of a blow by moving very slightly to absorb the force. Sutures are the most flexible in infants. Much of the early growth of the skull occurs at the sutural edges of the cranial bones.

Bones of the Head and Neck

The bones of the head and neck serve as a base during palpation of the soft tissues in the area. The bones also serve as markers when identifying the location of soft tissue lesions. These bones are also examined during a head and neck examination by a dental professional since they may be affected by a disease process. A dental professional must not only locate each of the head and neck bones but also recognize any abnormalities in the bony surface structure (discussed later in this chapter).

In order to recognize any bony abnormalities, the dental professional must know the normal anatomy of the bones of the head and neck. The knowledge of the normal anatomy includes locating the surface bony prominences, depressions, and articulations. It also includes knowing the openings in these bones and the blood vessels and nerves that travel through those openings. For effective study of the bones of the head and neck, it is helpful to use photographs and illustrations of these bones, as well as the skull model and palpation of a patient.

Bones also serve as landmarks when taking dental radiographs (see Chapter 1 for more information) and prior to administering a local anesthetic injection (see Chapter 9 for more information).

SKULL

There are 22 bones in the **skull** (skul) of a patient, not including the small bones of the middle ear (Table 3–1). The bones of the skull or braincase can be divided into the **cranium** (**kray**-nee-um) (the part housing the brain) with its **cranial bones** (**kray**-nee-al) and the **face** with its **facial bones.** The bones of the skull, whether facial or cranial, can be single or paired. These bones create the facial features, are involved in the temporomandibular joints, and participate with growth in the formation of dentition. Many texts use the alternative terms "neurocranium" for the cranial bones since they enclose the brain and "viscerocranium" for the facial bones.

Growth takes place in all the bones of the skull. Growth of the upper face takes place at the sutures between the maxillary bones and other bones, as well as at the bony surfaces. Growth in the lower face takes place at the bony surfaces of the mandible and at the head of its condyle. Inadequate or disproportionate growth of the upper face and mandible may leave inadequate room for the developing dentition and cause other occlusal problems. This failure of growth can be addressed by orthodontics and possibly endocrine therapy and surgery.

All the bones of the skull are immovable, except the mandible with its temporomandibular joint. The articulation of many of the bones in the skull is by sutures (Table 3–2). The skull also has a movable articulation with the bony vertebral column in the neck area.

Many of the skull bones have openings for important nerves and blood vessels of the head and neck (Table 3–3). Skull bones also have many associated processes that are involved in important structures of the face and head (Table 3–4).

It is easier to study the skull by first looking at its various views: the superior, anterior, lateral, and inferior views. Then the skull should be studied by looking at its individual bones and their various landmarks: the cranial bones, facial bones, and bones

Table 3–1

CRANIAL AND FACIAL BONES*

Cranial Bones	
Occipital bone	Single
Frontal bone	Single
Parietal bones	Paired
Temporal bones	Paired
Sphenoid bone	Single
Ethmoid bone	Single
Facial Bones	
Vomer	Single
Lacrimal bones	Paired
Nasal bones	Paired
Inferior nasal conchae	Paired
Zygomatic bones	Paired
Maxillary bones	Paired
Mandible	Single

*Note that there are single and paired bones.

Table 3–2

SUTURES OF THE SKULL AND THEIR BONY ARTICULATIONS

Suture	Bony Articulations
Coronal suture	Frontal and parietal bones
Sagittal suture	Parietal bones
Lambdoidal suture	Occipital and parietal bones
Squamosal suture	Temporal and parietal bones
Temporozygomatic suture	Zygomatic and temporal bones
Median palatine suture	Palatine bones
Transverse palatine suture	Maxillae and palatine bones

Table 3–3

BONY OPENINGS IN THE SKULL AND THEIR ASSOCIATED NERVES AND BLOOD VESSELS

Bony Opening	Location	Nerves and Vessels
Carotid canal	Temporal bone	Internal carotid artery
Cribiform plate with foramina	Ethmoid bone	Olfactory nerves
External acoustic meatus	Temporal bone	(Opening to tympanic cavity)
Foramen lacerum	Sphenoid, occipital, and temporal bones	(Cartilage)
Foramen magnum	Occipital bone	Spinal cord, vertebral arteries, and eleventh cranial nerve
Foramen ovale	Sphenoid bone	Mandibular division of the fifth cranial nerve
Foramen rotundum	Sphenoid bone	Fifth cranial nerve
Foramen spinosum	Sphenoid bone	Middle meningeal artery
Greater palatine foramen	Palatine bone	Greater palatine nerve and vessels
Hypoglossal canal	Occipital bone	Twelfth cranial nerve
Incisive foramen	Maxilla	Nasopalatine nerve and branches of the sphenopalatine artery
Inferior orbital fissure	Sphenoid bone and maxilla	Infraorbital and zygomatic nerves, infraorbital artery, and ophthalmic vein
Infraorbital foramen and canal	Maxilla	Infraorbital nerve and vessels
Internal acoustic meatus	Temporal bone	Seventh and eighth cranial nerves
Jugular foramen	Occipital and temporal bones	Internal jugular vein and ninth, tenth, and eleventh cranial nerves
Lesser palatine foramen	Palatine bone	Lesser palatine nerve and vessels
Mandibular foramen	Mandible	Inferior alveolar nerve and vessels
Mental foramen	Mandible	Mental nerve and vessels
Optic canal and foramen	Sphenoid bone	Optic nerve and ophthalmic artery
Petrotympanic fissure	Temporal bone	Chorda tympani nerve
Pterygoid canal	Sphenoid bone	Area nerves and vessels
Stylomastoid foramen	Temporal bone	Seventh cranial nerve
Superior orbital fissure	Sphenoid bone	Third, fourth, and sixth cranial nerves and ophthalmic nerve and vein

▼ Table 3–4

PROCESSES AND ASSOCIATED STRUCTURES OF THE SKULL BONES

Processes of the Skull	Skull Bones	Associated Structures
Alveolar process	Mandible	Contains roots of mandibular teeth
Alveolar process	Maxilla	Contains roots of maxillary teeth
Coronoid process	Mandible	Portion of ramus
Frontal process	Maxilla	Forms medial infraorbital rim
Frontal process	Zygomatic bone	Forms anterior lateral orbital wall
Greater wing	Sphenoid bone	Anterior process to sphenoid bone body
Lesser wing	Sphenoid bone	Posterolateral process to sphenoid bone body
Mastoid process	Temporal bone	Composed of mastoid air cells
Maxillary process	Zygomatic bone	Forms infraorbital rim and portion of anterior lateral orbital wall
Palatine process	Maxilla	Forms anterior hard palate
Postglenoid process	Temporal bone	Posterior to temporomandibular joint
Pterygoid process	Sphenoid bone	Consists of medial and lateral pterygoid plates
Styloid process	Temporal bone	Serves as attachment for muscles and ligaments
Temporal process	Zygomatic bone	Portion of zygomatic arch
Zygomatic process	Frontal bone	Lateral to orbit
Zygomatic process	Maxilla	Forms lateral portion of infraorbital rim
Zygomatic process	Temporal bone	Portion of zygomatic arch

of the neck. The features of the skull bones such as the fossae and paranasal sinuses should also be studied. This chapter follows this format of study.

▼ Superior View of the Skull

When the skull is viewed from above, four bones are visible (Figure 3–1). At the front of the skull is the single **frontal bone** (**frunt**-il). At the sides are the paired **parietal bones** (pah-**ri**-it-al). At the back of the skull is the single **occipital bone** (ok-**sip**-it-al).

▼ SUTURES ON SUPERIOR VIEW

Three sutures among these four skull bones are also visible on the superior view (Figure 3–2, Table 3–2). The suture extending across the skull, between the frontal and parietal bones, is the **coronal suture** (kor-**oh**-nahl). A second suture, the **sagittal suture** (**saj**-i-tel), extends from the front to the back of the skull, between the paired parietal bones. The third suture, located between the single occipital bone and the paired parietal bones, is the **lambdoidal suture** (lam-**doid**-al).

▼ Anterior View of the Skull

When the skull is viewed from the front, certain bones of the skull (or portions of these bones) are visible (Figure 3–3). These bones include the single frontal, ethmoid, vomer, and sphenoid bones and the mandible, and also the paired lacrimal, nasal, inferior nasal conchal, zygomatic, and maxillary bones.

▼ FACIAL BONES ON ANTERIOR VIEW

The **facial bones** visible on the anterior view of the skull include the **lacrimal bone** (**lak**-ri-mal), **nasal bone** (**nay**zil), **vomer** (**vo**-mer), **inferior nasal concha** (**nay**-zil **kong**-kah), **zygomatic bone** (zy-go-**mat**-ik), **maxilla** (mak-**sil**-ah), and **mandible** (**man**-di-bl) (Figure 3–4). The **palatine bones** (**pal**-ah-tine) are not visible on this view and are not strictly considered facial bones, but for

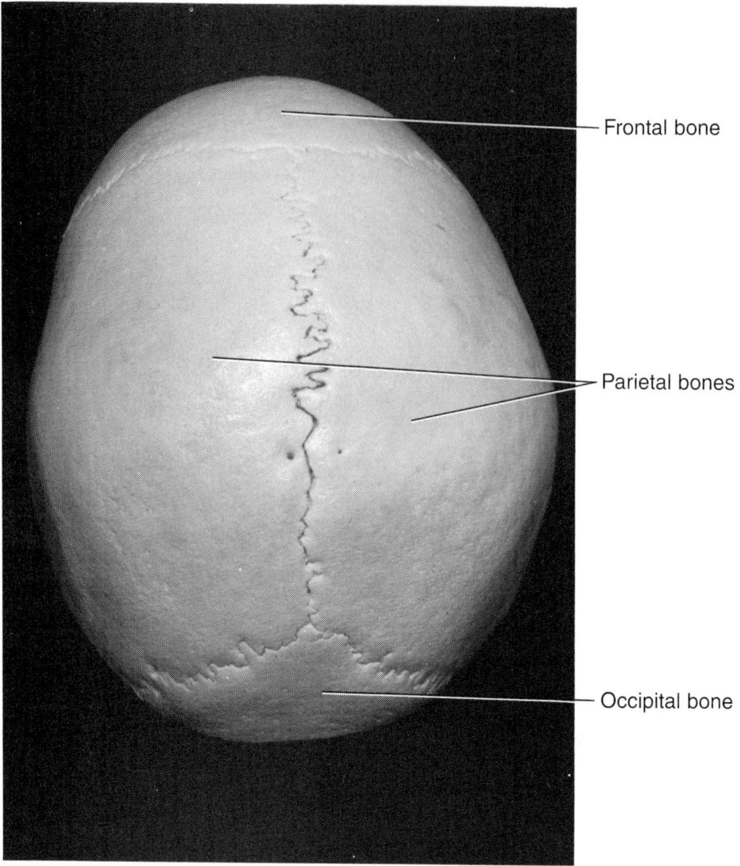

▼ **Figure 3–1**
Superior view of the bones of the skull.

ease of learning are included under the heading of facial bones. This main division of the bones of the skull is discussed further in this chapter.

▼ ORBIT AND ASSOCIATED STRUCTURES

The **orbit** (**or**-bit) or eye cavity, which contains and protects the eyeballs, is a prominent feature of the anterior view (Figure 3–5). Many bones of the skull form the walls and apex of the orbits (Table 3–5). The larger **orbital walls** (**or**-bit-al) are composed of the orbital plates of the frontal bone (making the roof or superior wall), ethmoid bone (forming the greatest portion of the medial wall), and lacrimal bone (at the anterior medial corner of the orbit) and the orbital surfaces of the maxilla (floor or inferior wall) and zygomatic bone (anterior part of the lateral wall). The orbital surface of the greater wing of the sphenoid bone is also included (the posterior part of the lateral wall).

The **orbital apex** (**or**-bit-al) or the deepest portion of the orbit is composed of the lesser wing of the sphenoid bone (forming the base) and the palatine bone (a small inferior portion) (Figure 3–6, Table 3–5). The round opening in the orbital apex is the **optic canal** (**op**-tik), which lies between the two roots of the

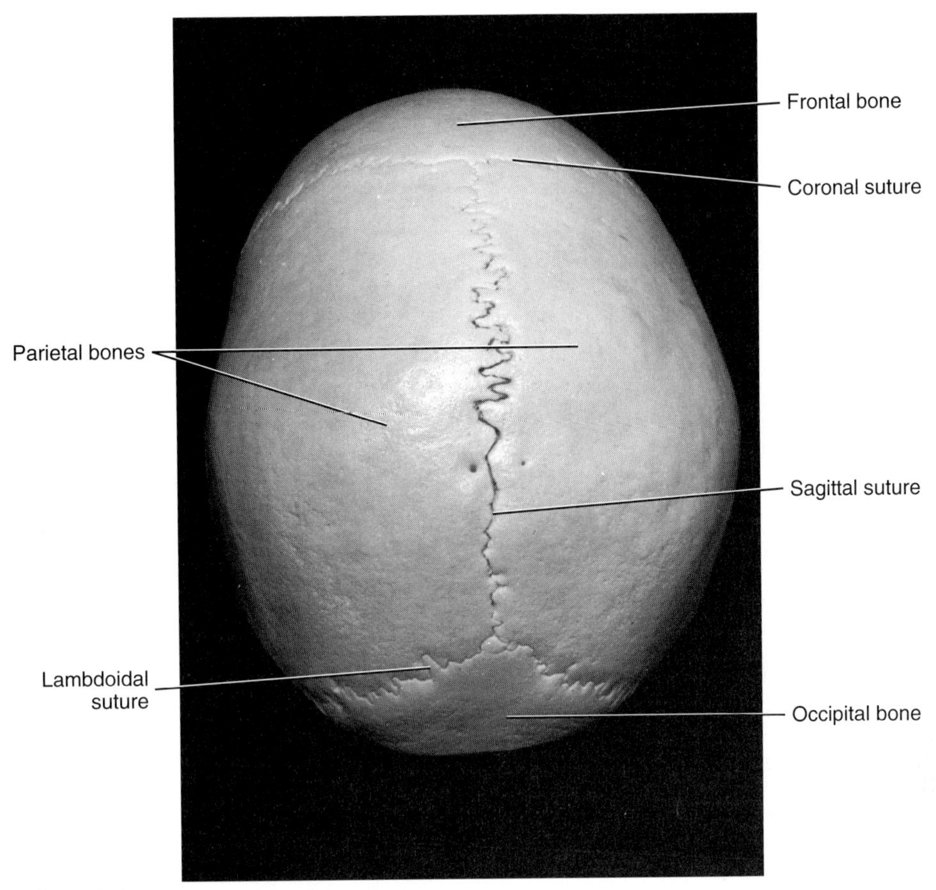

Figure 3–2
Superior view of the skull and its suture lines.

lesser wing of the sphenoid bone (Table 3–3). The optic nerve passes through the optic canal to reach the eyeball. The ophthalmic artery also extends through the canal to reach the eye.

There are two orbital fissures noted on the anterior aspect: the superior orbital fissure and inferior orbital fissure (Figure 3–7, Table 3–3). Lateral to the optic canal is the curved and slitlike **superior orbital fissure** (**or**-bit-al), between the greater and lesser wings of the sphenoid bone. Like the optic canal, the superior orbital fissure connects the orbit with the cranial cavity. The oculomotor or third cranial nerve, the trochlear or fourth cranial nerve, the abducens or sixth cranial nerve, and the ophthalmic nerve (from the trigeminal or fifth cranial nerve) and vein travel through this fissure.

The **inferior orbital fissure** (**or**-bit-al) can also be seen between the greater wing of the sphenoid bone and the maxilla. The inferior orbital fissure connects the orbit with the infratemporal and pterygopalatine fossae (discussed later in this chapter). The infraorbital and zygomatic nerves, branches of the maxillary nerve, and infraorbital artery enter the orbit through this fissure. The inferior ophthalmic vein travels through this fissure to join the pterygoid plexus of veins.

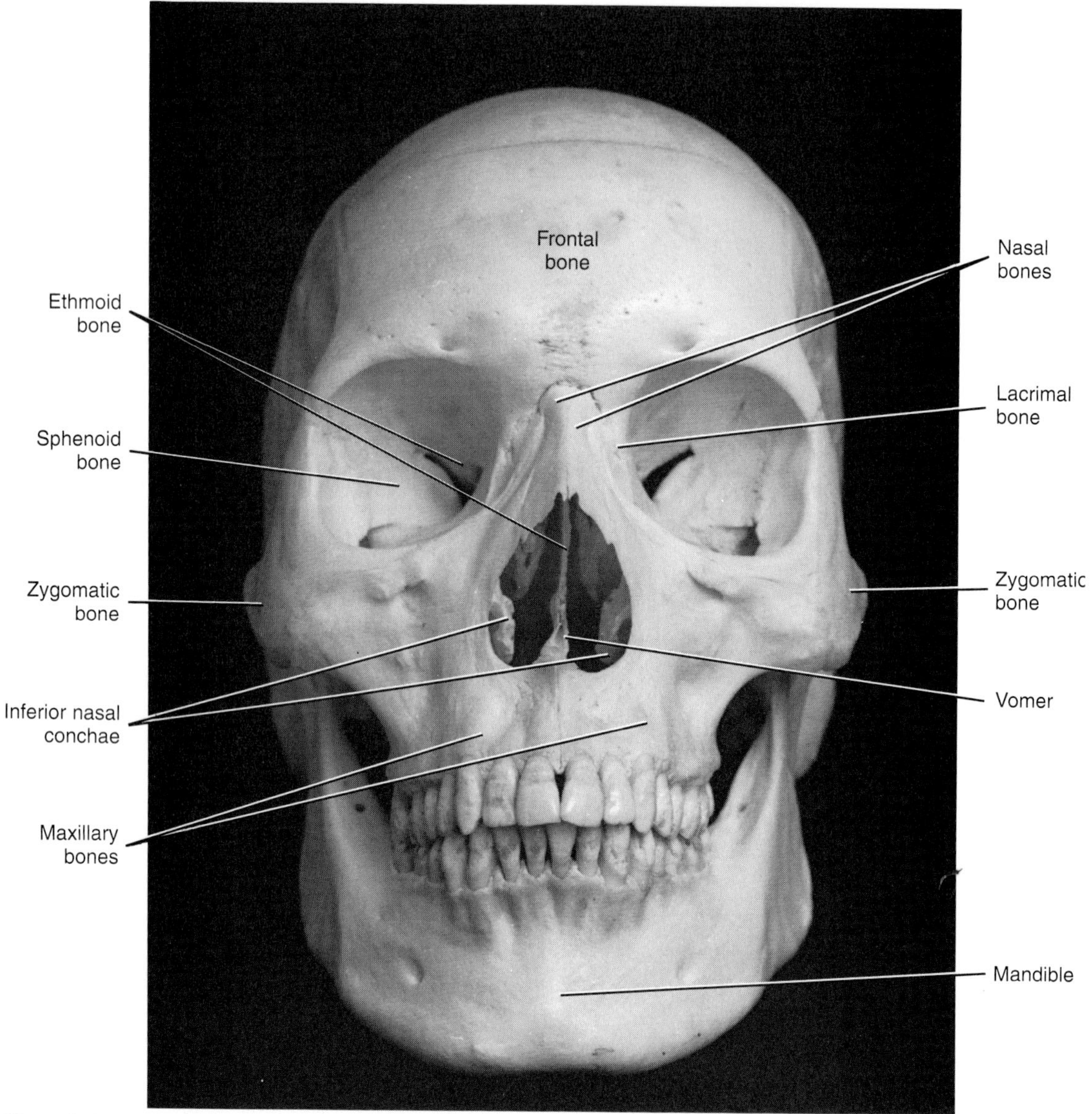

Figure 3–3
Anterior view of the bones of the skull.

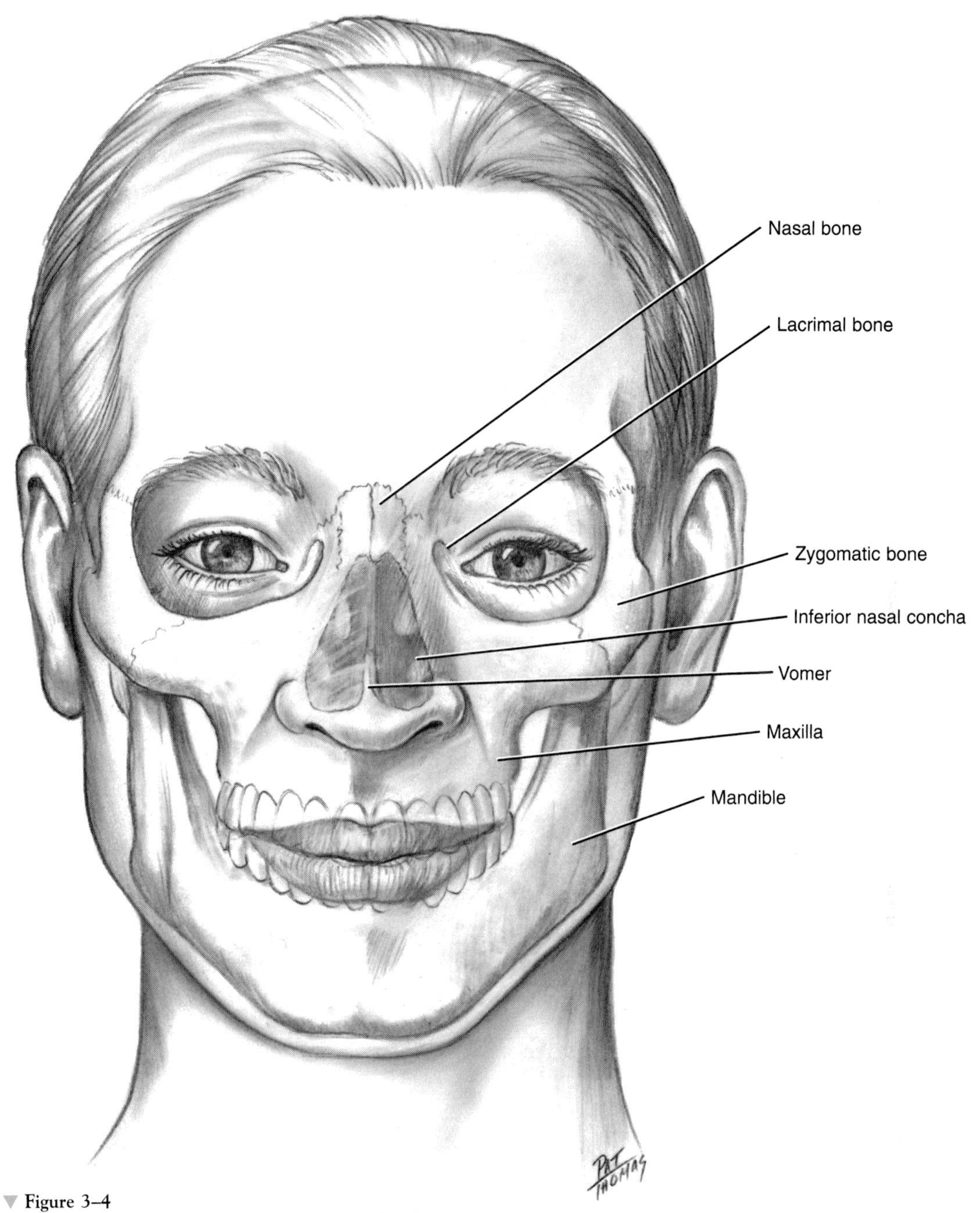

Figure 3–4
Anterior view of the facial bones and overlying facial tissues.

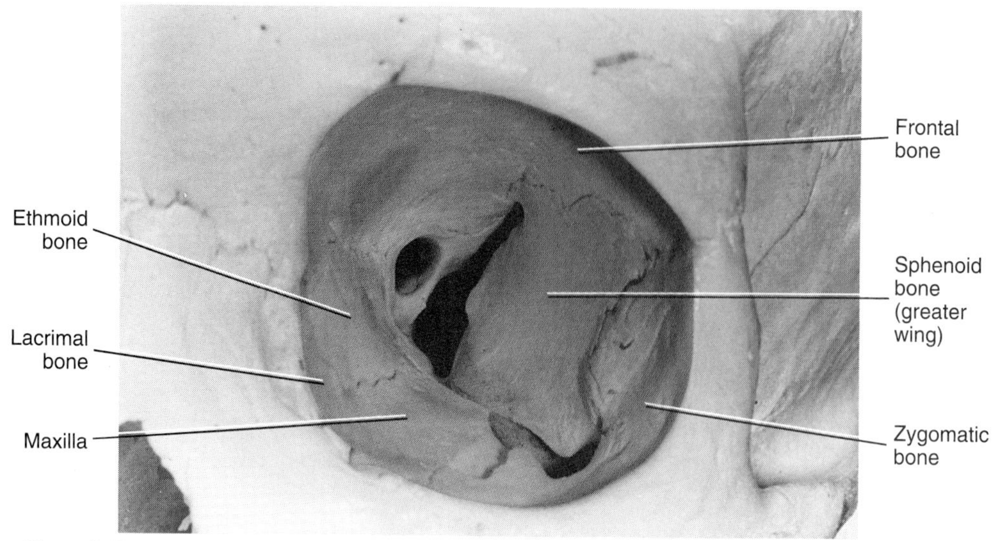

Figure 3–5
Anterior view of the left orbit of the skull and its walls.

NASAL CAVITY AND ASSOCIATED STRUCTURES

The **nasal cavity** (**nay**-zil **kav**-it-ee) can also be viewed from the anterior aspect (Figure 3–8). The **nasion** (**nay**-ze-on), a midpoint landmark, is located at the junction of the frontal and nasal bones. The anterior opening of the nasal cavity, the **piriform aperture** (**pir**-i-form), is large and triangular. The **bridge of the nose** is formed from the paired nasal bones. The lateral boundaries of the nasal cavity are formed by the maxillae.

Table 3–5

BONES OF THE SKULL THAT FORM THE ORBIT	
Portion of Orbit	**Skull Bones**
Roof or superior wall	Frontal bone
Medial wall	Ethmoid and lacrimal bones
Lateral wall	Zygomatic and sphenoid bones
Apex or base	Sphenoid and palatine bones

Each lateral wall of the nasal cavity has three projecting structures that extend inward from the maxilla, which are called the **nasal conchae** (**nay**-zil **kong**-kay) (Figure 3–8). These are the superior, middle, and inferior nasal conchae. Each extends scroll-like into the nasal cavity. The superior nasal concha and middle nasal concha are formed from the ethmoid bone. The inferior nasal concha is a separate facial bone. Beneath each concha is a groove known as a **nasal meatus** (**nay**-zil). Each of these meatus has openings through which the paranasal sinuses or nasolacrimal duct communicates with the nasal cavity.

The vertical partition of the nasal cavity, the **nasal septum** (**nay**-zil **sep**-tum), divides the nasal cavity into two portions (Figure 3–9). Anteriorly, the nasal septum is formed by both the nasal septal cartilage inferiorly and the perpendicular plate of the ethmoid bone superiorly. The posterior portions of the nasal septum are formed by the vomer.

Lateral View of the Skull

When viewed from the side, the skull shows both cranial bones and facial bones. A division between the

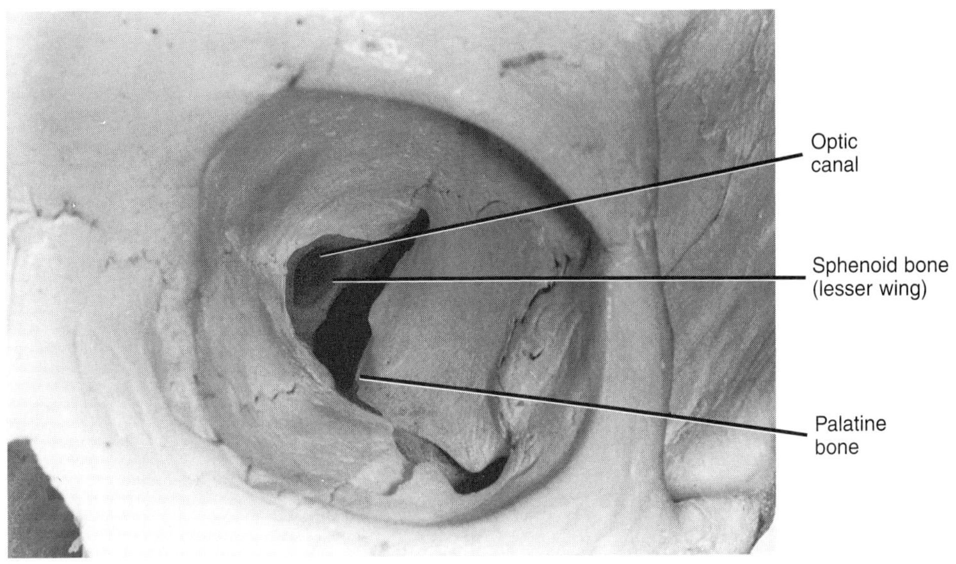

▼ **Figure 3–6**
Anterior view of the left orbit and the orbital apex.

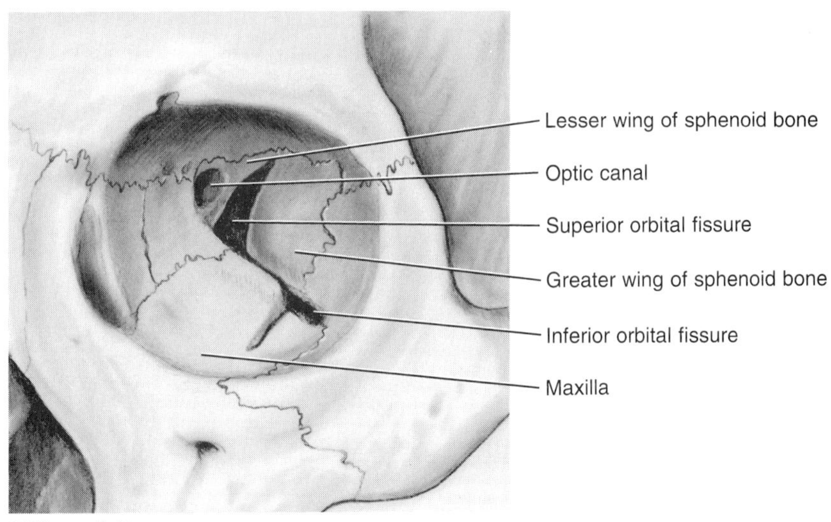

▼ **Figure 3–7**
Anterior view of the left orbit and the orbital fissures.

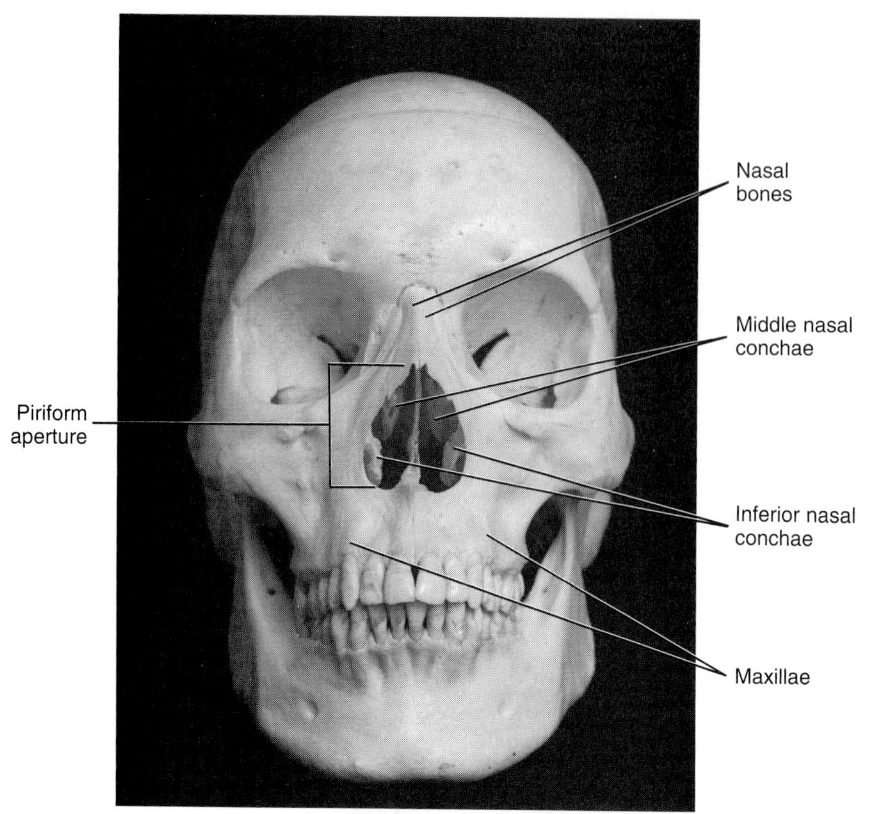

Figure 3–8
Anterior view of the skull and the nasal cavity.

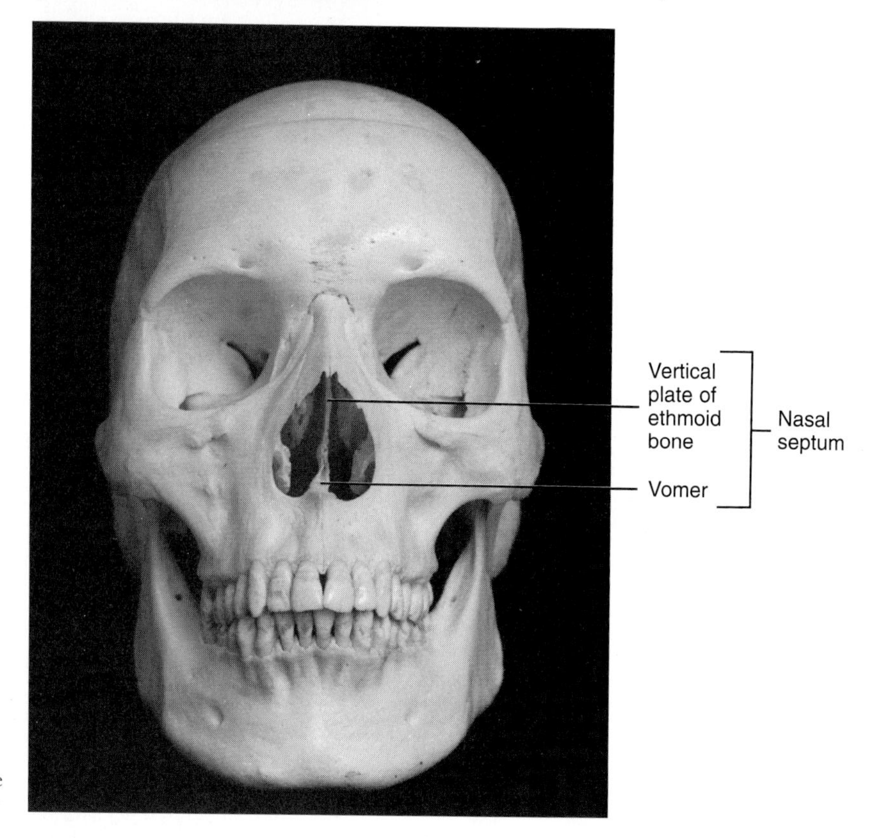

Figure 3–9
Anterior view of the skull and the nasal septum.

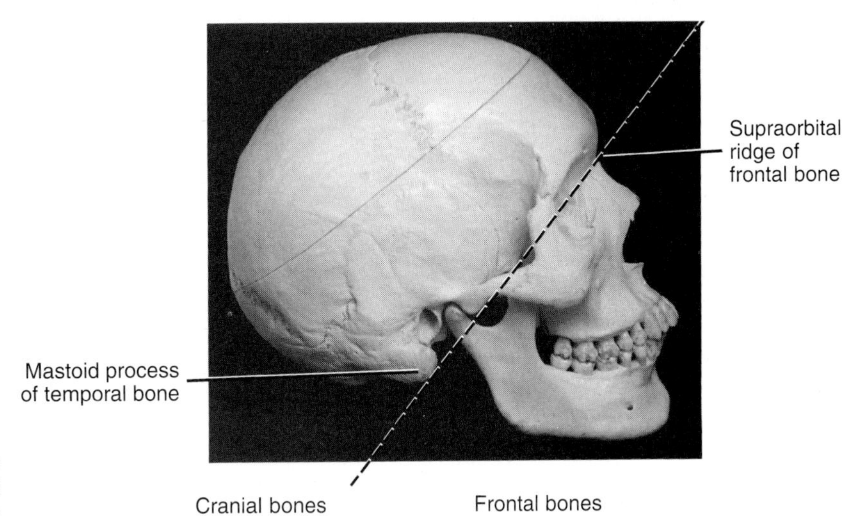

▼ **Figure 3–10**
Lateral view of the skull and an imaginary diagonal line separating the cranial bones and facial bones.

cranial bones and **facial bones** can be reinforced by making an imaginary diagonal line that passes downward and backward from the supraorbital ridge of the frontal bone to the tip of the mastoid process of the temporal bone (Figure 3–10). These two main divisions of the bones of the skull are discussed further in this chapter.

On the lateral surface of the skull are two separate parallel ridges or **temporal lines** (**tem**-poh-ral), crossing both the frontal and parietal bones (Figure 3–11). The superior ridge is the superior temporal line. The inferior ridge or inferior temporal line is the superior boundary of the temporal fossa and where the fan-shaped temporalis muscle attaches.

▼ CRANIAL BONES ON LATERAL VIEW

From the lateral view, the **cranium** (**kray**-nee-um) is easily seen, which includes the **cranial bones:** the **occipital** (ok-**sip**-it-al), **frontal** (**frunt**-il), **parietal** (pah-**ri**-it-al), **temporal** (**tem**-poh-ral), **sphenoid** (**sfe**-noid), and **ethmoid** bones (**eth**-moid). This main division of the bones of the skull is discussed further in this chapter.

Also present on the lateral view of the cranium are the associated sutures (Figure 3–12, Table 3–2). These sutures include the **coronal suture** (kor-**oh**-nahl), an articulation between the frontal and parietal bones, and the **lambdoidal suture** (lam-**doid**-al), an articulation

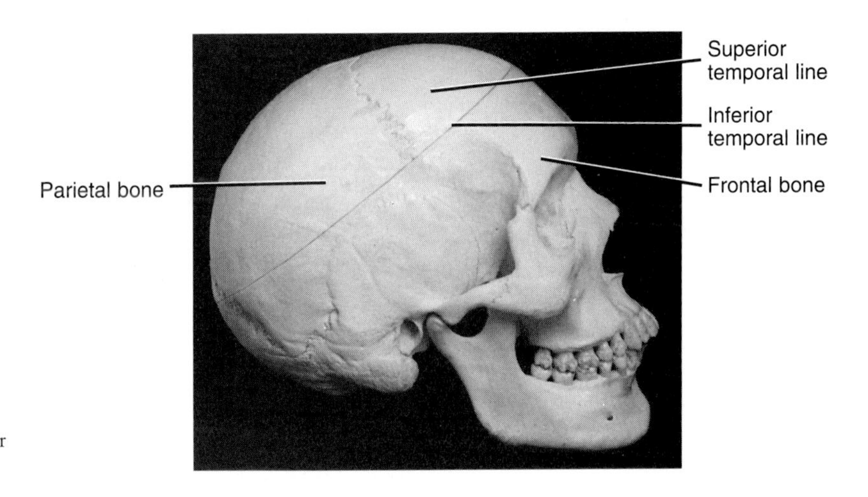

▼ **Figure 3–11**
Lateral view of the skull and superior and inferior temporal lines.

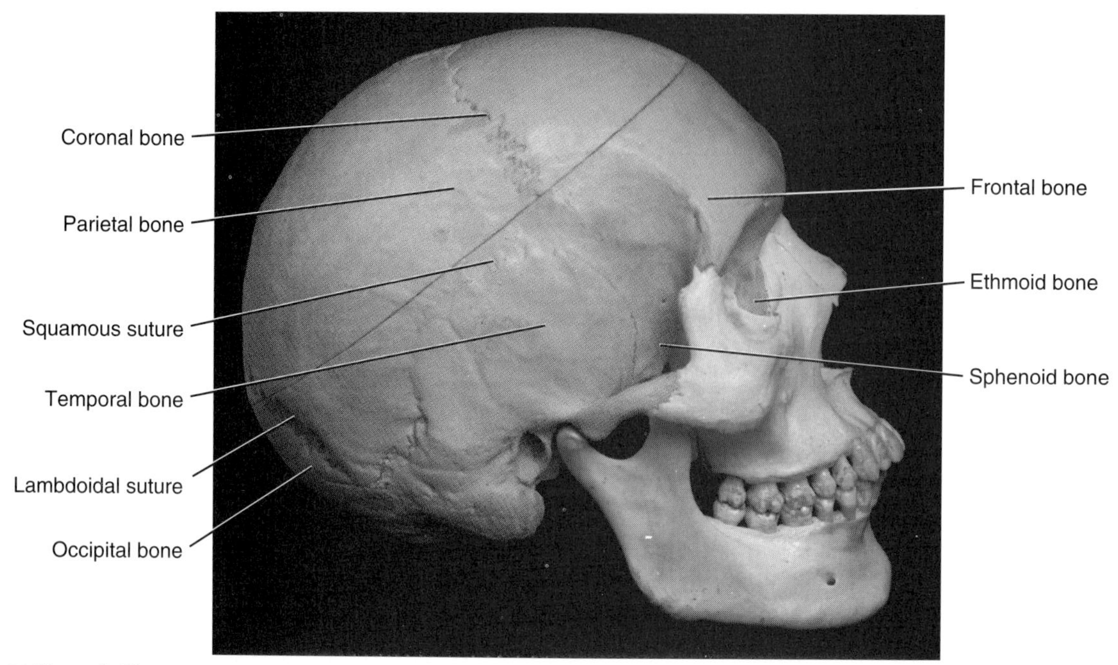

▼ **Figure 3–12**
Lateral view of the cranial bones and their suture lines.

between the parietal and occipital bones. Also present is the arched **squamosal suture** (**skway**-mus-al), between the temporal and parietal bones.

▼ FOSSAE ON LATERAL VIEW

The **temporal fossa** (**tem**-poh-ral) is easily seen on the lateral aspect of the skull (Figure 3–13). The temporal fossa is formed by several bones of the skull and contains the body of the temporalis muscle. Inferior to the temporal fossa is the **infratemporal fossa** (in-frah-**tem**-poh-ral). Deep to the infratemporal fossa and harder to see is the **pterygopalatine fossa** (**teh**-ri-go-**pal**-ah-tine). The temporal, infratemporal, and pterygopalatine fossae are discussed later in this chapter and contain many important head and neck structures.

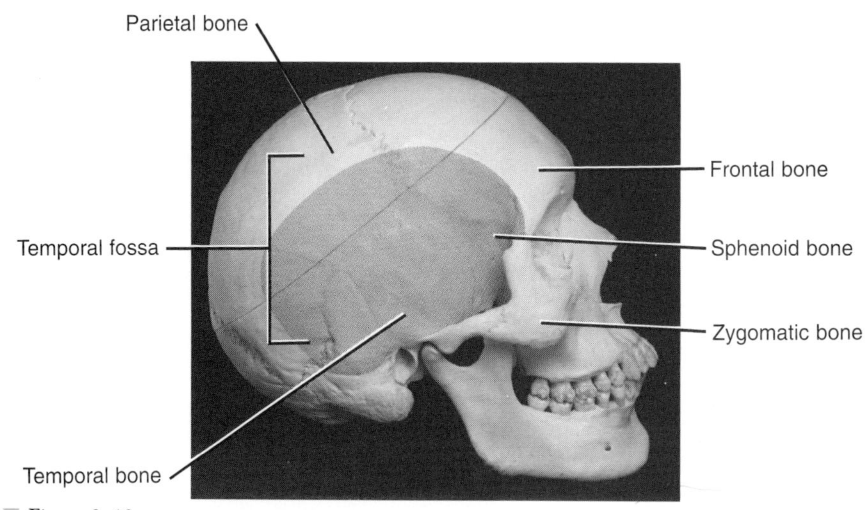

▼ **Figure 3–13**
Lateral view of the skull and temporal fossa.

▼ ZYGOMATIC ARCH AND TEMPOROMANDIBULAR JOINT ON LATERAL VIEW

Farther inferior on the lateral aspect are many important bony landmarks (Figure 3–14). The **zygomatic arch** (zi-go-**mat**-ik) or cheek bone is visible, formed by the union of the broad temporal process of the zygomatic bone and the slender zygomatic process of the temporal bone (Table 3–4). The suture between these two bones is the **temporozygomatic suture** (tem-por-oh-zi-go-**mat**-ik). The zygomatic arch serves as the origin for the masseter muscle.

The **temporomandibular joint** (tem-poh-ro-man-**dib**-you-lar) is also noted and is a movable articulation between the temporal bone and the mandible. The specifics of this joint are discussed in Chapter 5. Further landmarks of these bones are covered later in this chapter.

▼ Inferior View of the External Surface of the Skull

Most of the structures of the inferior aspect of the skull surface are more easily viewed on the skull model if the mandible is temporarily removed. The **maxillary, zygomatic, vomer, temporal, sphenoid, occipital,** and **palatine bones** are visible on this inferior view of the skull's external surface (Figure 3–15).

▼ HARD PALATE AND ASSOCIATED STRUCTURES

At the anterior portion of the skull's inferior aspect is the **hard palate** (**pal**-it), bordered by the **alveolar process of the maxilla** (al-**ve**-o-lar, mak-**sil**-ah) with its **maxillary teeth** (**mak**-sil-lare-ee) (Figure 3–16, Table 3–4). The hard palate is formed by the two palatine processes of the maxillae and the two horizontal plates of the palatine bones.

Two prominent sutures are present on the hard palate (Table 3–2). One suture is the **median palatine suture** (**pal**-ah-tine), a midline articulation between the two palatine processes of the maxillae anteriorly and the two horizontal plates of the palatine bones posteriorly. The other suture is the **transverse palatine suture** (**pal**-ah-tine), an articulation between the two palatine processes of the maxillae and the two horizontal plates of the palatine bones.

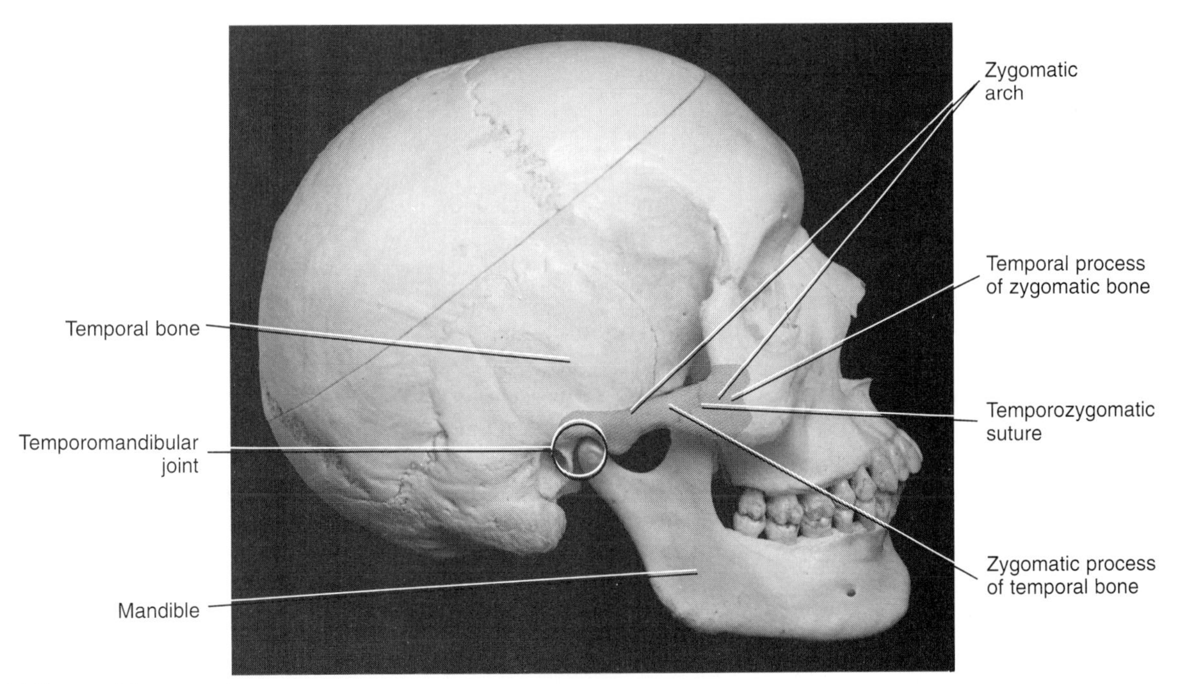

▼ **Figure 3–14**
Lateral view of the skull showing the zygomatic arch and the temporomandibular joint.

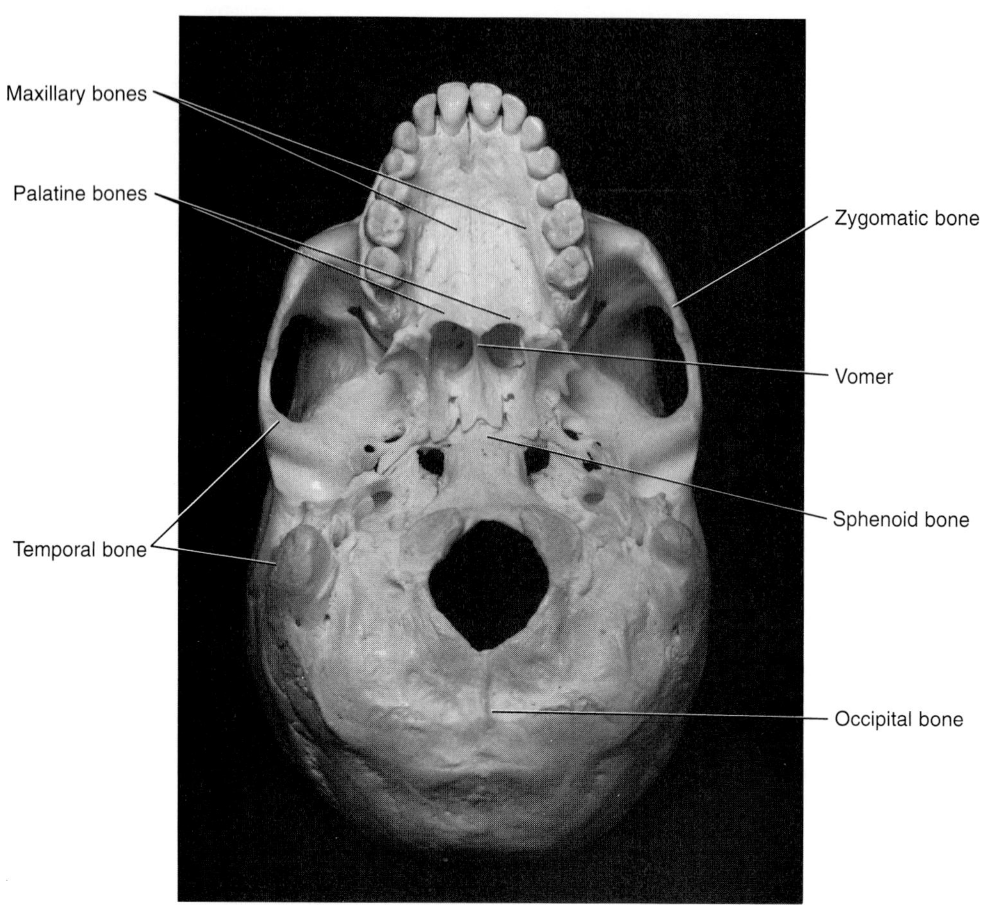

▼ **Figure 3–15**
Inferior view of the external surface of the skull.

The hard palate forms the floor of the nasal cavity, as well as the roof of the mouth (Figure 3–16). The posterior edge of the hard palate forms the inferior border of two funnel-shaped cavities, the **posterior nasal apertures** (**nay**-zil) or choanae (ko-**a**-nay). The superior border of each aperture is formed by the vomer and the sphenoid bone. The posterior edge of the vomer forms the medial border of the posterior nasal apertures. The posterior nasal apertures are the posterior openings of the nasal cavity.

Near the superior border of each posterior nasal aperture is a small canal, the **pterygoid canal** (**teh**-ri-goid) (Figure 3–16, Table 3–3). The pterygoid canal extends to open into the **pterygopalatine fossa** (**teh**-ri-go-**pal**-ah-tine) and carries the pterygoid nerve and blood vessels (the fossa is discussed later in this chapter).

▼ MIDDLE PORTION OF THE EXTERNAL SKULL SURFACE

The middle portion of the inferior aspect of the external skull surface has many important surface prominences and depressions on the **sphenoid bone** (Figure 3–17). The lateral borders of the posterior nasal apertures are formed on each side by the **pterygoid process** (**teh**-ri-goid) of the sphenoid bone (Table 3–4).

Each pterygoid process consists of a thin **medial pterygoid plate** (**teh**-ri-goid) and a flattened **lateral pterygoid plate.** The depression between the medial and lateral plates is called the **pterygoid fossa.** At the inferior portion of the medial plate of the pterygoid process is a thin curved process, the **hamulus** (**ha**-mu-lis). The sphenoid bone and its landmarks are covered later in this chapter.

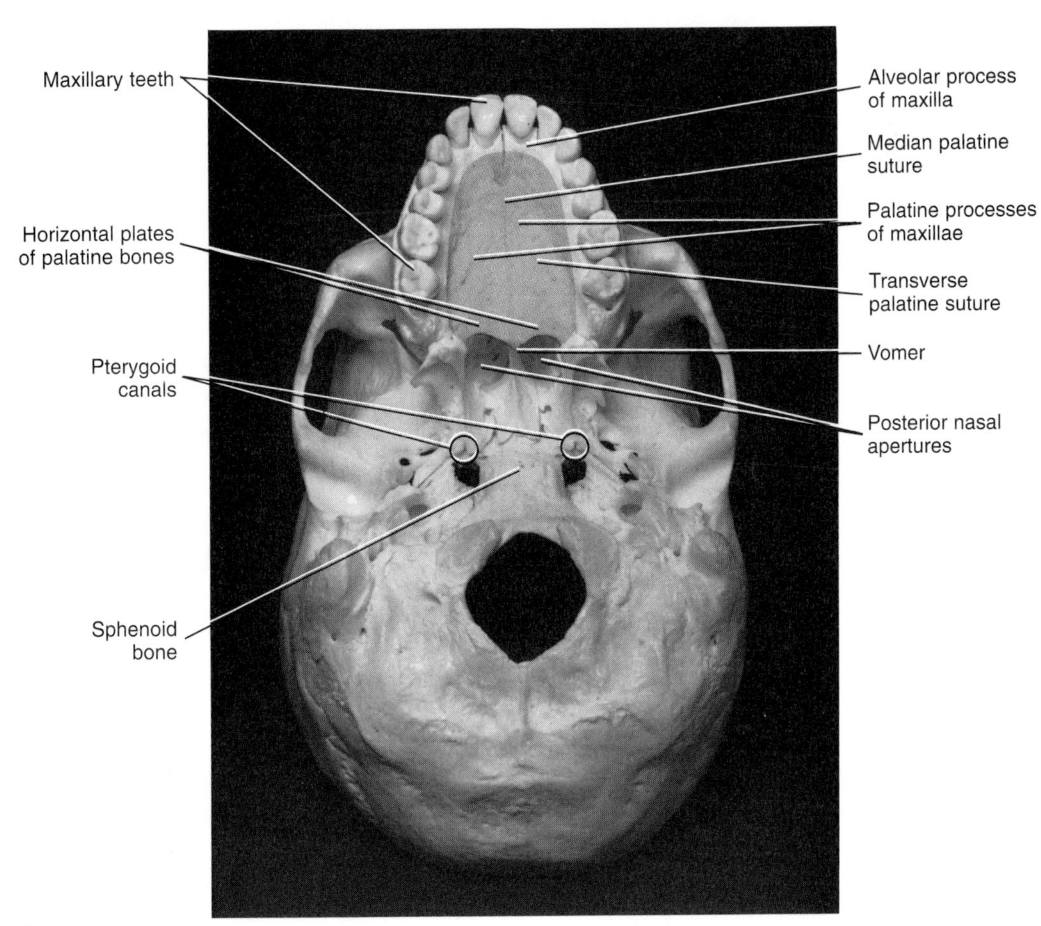

Figure 3–16
External surface of the skull with the hard palate highlighted.

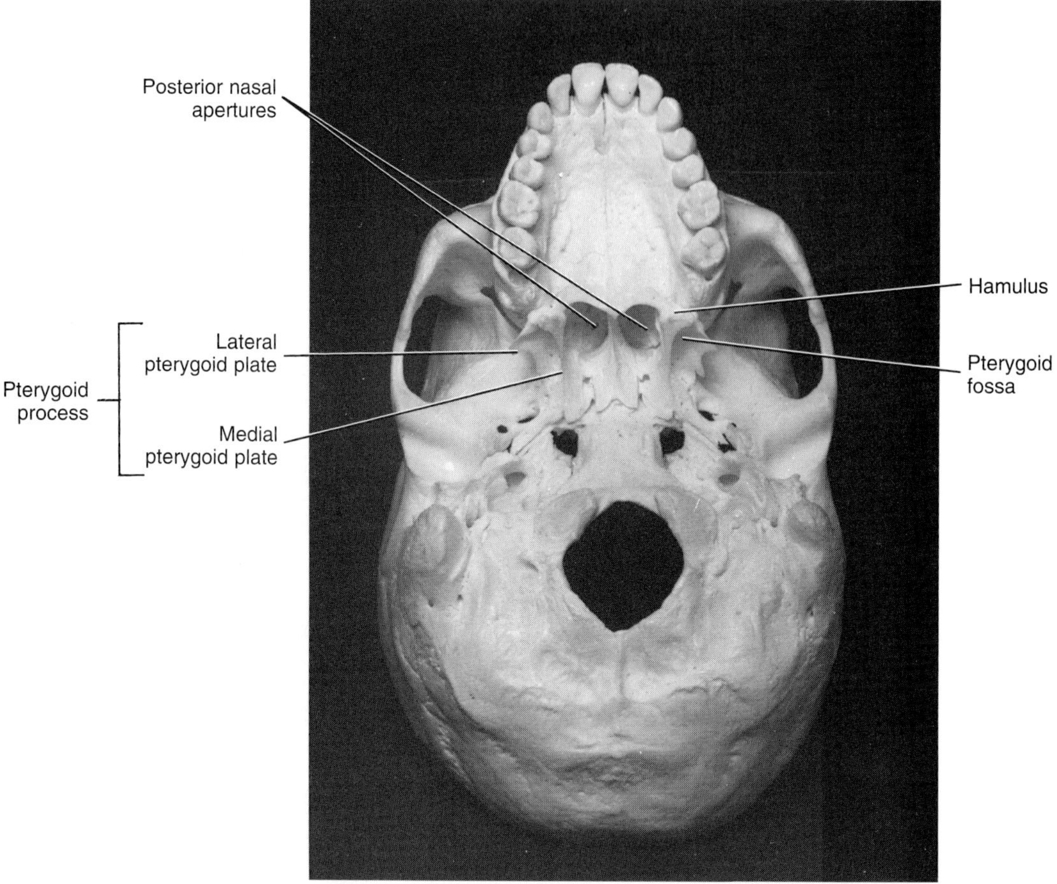

▼ **Figure 3–17**
Structures on the middle portion of the external skull surface.

▼ FORAMINA OF THE EXTERNAL SKULL SURFACE

The underside of the skull has a large number of foramina (Figure 3–18, Table 3–3). These openings provide entrance and exit for the arteries and veins that supply the brain and facial tissues (see Chapter 6 for more information on the blood vessels and associated foramina and canals). They are also the way the cranial nerves pass to and from the brain (see Chapter 8 for more information on the cranial nerves and their associated foramina and canals).

The larger anterior oval opening on the sphenoid bone is the **foramen ovale** (**ova**-lee), for the mandibular division of the trigeminal or fifth cranial nerve. The smaller and more posterior opening is the **foramen spinosum** (**spine**-o-sum), which carries the middle meningeal artery into the cranial cavity. The foramen spinosum receives its name from the nearby **spine of the sphenoid bone,** which is at the posterior extremity of the sphenoid bone.

Also on the external surface of the skull is the large irregularly shaped **foramen lacerum** (lah-**ser**-um), which in life is filled with cartilage (Figure 3–18). Posterolateral to the foramen lacerum is a round opening in the petrous portion of the temporal bone, the **carotid canal** (kah-**rot**-id). The carotid canal carries the internal carotid artery and sympathetic carotid plexus. A pointed bony projection, the **styloid process** (**sty**-loid), is visible lateral and posterior to the carotid canal (Table 3–4). Immediately posterior to the styloid process is the **stylomastoid foramen** (sty-lo-**mas**-toid), an opening through which the facial or seventh cranial nerve exits from the skull to the face.

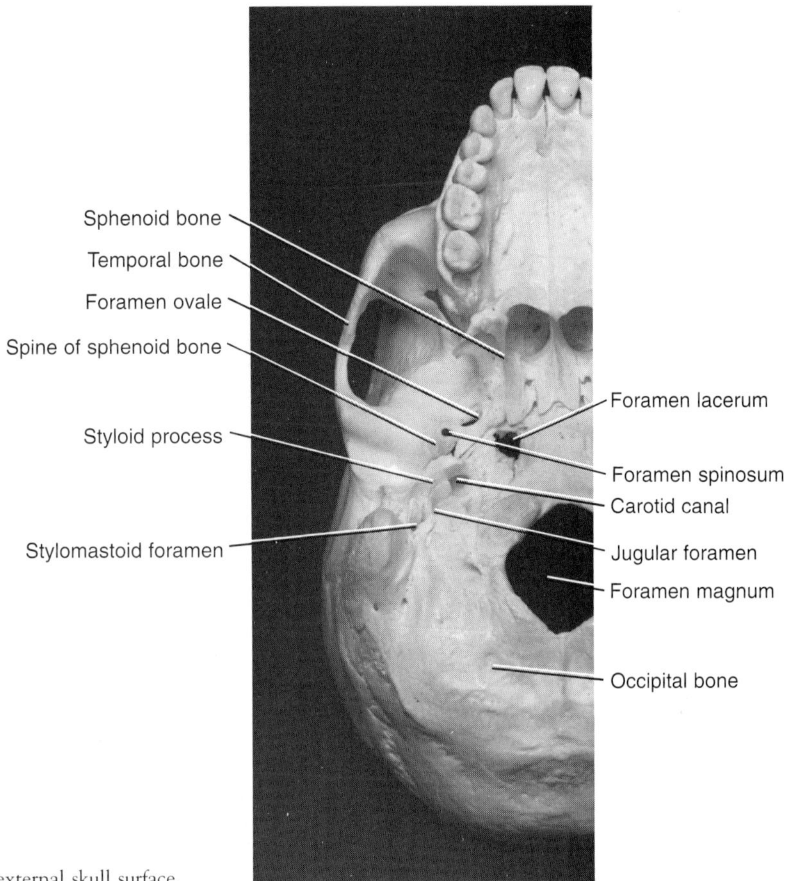

▼ **Figure 3–18**
Foramina and associated structures of the external skull surface.

The **jugular foramen** (**jug**-you-lar), just medial to the styloid process, is more easily seen if the skull model is tilted to one side. The jugular foramen is the opening through which pass the internal jugular vein and three cranial nerves: the glossopharyngeal or ninth cranial nerve, vagus or tenth cranial nerve, and spinal accessory or eleventh cranial nerve.

The largest opening on the inferior view is the **foramen magnum** (**mag**-num) of the occipital bone, through which pass the spinal cord, vertebral arteries, and accessory or eleventh cranial nerve.

▼ Superior View of the Internal Surface of the Skull

The internal surface of the skull is viewed by carefully removing the top half of the skull model. The **frontal, ethmoid, sphenoid, temporal, occipital,** and **parietal bones** are visible from this view of the internal surface of the skull (Figure 3–19).

▼ FORAMINA OF THE INTERNAL SKULL SURFACE

Also present are the inside openings of the **optic canal, superior orbital fissure, foramen ovale, foramen spinosum, carotid canal, jugular foramen,** and **foramen magnum,** as discussed before when viewing the external skull surface (Figure 3–19, Table 3–3). Additionally, many other foramina are present on the internal surface of the skull. The perforated **cribriform plate** (**krib**-ri-form), with foramina for the olfactory nerves, and the **foramen rotundum** (row-**tun**-dum), for the maxillary division of the fifth cranial or trigeminal nerve, are also seen from this view. Finally, also present are the **hypoglossal canal** (hi-poh-**gloss**-al), for the twelfth cranial or hypoglossal nerve, and the **internal acoustic meatus** (ah-**koos**-tik), for the seventh cranial or facial nerve and the eighth cranial or vestibulocochlear nerve.

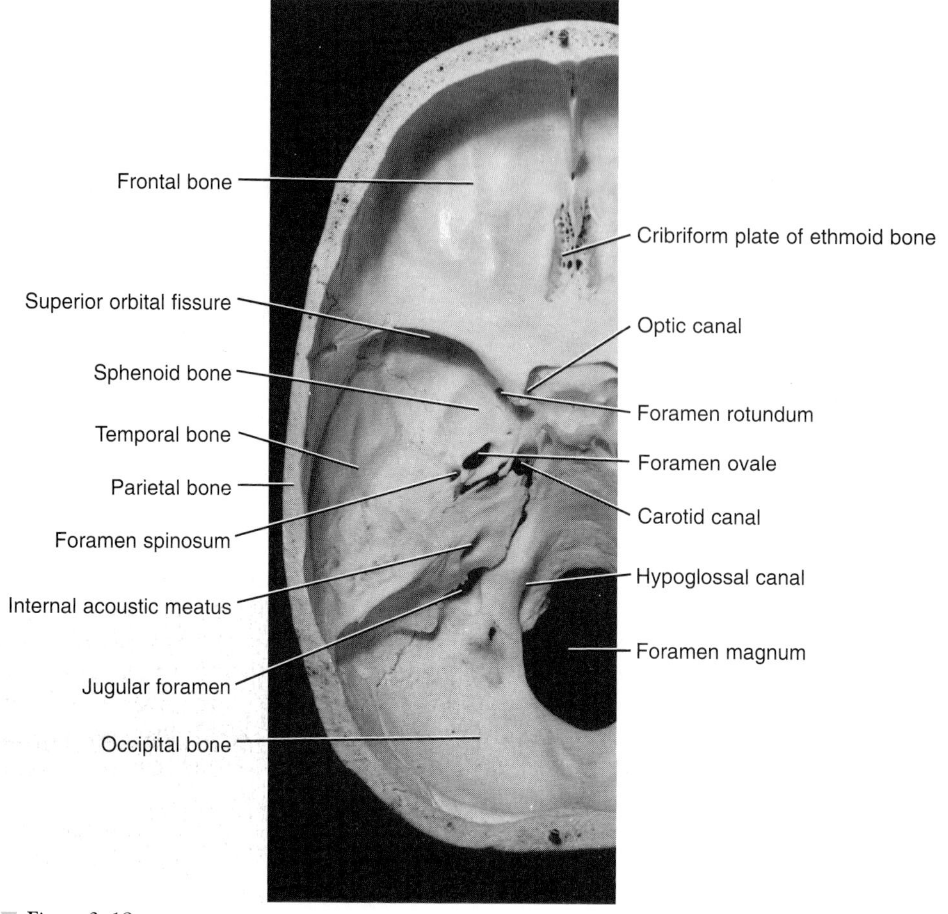

Figure 3–19
Superior view of the internal surface of the skull showing foramina.

CRANIAL BONES

The **cranium** (**kray**-nee-um) is formed from the **cranial bones** (**kray**-nee-al). The cranial bones include the single occipital, frontal, sphenoid, and ethmoid bones and the paired parietal and temporal bones (Figure 3–20).

Occipital Bone

The **occipital bone** (ok-**sip**-it-al) is a single cranial bone located in the most posterior portion of the skull (Figure 3–21). The occipital bone articulates with the parietal, temporal, and sphenoid bones of the skull. The occipital bone can easily be studied from an inferior view of its external surface.

INFERIOR VIEW OF EXTERNAL SURFACE OF THE OCCIPITAL BONE

On the external surface of the occipital bone from an inferior view, it can be seen that the **foramen magnum** (**mag**-num) is completely formed by this bone (Figure 3–22, Table 3–3). Lateral and anterior to the foramen magnum are the paired **occipital condyles** (ok-**sip**-it-al), curved and smooth projections. The occipital condyles have a movable articulation with the atlas, the first cervical vertebra of the vertebral column (discussed later in this chapter). On the stout basilar portion (**bas**-i-lar), a four-sided plate anterior to the foramen magnum, is a midline projection, the pharyngeal tubercle (fah-**rin**-je-al).

When tilting the skull model, the openings anterior and lateral to the foramen magnum are visible

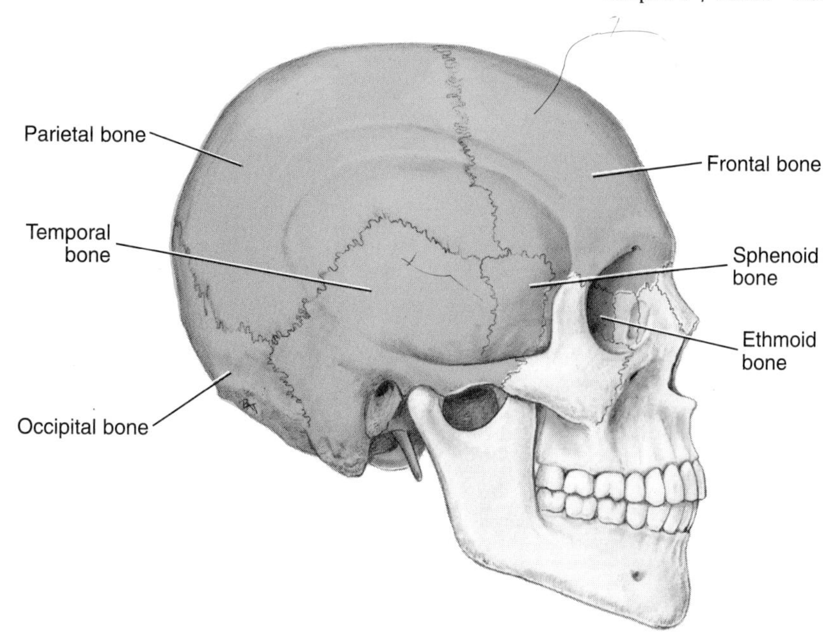

▼ **Figure 3–20**
Lateral view of the skull and the cranial bones.

on the inferior view of the occipital bone (Table 3–3). These openings are the paired **hypoglossal canals** (hi-poh-**gloss**-al). The hypoglossal or twelfth cranial nerve is transmitted through the hypoglossal canals. The **jugular notch of the occipital bone** (**jug**-you-lar), the medial portion of the two bones that forms the jugular foramen, is also present (a portion of the temporal bone is the other).

▼ Frontal Bone

The **frontal bone** (**frunt**-il) is a single cranial bone that forms both the forehead and the superior portion of the orbits (Figure 3–23). The frontal bone articulates with the parietal bones, sphenoid bone, lacrimal bones, nasal bones, ethmoid bone, zygomatic bones, and maxillae. The frontal bone's portion of the superior

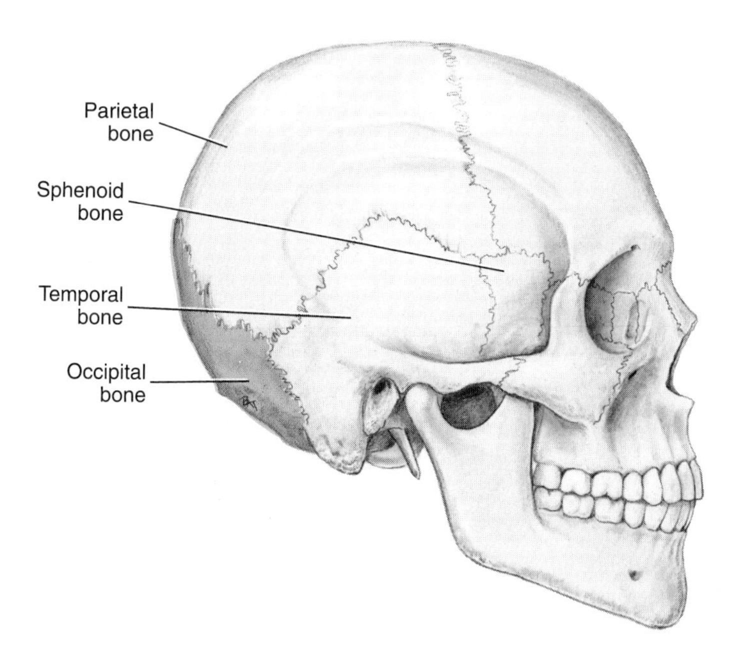

▼ **Figure 3–21**
Lateral view of the skull with the occipital bone and its bony articulations.

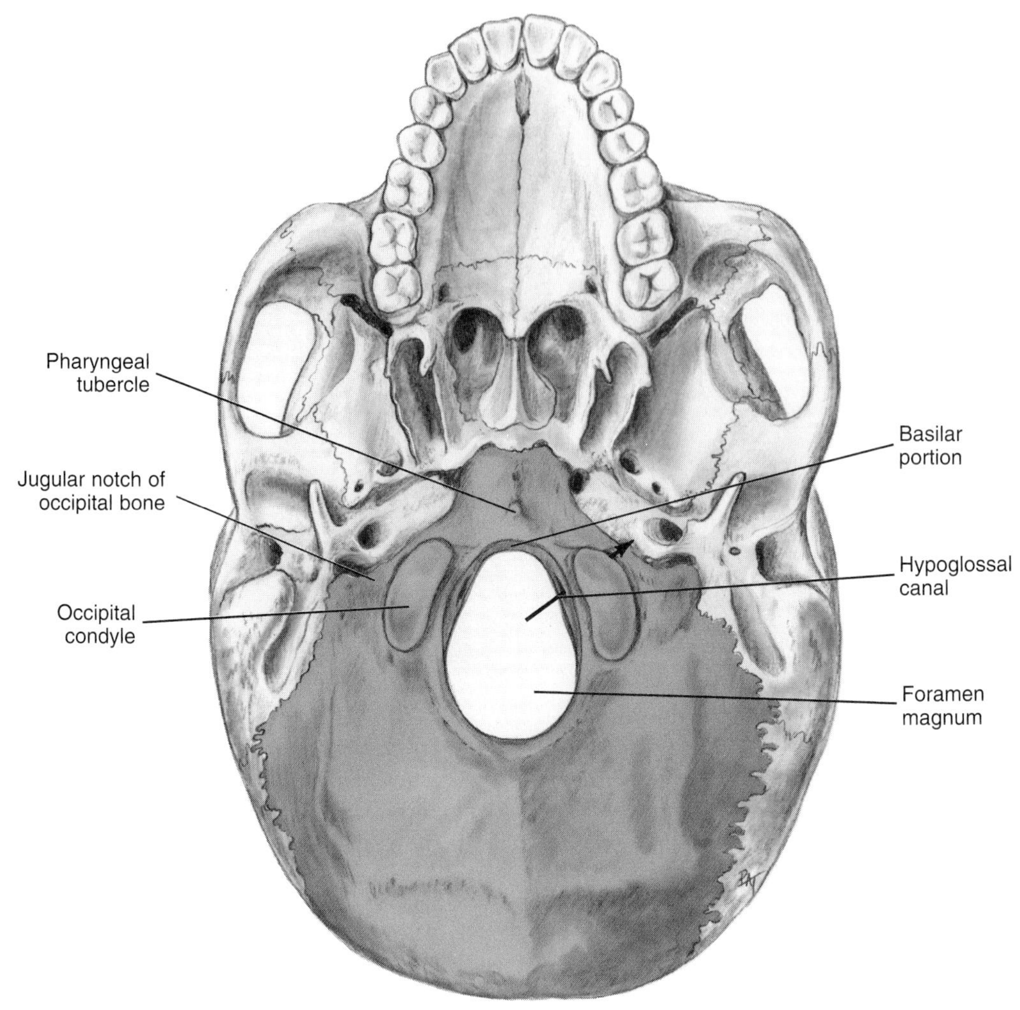

▼ Figure 3–22
Inferior view of the external surface of the skull with the occipital bone highlighted.

temporal line and inferior temporal line is visible when the bone is viewed from the lateral aspect. Internally, the frontal bone contains the paired paranasal sinuses, the **frontal sinuses** (**sy**-nus-es) (discussed later in this chapter). The frontal bone can also be studied from anterior and inferior views.

▼ ANTERIOR VIEW OF FRONTAL BONE

On the anterior aspect (Figure 3–24), certain landmarks are visible on the frontal bone. The orbital plates of the frontal bone create the superior wall or orbital roof. The curved elevations over the superior portion of the orbit are the **supraorbital ridges** (soo-prah-**or**-bit-al), subjacent to the eyebrows. The supraorbital ridges are more prominent in adult males. The **supraorbital notch** is located on the medial portion of the supraorbital ridge and is where the supraorbital artery and nerve travel from the orbit to the forehead. The supraorbital notch is located about an inch from the midline and when palpated with pressure can produce patient discomfort.

Between the supraorbital ridges is the **glabella** (glah-**bell**-ah), the smooth elevated area between the eyebrows, which tends to be flat in children and adult females and forms a rounded prominence in adult males. Lateral to the orbit there is a projection, the orbital surface of the **zygomatic process of the frontal bone** (zi-go-**mat**-ik) (Table 3–4).

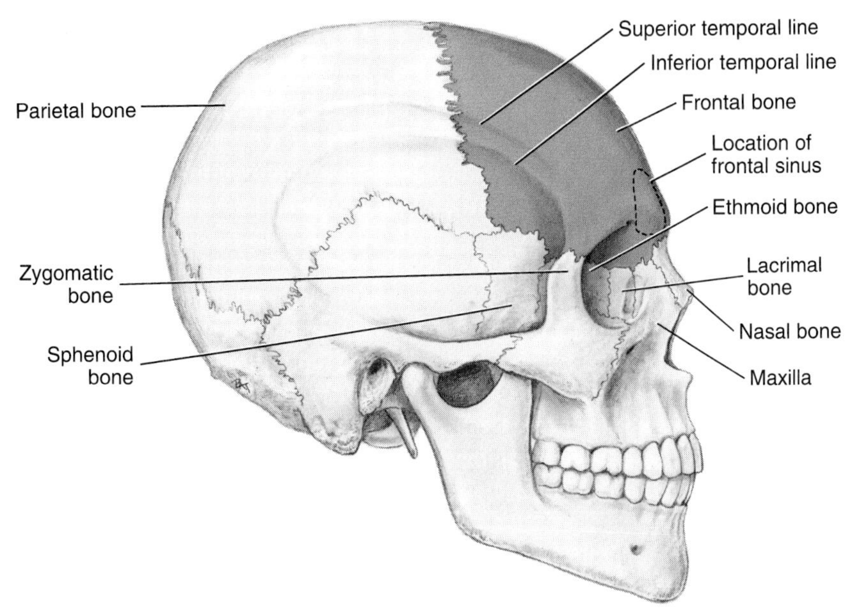

▼ **Figure 3–23**
Lateral view of the skull with the frontal bone and its bony articulations.

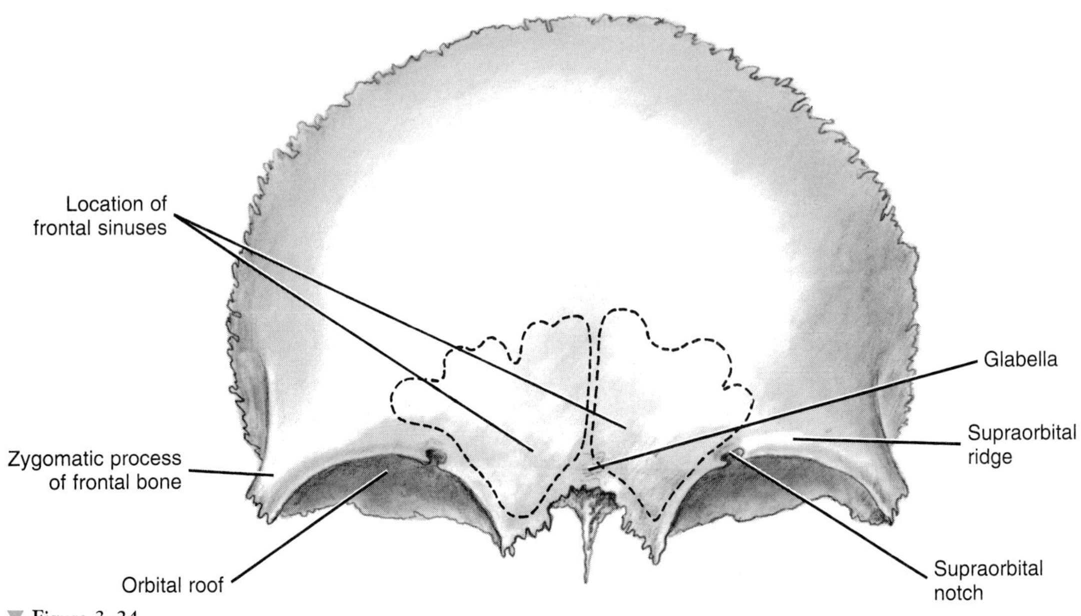

▼ **Figure 3–24**
Anterior view of the frontal bone.

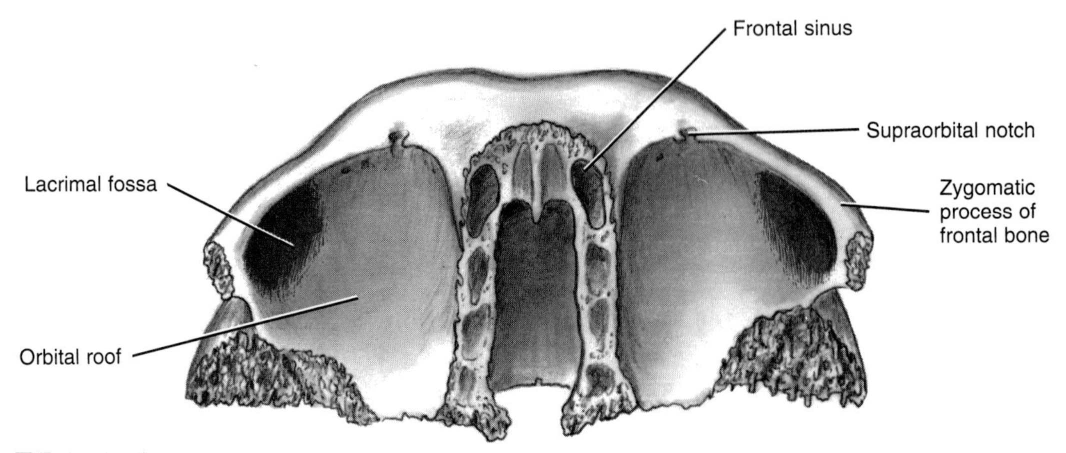

Figure 3–25
Inferior view of the frontal bone with the lacrimal fossa highlighted.

▼ INFERIOR VIEW OF FRONTAL BONE

From the inferior view of the frontal bone, each **lacrimal fossa** (**lak**-ri-mal) is visible (Figure 3–25). The lacrimal fossa is located just inside the lateral portion of the supraorbital ridge. This fossa contains the **lacrimal gland,** which produces lacrimal fluid or tears. After lubricating the eye, the lacrimal fluid empties into the nasal cavity through the nasolacrimal duct.

▼ Parietal Bones

The **parietal bones** (pah-**ri**-it-al) are paired cranial bones and articulate with each other at the **sagittal suture** (**saj**-i-tel) (Figure 3–26, Table 3–2). The

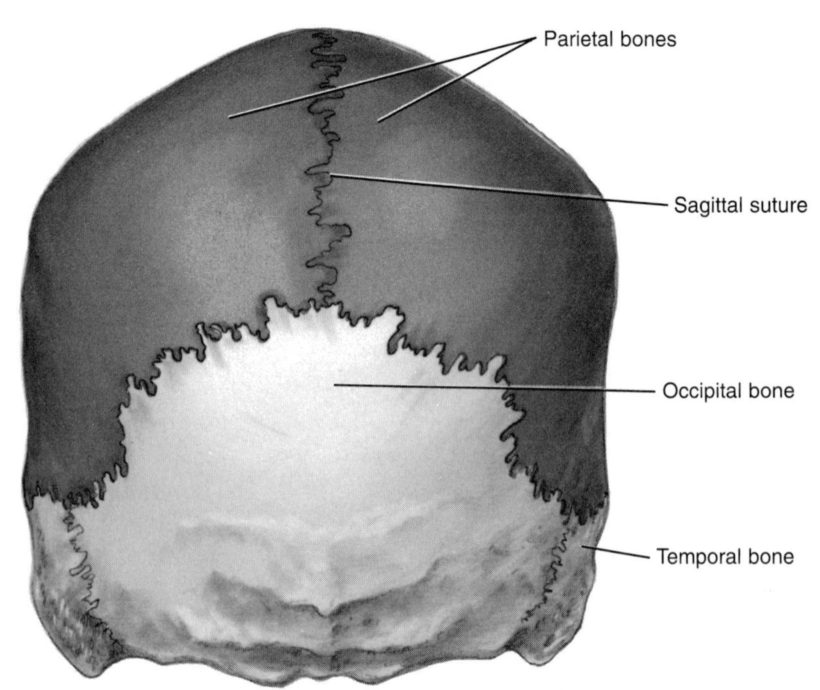

Figure 3–26
Posterior view of the skull with the parietal bone and some of its bony articulations.

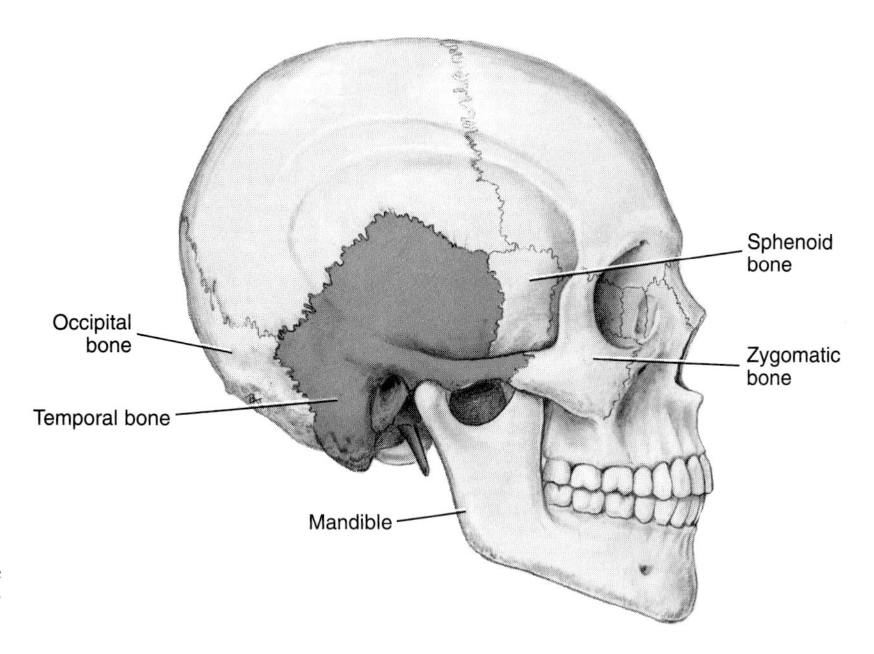

▼ **Figure 3–27**
Lateral view of the skull with the temporal bone and its bony articulations.

parietal bones also articulate with the occipital, frontal, temporal, and sphenoid bones.

▼ Temporal Bones

The **temporal bones** (**tem**-poh-ral) are paired cranial bones that form the lateral walls of the skull (Figure 3–27). Each temporal bone articulates with one zygomatic and one parietal bone, the occipital and sphenoid bones, and the mandible. Each temporal bone is composed of three portions: the squamous, tympanic, and petrous portions.

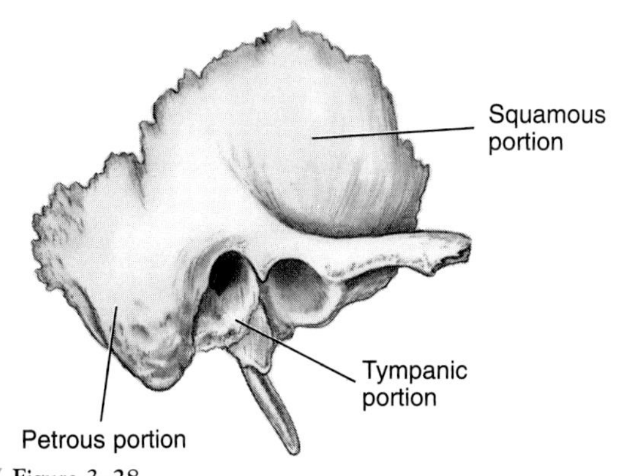

▼ **Figure 3–28**
Lateral view of the portions of the temporal bone: squamous portion, tympanic portion, and petrous portion.

▼ PORTIONS OF TEMPORAL BONE

The portions of the temporal bone can be viewed from the lateral aspect of the skull (Figure 3–28). The large fan-shaped flat portion on each of the temporal bones is the **squamous portion of the temporal bone** (**skwa**-mus). The second portion is the small, irregularly shaped **tympanic portion of the temporal bone** (tim-**pan**-ik), which is associated with the ear canal. The third portion is the **petrous portion of the temporal bone** (pe-**tros**), which is inferiorly located and helps form the cranial floor.

▼ SQUAMOUS PORTION OF TEMPORAL BONE

In addition to helping form the braincase, the **squamous portion of the temporal bone** forms the zygomatic process of the temporal bone (zi-go-**mat**-ik), which forms part of the zygomatic arch (Figure 3–29, Table 3–4). This portion of the temporal bone also forms the cranial portion of the temporomandibular joint. This joint is discussed in detail in Chapter 5. On the inferior surface of the zygomatic process of the temporal bone is the **articular fossa** (ar-**tik**-you-ler) or mandibular fossa.

Anterior to the articular fossa is the **articular eminence** (ar-**tik**-you-ler), and posterior is the **postglenoid process** (post-**gle**-noid) (Table 3–4). The articular fossa and eminence are portions of the tempo-

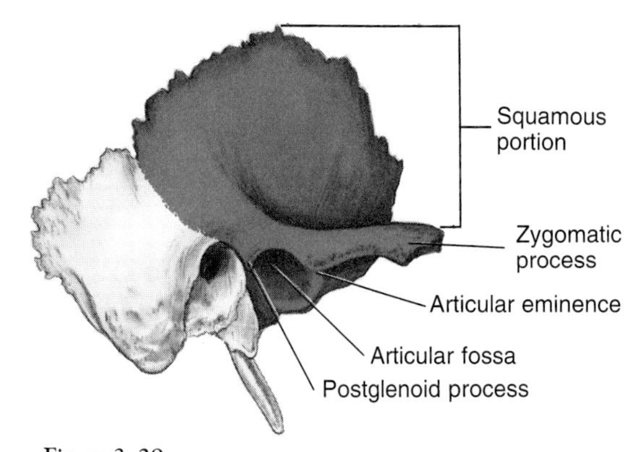

Figure 3–29
Lateral view of the temporal bone with the squamous portion highlighted.

ral bone that articulate with the mandible at the temporomandibular joint.

▼ TYMPANIC PORTION OF TEMPORAL BONE

The **tympanic portion of the temporal bone** forms most of the **external acoustic meatus** (ah-**koos**-tik), a short canal leading to the tympanic cavity, located posterior to the articular fossa (Figure 3–30, Table 3–3). Also posterior to the articular fossa, the tympanic portion is separated from the petrosal portion by a fissure, the **petrotympanic fissure** (pe-troh-tim-**pan**-ik), through which the chorda tympani nerve emerges.

▼ PETROUS PORTION OF TEMPORAL BONE

On the inferior aspect of the **petrous portion of the temporal bone** (pe-**tros**), posterior to the external acoustic meatus, is a large roughened projection, the **mastoid process** (**mass**-toid) (Figure 3–31, Table 3–4). The mastoid process is composed of air spaces or **mastoid air cells** that communicate with the middle ear cavity. The mastoid process also serves as the site for attachment of the large muscles of the neck, such as the sternocleidomastoid muscle.

Medial to the mastoid process is the **mastoid notch** (**mass**-toid) (Figure 3–31A and B). Inferior and medial to the external acoustic meatus is a long pointed bony projection, the **styloid process** (**sty**-loid), a structure that serves for the attachment of muscles and ligaments (Table 3–4). The **stylomastoid foramen** (sty-lo-**mass**-toid) carries the facial or seventh cranial nerve and is named for its location between the styloid process and mastoid process (Table 3–3). When the skull model is tilted, the **jugular notch of the temporal bone** (**jug**-you-lar) is visible (Table 3–3), the lateral portion of the two bones that form the jugular foramen (the other bone is the occipital bone).

On the intracranial surface is the **internal acoustic meatus** (ah-**koos**-tik), which carries the vestibulocochlear or eighth cranial nerve and the facial or seventh cranial nerve (Figure 3–21, Table 3–3). Both of these cranial nerves enter the skull from the brain. The vestibulocochlear nerve remains inside the petrous portion of the temporal bone, which houses the inner ear. In contrast, the facial nerve takes a convoluted path through the bone, eventually emerging at the stylomastoid foramen.

▼ Sphenoid Bone

The next cranial bone of the skull to be considered is the single **sphenoid bone** (**sfe**-noid) (Figures 3–32 and 3–33). The sphenoid bone is a midline bone that articulates with the frontal, parietal, ethmoid, tempo-

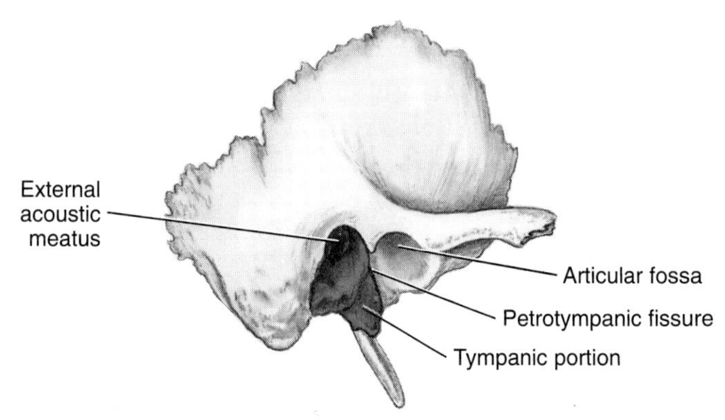

▼ **Figure 3–30**
Lateral view of the temporal bone with the tympanic portion highlighted.

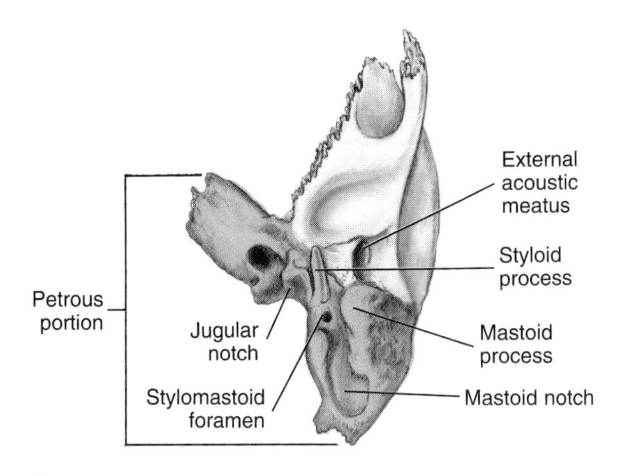

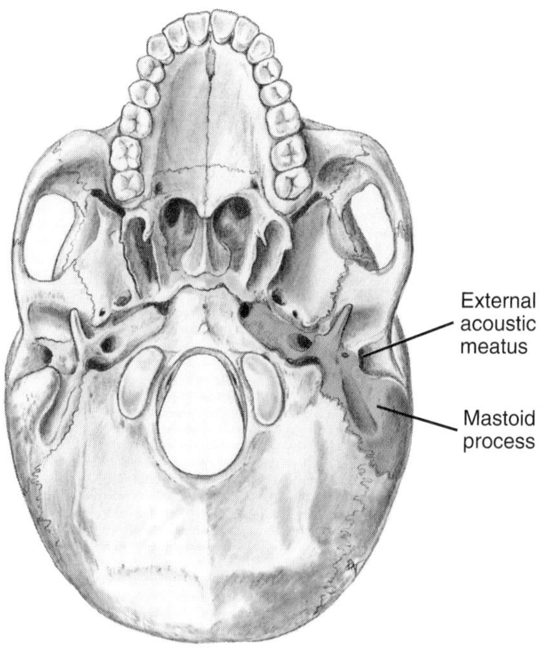

▼ **Figure 3–31**
A: Inferior view of the temporal bone with the petrous portion highlighted.
B: Location of the petrous portion of the temporal bone demonstrated on an inferior view of the skull.

ral, zygomatic, maxillary, palatine, vomer, and occipital bones. This bone is very complex, with some portions of it encountered in almost every significant area of the skull. The sphenoid bone has important foramina and different portions: the body and its processes.

▼ FORAMINA OF THE SPHENOID BONE

Many foramina or fissures are located in the sphenoid bone, such as the **superior orbital fissure, foramen ovale, foramen rotundum,** and **foramen spinosum,** which carry important nerves and blood vessels of the head and neck (Figures 3–32 and 3–33, Table 3–3).

▼ BODY OF THE SPHENOID BONE

The middle portion of the sphenoid bone is the **body of the sphenoid bone** (**sfe**-noid), which articulates on its anterior surface with the ethmoid bone (Figures 3–32 and 3–33). The body of sphenoid bone articulates posteriorly with the basilar portion of the occipital bone. The body contains the paired paranasal sinuses, the **sphenoid sinuses** (**sfe**-noid **sy**-nus-es) (discussed later in this chapter).

▼ PROCESSES OF THE SPHENOID BONE

The body of the sphenoid bone has three paired processes that arise from it: the lesser wing, greater wing, and pterygoid process (Figures 3–32 to 3–34, Table 3–4). The anterior process is the **lesser wing of the sphenoid bone,** which makes up the base of the orbital apex. The posterolateral process is the **greater wing of the sphenoid bone.**

Inferior to the greater wing of the sphenoid bone is the **pterygoid process** (**teh**-ri-goid), an area for the attachment of some of the muscles of mastication. The pterygoid process consists of two plates, the flattened **lateral pterygoid plate** and thinner **medial pterygoid plate,** with the **pterygoid fossa** between them. The **hamulus** (**ha**-mu-lis), a thin curved process, is the inferior termination of the medial pterygoid plate.

A sharp pointed area, the **spine of the sphenoid bone,** is located at the posterior corner of each greater wing of the sphenoid bone. Each greater wing is divided into two smaller surfaces by the **infratemporal crest** (in-frah-**tem**-poh-ral), the temporal and infratemporal surfaces.

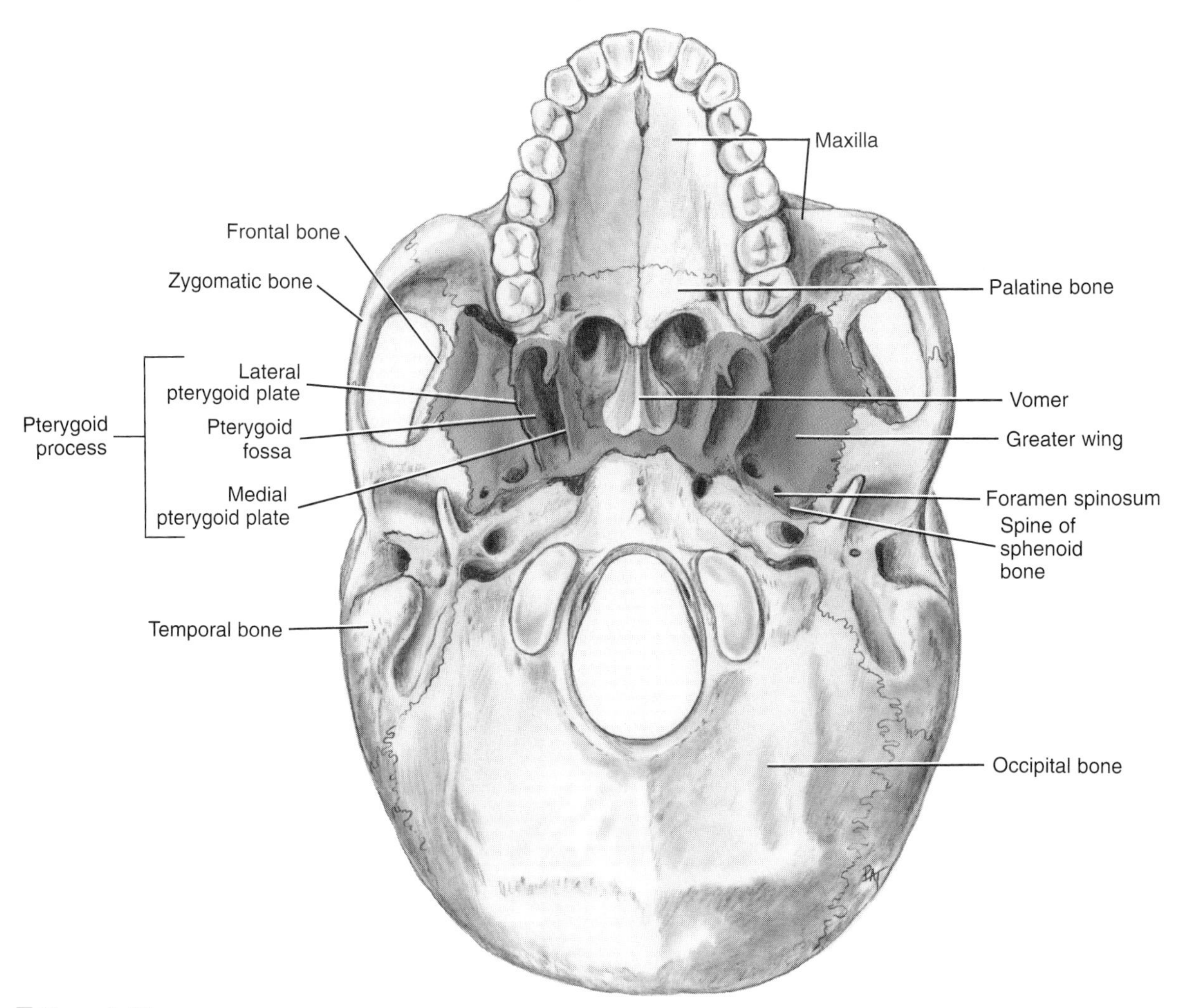

Figure 3–32
Inferior view of the external surface of the skull with the sphenoid bone highlighted.

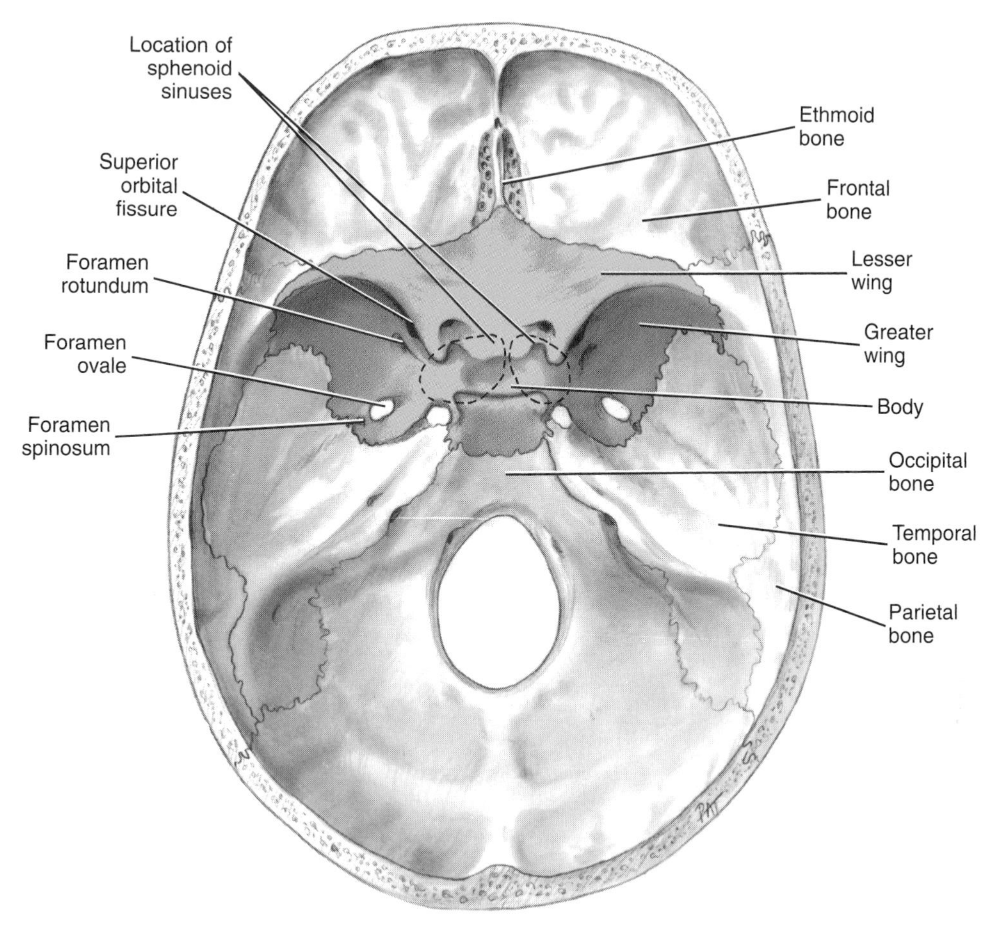

▼ **Figure 3–33**
Superior view of the internal surface of the skull with the sphenoid bone highlighted.

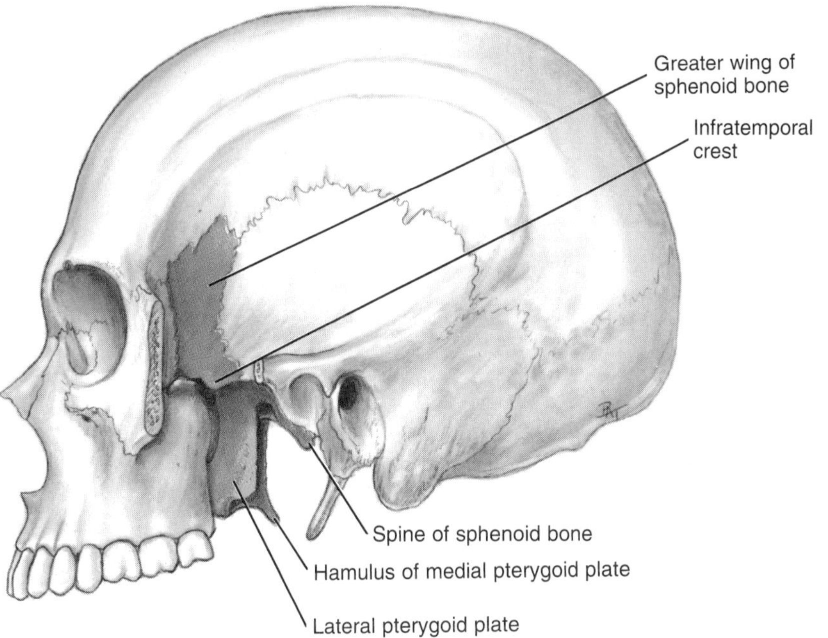

▼ **Figure 3–34**
Cutaway view of the lateral aspect of the upper portion of the skull with the sphenoid bone highlighted.

Ethmoid Bone

The **ethmoid bone** (**eth**-moid) is a single midline cranial bone of the skull (Figure 3–35). The ethmoid articulates with the frontal, sphenoid, lacrimal, and maxillary bones and adjoins the vomer at its inferior and posterior border.

PLATES OF THE ETHMOID BONE AND ASSOCIATED STRUCTURES

The ethmoid bone has two unpaired plates, the midline vertical **perpendicular plate** (per-pen-**dik**-you-lar) and the horizontal **cribriform plate** (**krib**-ri-form), which it crosses. The perpendicular plate is easily seen in the nasal cavity and aids the vomer and nasal septal cartilage in forming the nasal septum (Figure 3–35). The cribriform plate, visible from the inside of the cranial cavity and present on the superior aspect of the bone, is perforated by foramina to allow the passage of olfactory nerves for the sense of smell (Figures 3–36 and 3–37).

The **ethmoid sinuses** (**eth**-moid **sy**-nus-es) or ethmoid air cells are a variable number of small cavities in the lateral mass of the ethmoid bone and are discussed later in this chapter. A vertical midline continuation of the perpendicular plate into the cranial cavity is the **crista galli** (**kris**-tah **gal**-lee). The crista galli serves as an attachment for layers covering the brain.

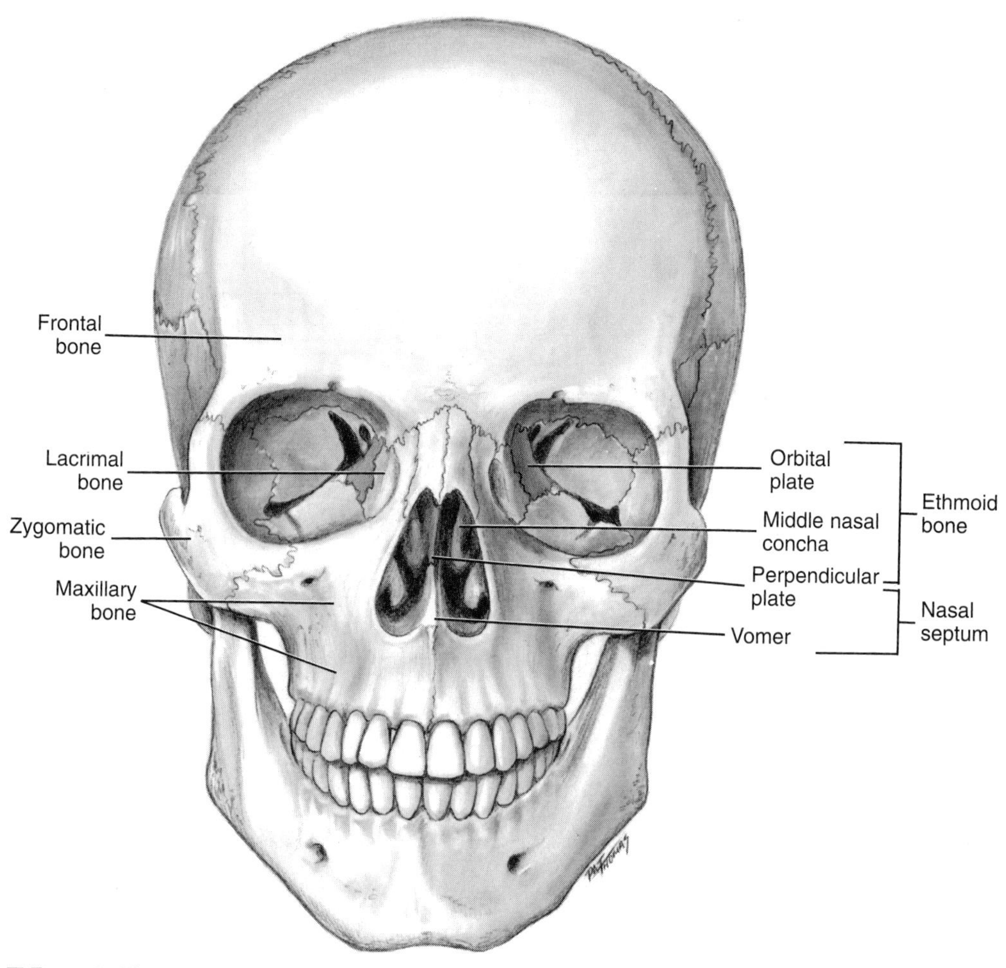

Figure 3–35
Anterior view of the skull with the ethmoid bone highlighted.

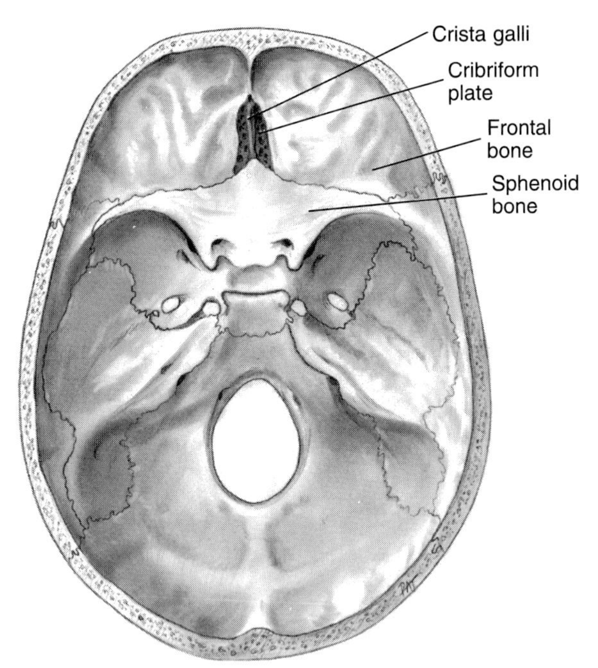

▼ **Figure 3–36**
Superior view of the internal surface of the skull with the ethmoid bone highlighted.

▼ NASAL CONCHAE AND ASSOCIATED STRUCTURES

The lateral portions of the ethmoid bone form the **superior nasal conchae** (**nay**-zil **kong**-kay) and **middle nasal conchae** in the nasal cavity and the paired orbital plates (Figures 3–37 and 3–38). The **orbital plate of the ethmoid bone** (**or**-bit-al) forms the medial orbital wall. Between the orbital plate and the conchae are the **ethmoid sinuses,** which consist of air-filled spaces (discussed later in this chapter).

▼ FACIAL BONES

The **facial bones** create the facial features and serve as a base for dentition. The facial bones include the single vomer and mandible and the paired lacrimal, nasal, inferior nasal conchal, zygomatic, and maxillary bones (Figure 3–4). For ease of learning, the palatine bones will be considered under the heading of facial bones, but they are not strictly considered facial bones.

Many bones of the face are shared by two or more soft tissue components of the face. For example, the frontal bone forms both the forehead and the areas around the eyes. This is important to remember since an abnormality in one facial bone often involves many soft tissue components (discussed later in this chapter).

▼ Vomer

The **vomer** (**vo**-mer) is a single facial bone of the skull that forms the posterior portion of the nasal septum. It is located in the midsagittal plane inside the nasal cavity. The articulations of the vomer are easily seen on a lateral view of the bone (Figure 3–39). The vomer articulates with the ethmoid bone on its anterosuperior border, the nasal cartilage anteriorly, the palatine bones and maxillae inferiorly, and the sphenoid bone on its posterosuperior border. The posteroinferior border is free of any bony articulation. The vomer also has no muscle attachments.

▼ Lacrimal Bones, Nasal Bones, and Inferior Nasal Conchae

The paired **lacrimal bones** (**lak**-ri-mal) are irregular thin plates of bone that form a small portion of the

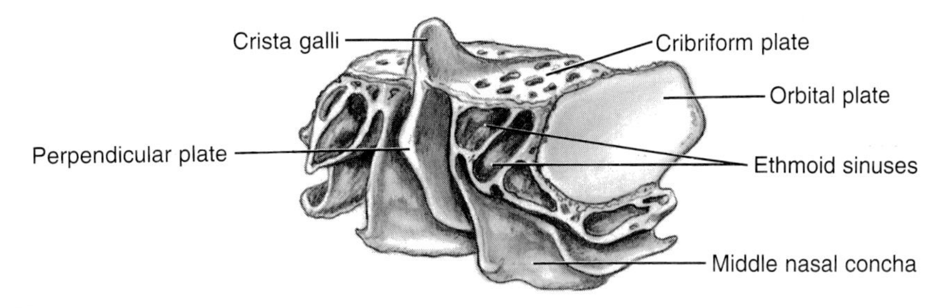

▼ **Figure 3–37**
Oblique anterior view of the ethmoid bone with its perpendicular, cribriform, and orbital plates.

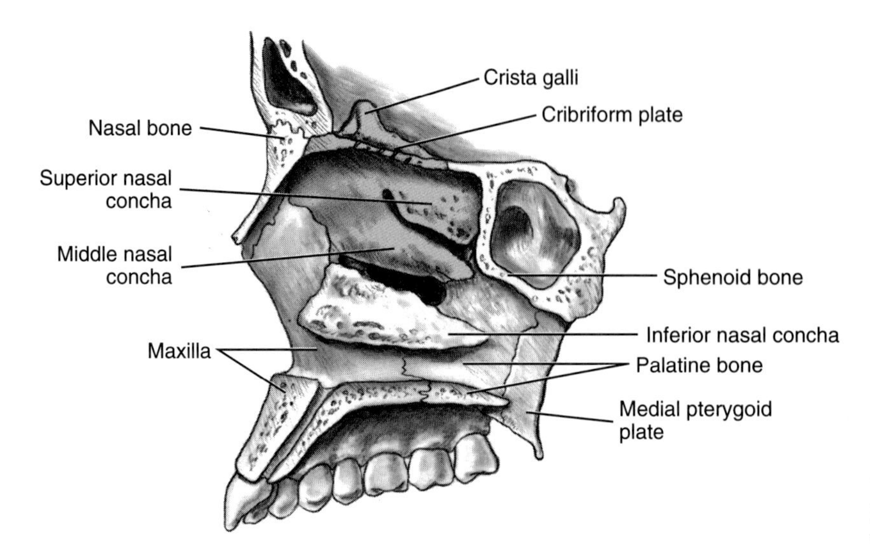

Figure 3–38
Lateral wall of the right nasal cavity with the ethmoid bone highlighted.

anterior medial wall of the orbit (Figure 3–40). Each lacrimal bone articulates with the ethmoid, frontal, and maxillary bones. The **nasolacrimal duct** (nay-so-**lak**-rim-al dukt) is formed at the junction of the lacrimal and maxillary bones. Lacrimal fluid or tears from the lacrimal gland are drained through this duct into the inferior nasal meatus.

The **nasal bones** (**nay**-zil) are paired facial bones that form the bridge of the nose, articulating with each other in the midline above the piriform aperture. The nasal bones fit between the frontal processes of the maxillae and thus articulate with the frontal bone superiorly and the maxillae laterally (Figure 3–40).

The **inferior nasal conchae** (**nay**-zil **kong**-kee) are paired facial bones that project off the maxilla to form part of the lateral walls of the nasal cavity (Figure 3–40). Unlike the superior and middle nasal conchae that also project off the maxillae, the inferior nasal conchae are separate facial bones. Each inferior nasal concha is composed of thin spongy bone curved upon itself like a scroll. These bones articulate with the ethmoid, lacrimal, palatine, and maxillary bones. The inferior nasal conchae do not have any muscle attachments.

Zygomatic Bones

The **zygomatic bones** (zi-go-**mat**-ik) are paired facial bones of the skull that form the cheek bones or malar

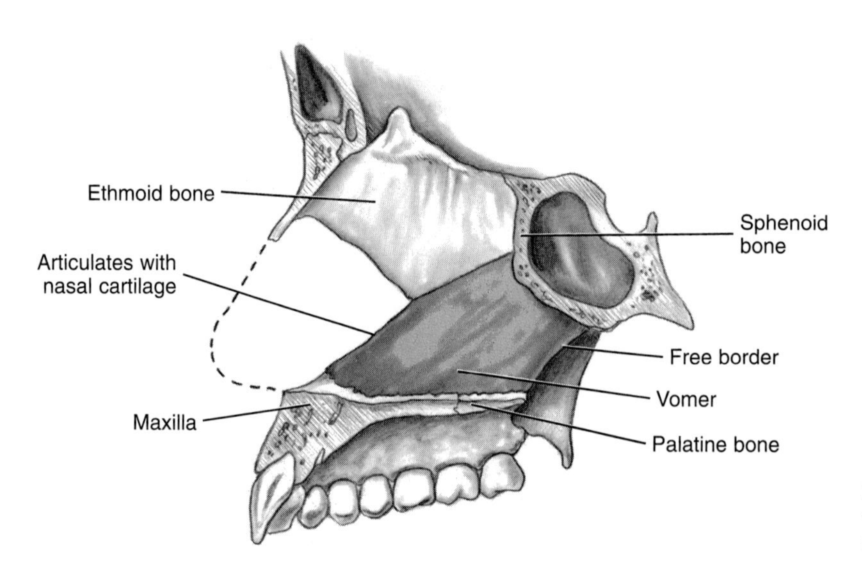

Figure 3–39
Medial wall of the left nasal cavity with the vomer highlighted (outline of nasal cartilage).

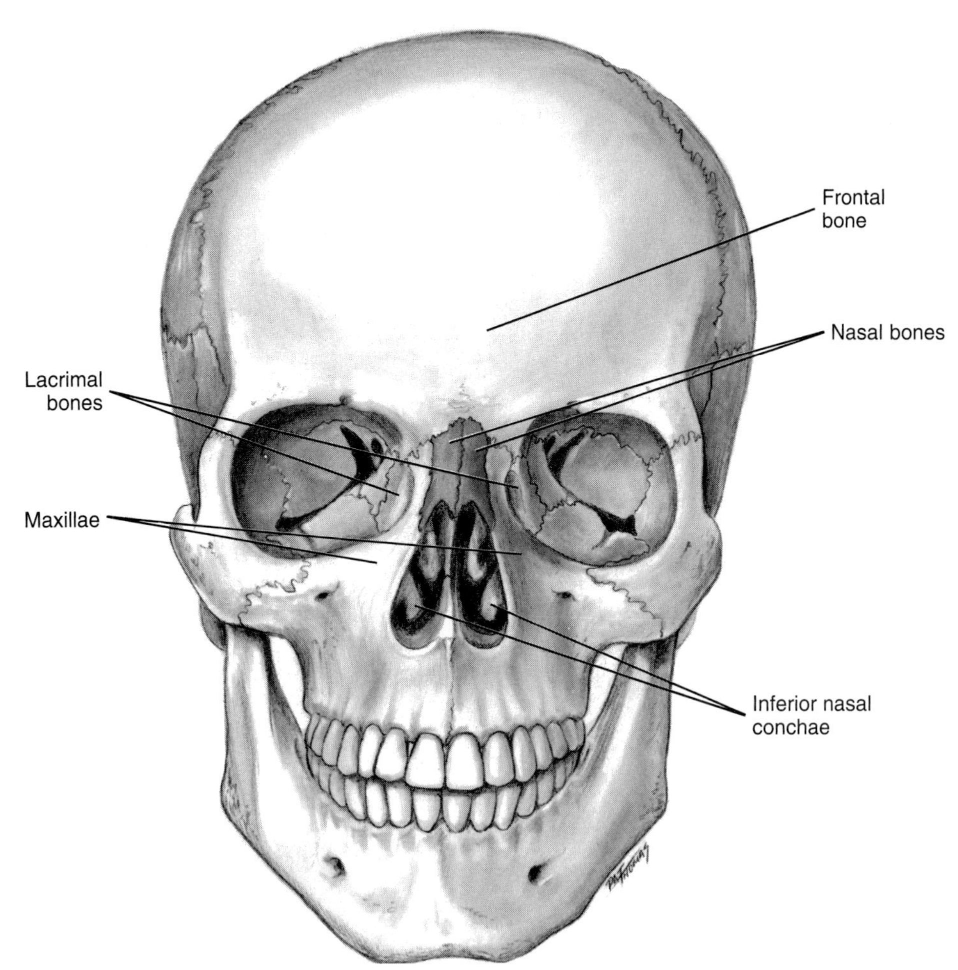

▼ **Figure 3–40**
Anterior view of the skull with the nasal bones, lacrimal bones, and inferior nasal conchae highlighted.

surfaces (Figures 3–41 and 3–42). The zygomatic bones articulate with the frontal, temporal, sphenoid, and maxillary bones. Each zygomatic bone has a diamond shape composed of three processes with similarly named associated bony articulations: the frontal, temporal, and maxillary processes.

▼ PROCESSES OF THE ZYGOMATIC BONE

Each process of the zygomatic bone forms important structures of the skull (Figures 3–41 and 3–42, Table 3–4). The orbital surface of the **frontal process of the zygomatic bone** (**frunt**-il) forms the anterior lateral orbital wall. The **temporal process of the zygomatic bone** (**tem**-poh-ral) forms the zygomatic arch together with the zygomatic process of the temporal bone. The orbital surface of the **maxillary process of the zygomatic bone** (**mak**-sil-lare-ee) forms a portion of the **infraorbital rim** (in-frah-**or**-bit-al rim) and a small portion of the anterior part of the lateral orbital wall.

▼ Palatine Bones

The **palatine bones** (**pal**-ah-tine) are paired bones of the skull that are not strictly considered facial bones, but they will be considered under this heading for ease of learning. Each palatine bone consist of two plates, the horizontal and vertical plates.

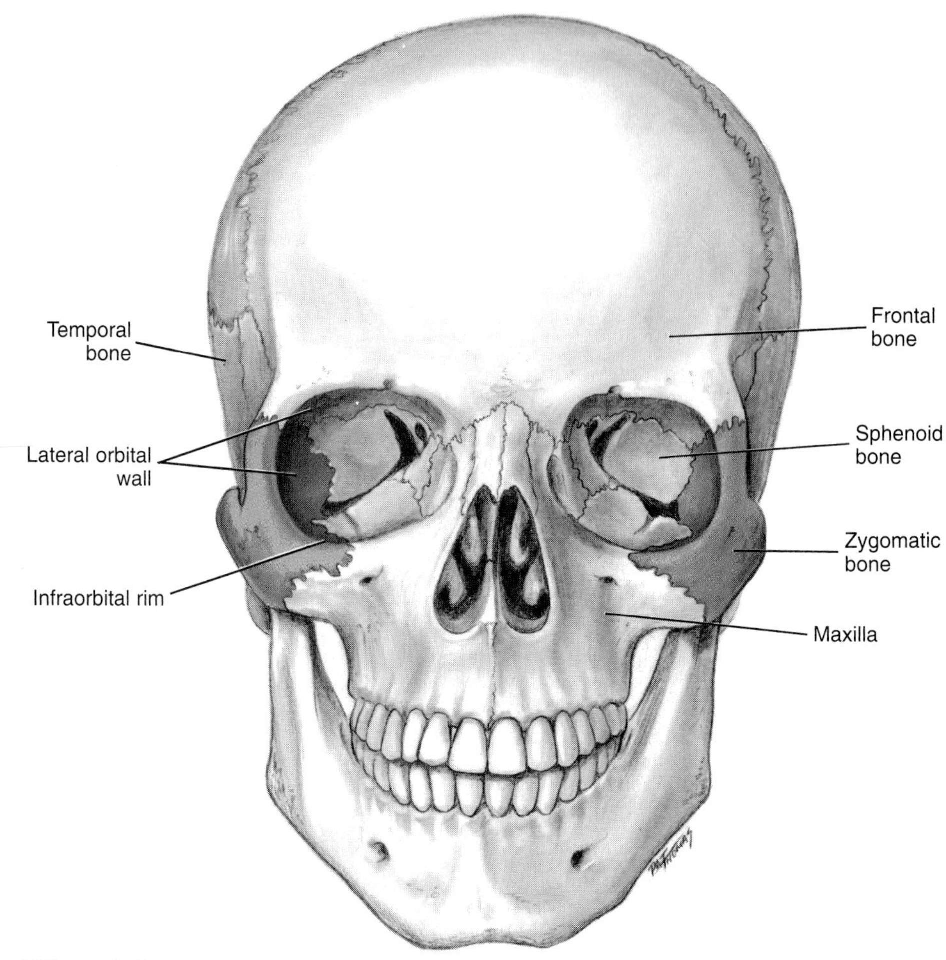

▼ Figure 3–41
Anterior view of the skull with the zygomatic bones highlighted.

▼ PLATES OF THE PALATINE BONE

Both the horizontal and vertical plates can be seen from a posterior view of a palatine bone (Figure 3–43). The **horizontal plates of the palatine bones** form the posterior portion of the hard palate. The **vertical plates of the palatine bones** form part of the lateral walls of the nasal cavity, and each plate contributes a small lip of bone to the orbital apex.

▼ SUTURE OF THE PALATINE BONES

The palatine bones serve as a link between the maxillae and the sphenoid bone with which they articulate, as well as each other. The two horizontal plates articulate with each other at the posterior portion of the **median palatine suture** (**pal**-ah-tine) (Figure 3–44, Table 3–2).

▼ FORAMINA OF THE PALATINE BONES

There are two important foramina in the palatine bones that transmit nerves and blood vessels to this region, the greater and lesser palatine foramina (Figure 3–44, Table 3–3). The **greater palatine foramen** (**pal**-ah-tine) is located in the posterolateral region of each of the palatine bones, usually distal to the third maxillary molar. The greater palatine foramen trans-

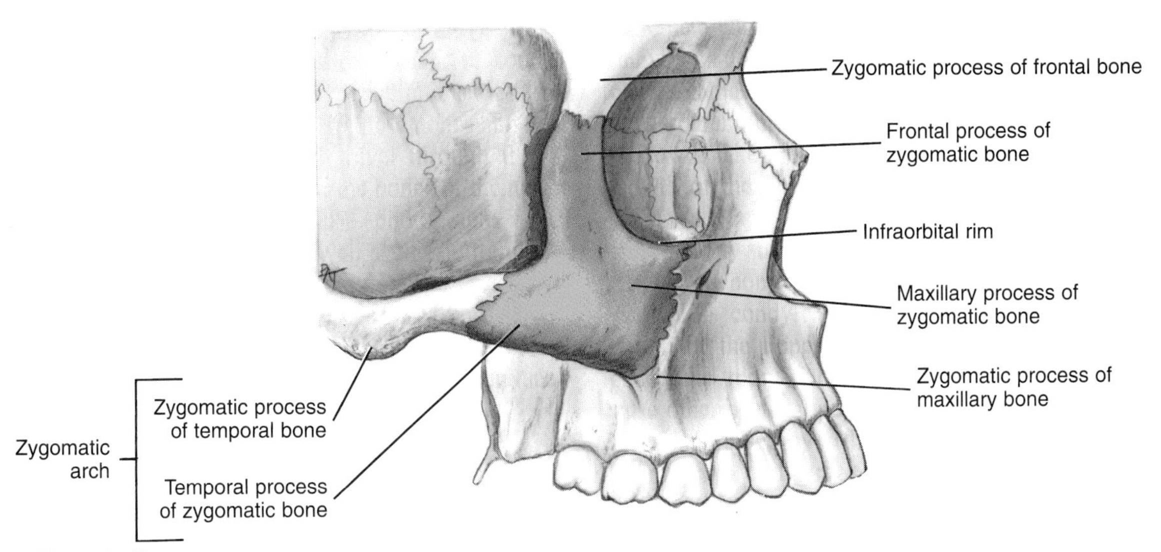

▼ **Figure 3–42**
Lateral view of the upper portion of the skull with the zygomatic bone highlighted.

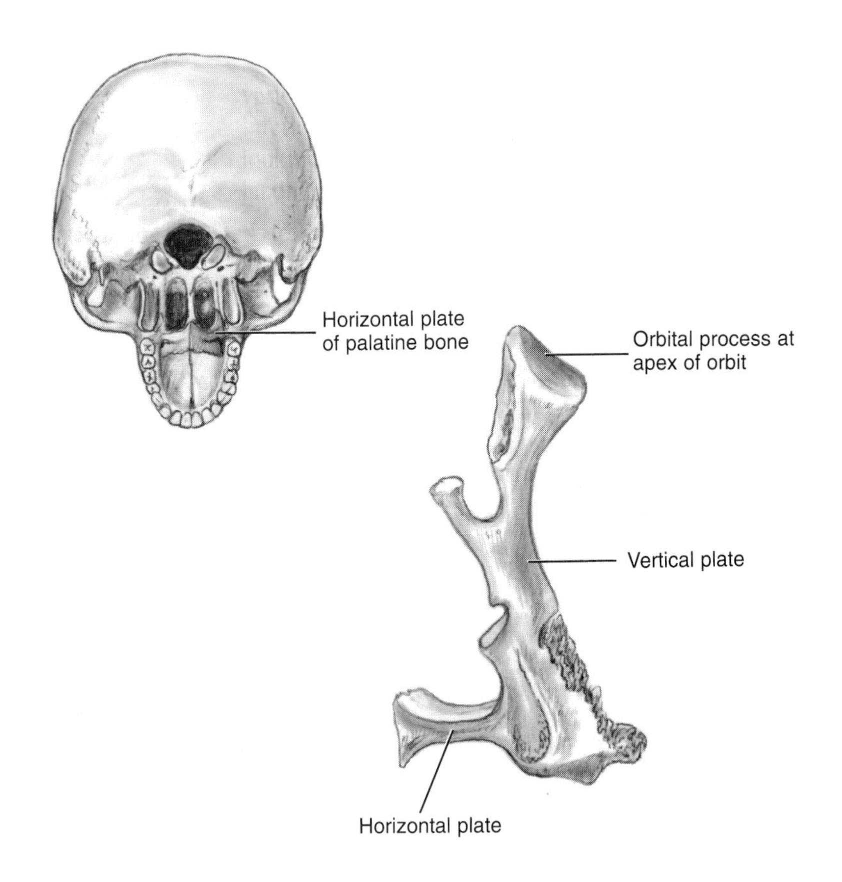

▼ **Figure 3–43**
Posterior view of the right palatine bone with its location demonstrated on a posterior-inferior view of the skull.

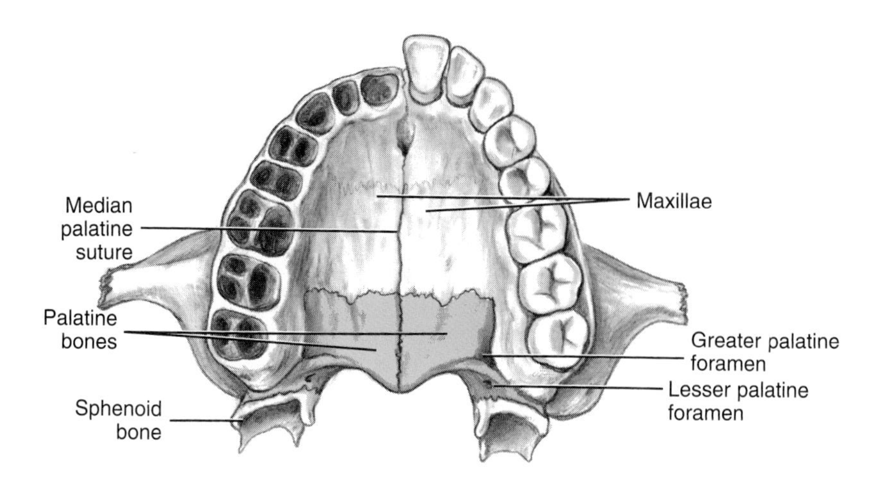

▼ Figure 3–44
Inferior view of the hard palate with the palatine bones highlighted.

mits the greater palatine nerve and blood vessels and is a landmark for the administration of a local anesthetic for the greater palatine nerve block (see Chapter 9).

A smaller opening nearby, the **lesser palatine foramen** (**pal**-ah-tine), transmits the lesser palatine nerve and blood vessels to the soft palate and tonsils. Both foramina are openings of the pterygopalatine canal that carries the descending palatine nerves and blood vessels from the pterygopalatine fossa to the palate.

▼ Maxilla

The upper jaw or **maxilla** (mak-**sil**-ah) consists of two maxillary bones or **maxillae** (mak-**sil**-lay) that are fused together (Figure 3–45). The maxillae articulate with the frontal, lacrimal, nasal, inferior nasal conchal, vomer, sphenoid, ethmoid, palatine, and zygomatic bones. Each maxilla includes a body and four processes: the frontal, zygomatic, palatine, and alveolar processes.

The **body of the maxilla** has orbital, nasal, infratemporal, and facial surfaces. The bodies contain air-filled spaces or paranasal sinuses, the **maxillary sinuses** (**mak**-sil-lare-ee **sy**-nuses) (discussed later in this chapter). The maxilla can be studied from its three views: the anterior, lateral, and inferior views.

▼ ANTERIOR VIEW OF THE MAXILLA

The **frontal process of the maxilla** (**frunt**-il) articulates with the frontal bone and forms the medial orbital rim with the lacrimal bone on its anterior surface (Figure 3–45, Table 3–4). Each maxilla's orbital surface is separated from the sphenoid bone by the **inferior orbital fissure** (**or**-bit-al) (Table 3–3). The inferior orbital fissure carries the infraorbital and zygomatic nerves, infraorbital artery, and inferior ophthalmic vein. The groove in the floor of the orbital surface is the **infraorbital sulcus** (in-frah-**or**-bit-al).

The infraorbital sulcus becomes the **infraorbital canal** (in-frah-**or**-bit-al) and then terminates on the facial surface of the maxilla as the **infraorbital foramen** (Table 3–3). This foramen transmits the infraorbital nerve and blood vessels. The infraorbital foramen is a landmark for the administration of a local anesthetic for the infraorbital block (see Chapter 9). Palpation of the infraorbital foramen will cause a mild aching in a patient. Inferior to the infraorbital foramen is an elongated depression, the **canine fossa** (**kay**-nine). The canine fossa is just posterosuperior to the roots of the maxillary canine teeth.

Each tooth of the maxillary arch is covered by a prominent facial ridge of bone, a portion of the **alveolar process of the maxilla** (al-**ve**-o-lar) (Table 3–4). The facial ridge over the maxillary canine, the **canine eminence** (**kay**-nine), is especially promi-

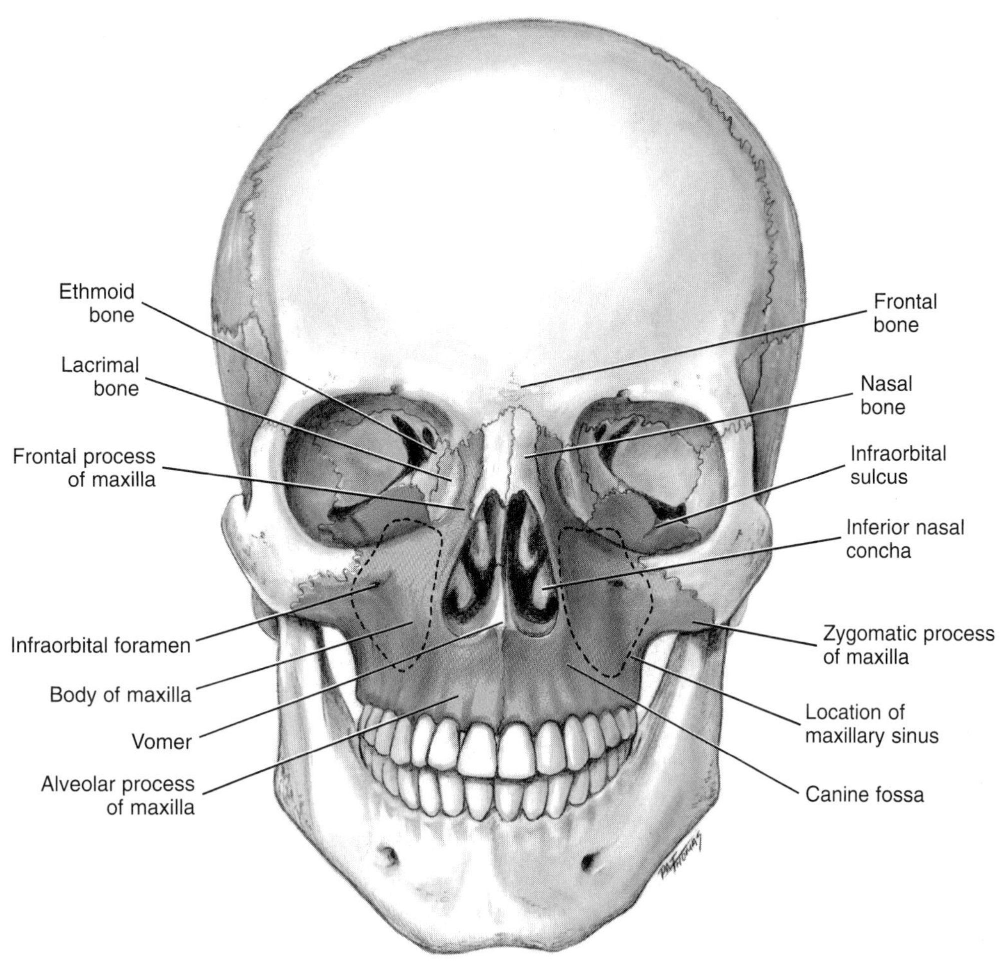

▼ Figure 3–45
Anterior view of the skull with the maxilla and its bony articulations (articulation of the maxilla with the pterygoid process of the sphenoid bone and palatine bones cannot be seen in this view).

nent. The maxillary bone over the facial surface of the maxillary teeth is less dense than the mandible over similar teeth. This allows a greater incidence of clinically adequate local anesthesia for the maxillary teeth when the drug is administered as a local infiltration (see Chapter 9).

▼ LATERAL VIEW OF THE MAXILLA

The **zygomatic process of the maxilla** (zy-go-**mat**-ik) articulates with the zygomatic bone laterally, completing the **infraorbital rim** (in-frah-**or**-bit-al) (Figure 3–46, Table 3–4). Some of the landmarks noted in the anterior view of the maxilla are also present.

▼ INFERIOR VIEW OF THE MAXILLAE

On the inferior surface, each **palatine process of the maxilla** (**pal**-ah-tine) articulates with the other to form the anterior, major portion of the hard palate (Figure 3–47, Table 3–4). The suture between these two palatine processes of the maxillae is the anterior portion of the **median palatine suture** (Table 3–2). In the patient, this is covered by the median palatine raphe, a midline fibrous band of tissue.

In the anterior midline portion of the palatine process, just posterior to the maxillary central incisors, is the **incisive foramen** (in-**sy**-ziv) (Table 3–3). This foramen carries the branches of the right and left

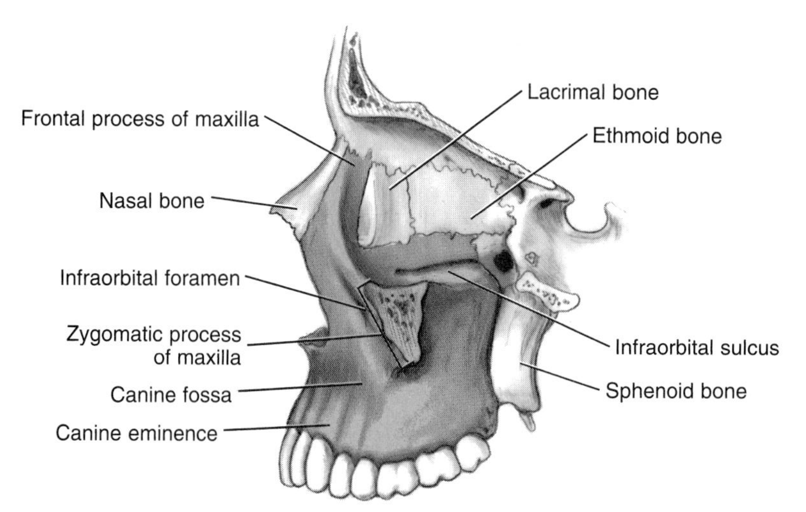

▼ **Figure 3–46**
Cutaway view of the lateral aspect of the skull with the maxilla highlighted.

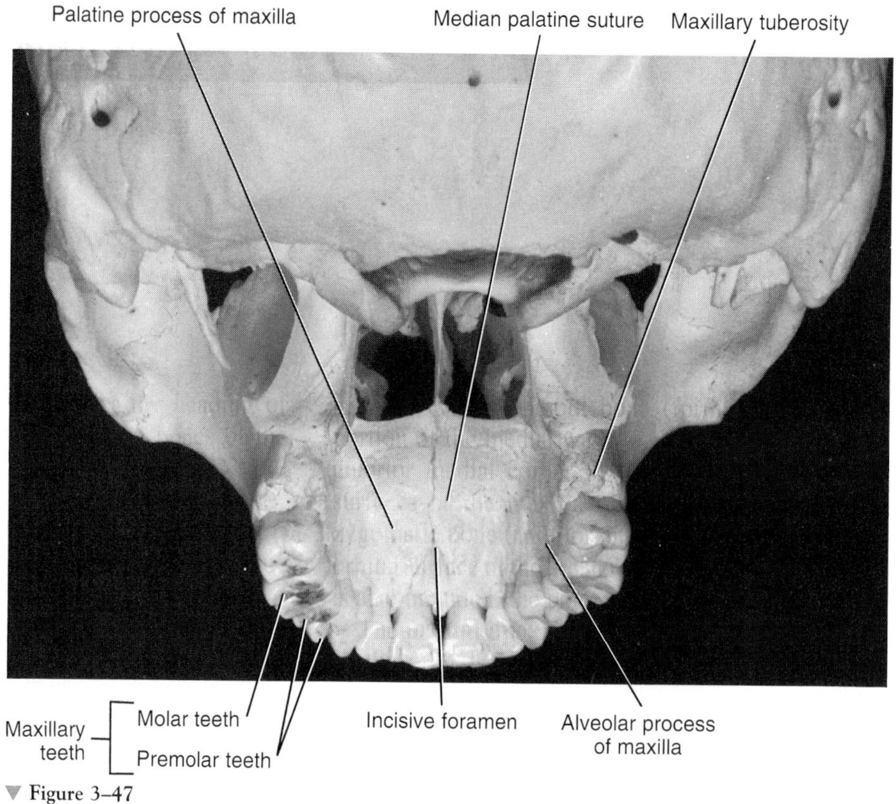

▼ **Figure 3–47**
Posterior-inferior view of the maxillae and hard palate.

nasopalatine nerves and blood vessels from the nasal cavity to the palate. The incisive foramen is a landmark for the administration of a local anesthetic for the nasopalatine nerve block (see Chapter 9). The soft tissue bulge over the incisive foramen in the patient is called the **incisive papilla.**

The **alveolar process of the maxilla** (al-**ve**-o-lar) usually contains the roots of the maxillary teeth (Figure 3–47, Table 3–4). The alveolar process of the maxilla can become resorbed in a patient who is completely edentulous in the maxillary arch (resorption occurs to a lesser extent in partially edentulous cases), possibly leading to problems with the maxillary sinuses (discussed later in this chapter). The body of the maxilla is not resorbed with tooth loss, but its walls may become thinner in this case.

The density of the maxillary bone in an area determines the route that a dental infection takes with abscess and fistula formation (see Chapter 12). Finally, the differences in alveolar process density determine the easiest and most convenient areas of bony fracture used during tooth extraction. Thus the maxillary teeth are mechanically easier to remove by fracturing the thinner facial surface rather than the thicker lingual surface.

On the posterior portion aspect of the body of the maxilla is a rounded roughened elevation, the **maxillary tuberosity** (**mak**-sil-lare-ee), just posterior to the most distal molar of the maxillary dentition. The maxillary tuberosity is perforated by one or more **posterior superior alveolar foramina** (al-**ve**-o-lar), where the posterior superior alveolar nerve and blood vessel branches enter the bone from the back. The maxillary tuberosity is a landmark for the administration of a local anesthetic for the posterior superior alveolar block (see Chapter 9).

▼ Mandible

The lower jaw or **mandible** (**man**-di-bl) is a single facial bone that is the only freely movable bone of the skull (Figure 3–48). This bone is also the largest and strongest facial bone. The mandible has a movable articulation with the temporal bones at each temporomandibular joint (see Chapter 5 for more information). The mandible is more effectively studied when it is temporarily removed from the skull model. The mandible can be studied from three views: the anterior, lateral, and medial views.

▼ ANTERIOR VIEW OF THE MANDIBLE

On the anterior surface of the mandible are many important landmarks (Figure 3–49). The **mental protuberance** (**ment**-il pro-**too**-ber-ins), the bony prominence of the chin, is located beneath the roots of the mandibular incisors. The mental protuberance is more pronounced in males but can be visualized and palpated in females. In the midline on the surface of

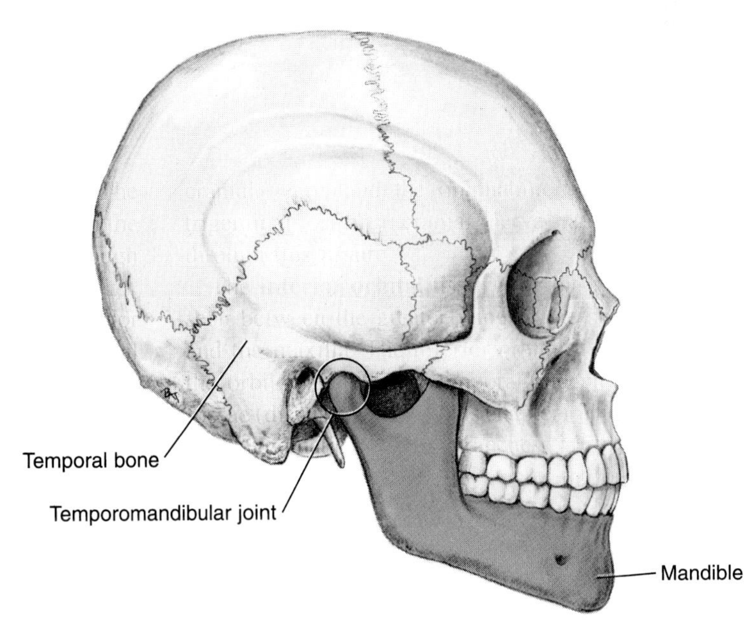

▼ **Figure 3–48**
Lateral view of the skull showing the mandible and the temporomandibular joint.

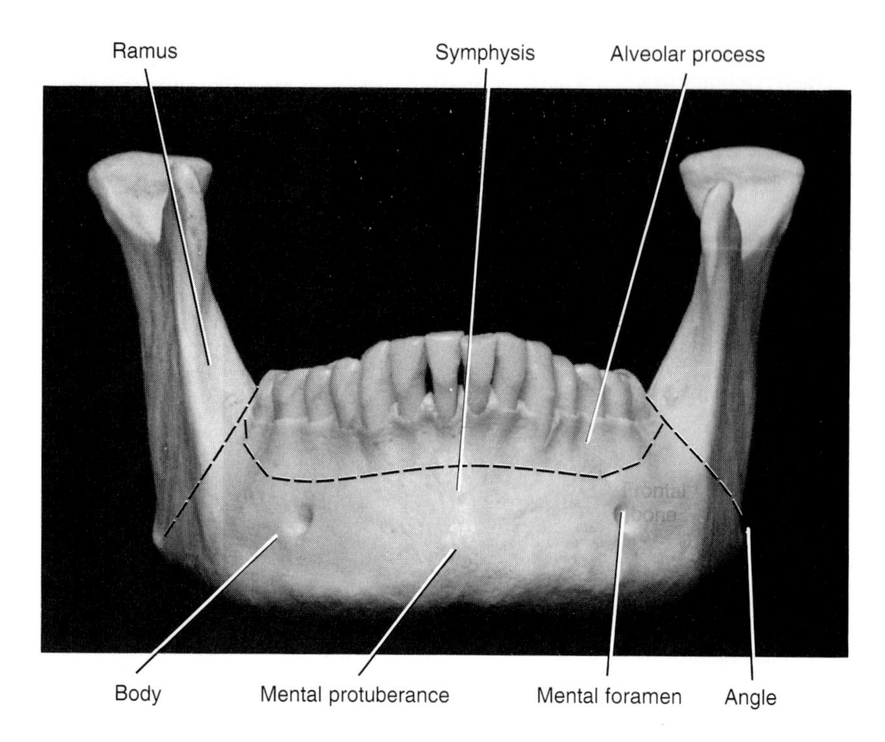

▼ **Figure 3–49**
Anterior view of the mandible and its associated landmarks.

the mandible is a faint ridge, an indication of the mandibular **symphysis** (**sim**-fi-sis), where the bone is formed by the fusion of right and left processes. Like other symphyses in the body, this is a midline articulation where the bones are joined by fibrocartilage, but this articulation fuses together in early childhood.

Farther posteriorly on the surface of the mandible, typically between the apices of the first and second mandibular premolars, is an opening, the **mental foramen** (**ment**-il) (Table 3–3). As the mandibular growth proceeds in young children, the mental foramen alters direction from anterior to posterosuperior. The mental foramen allows the entrance of the mental nerve and blood vessels into the mandibular canal (discussed later in this chapter).

The mental foramen's posterosuperior opening in adults signifies the changed direction of the emerging mental nerve. This is an important landmark to note on an oral radiograph prior to the administration of a local anesthetic for a mental or incisive block (Chapter 9). It is also important not to confuse the mental foramen on a radiograph with a periapical lesion related to the teeth or other oral lesions.

The heavy horizontal portion of the lower jaw below the mental foramen is called the **body of the mandible** (Figure 3–49). Superior to this, the portion of the lower jaw usually containing the roots of the mandibular teeth is the **alveolar process of the mandible** (al-**ve**-o-lar) (Table 3–4). In children, the body of the mandible, along with the alveolar process, elongates to provide space for the additional teeth as one nears adulthood.

If a patient becomes completely edentulous in the mandibular arch (or to some extent in partial edentulous cases), the lower alveolar process can become resorbed. This resorption can occur to such an extent that the mental foramen is virtually on the superior border of the mandible, instead of opening on the anterior surface, changing its relative position. The body of the mandible is not affected and remains thick and rounded.

The alveolar process of the mandibular incisors is less dense than the body of the mandible in that area and even less dense than the alveolar process of the posterior teeth, allowing local infiltration of the mandibular incisors with a local anesthetic with varying degrees of success (see Chapter 9). The density of the mandibular bone in an area also determines the route that a dental infection takes with abscess and fistula formation (see Chapter 12). Finally, the differences in alveolar process density determine the easiest and most convenient areas of bony fracture used during tooth extraction. Thus the mandibular third molar is mechanically easier to remove by fracturing the thinner lingual surface (being careful of the nearby lingual nerve) rather than the thicker buccal surface.

▼ LATERAL VIEW OF THE MANDIBLE

On the lateral aspect of the mandible, the stout, flat plate of the **ramus** (**ray**-mus) extends upward and backward from the **body of the mandible** on each side (Figure 3–50). During growth of the body, the body of the mandible and alveolar process elongates behind the mental foramen, providing space for three additional permanent teeth. The ramus, which serves as the primary area for the attachment of the muscles of mastication, grows upward and backward, displacing the mental protuberance of the chin downward and forward as one nears adulthood.

The anterior border of the ramus is a thin, sharp margin that terminates in the **coronoid process** (**kor**-ah-noid) (Table 3–4). The main portion of the anterior border of the ramus forms a concave forward curve called the **coronoid notch.** The coronoid notch is a landmark for the administration of a local anesthetic for the inferior alveolar block (see Chapter 9). Inferior to the coronoid notch, the anterior border of the ramus becomes the **external oblique line** (**ob**-leek). The external oblique line is a crest where the ramus joins the body of the mandible.

The posterior border of the ramus is thickened and extends from the **angle of the mandible** to a projection, the **condyle of the mandible** with its neck (Figure 3–50). The **articulating surface of the condyle** (ar-**tik**-you-late-ing) is an oval head involved in the temporomandibular joint. Between the coronoid

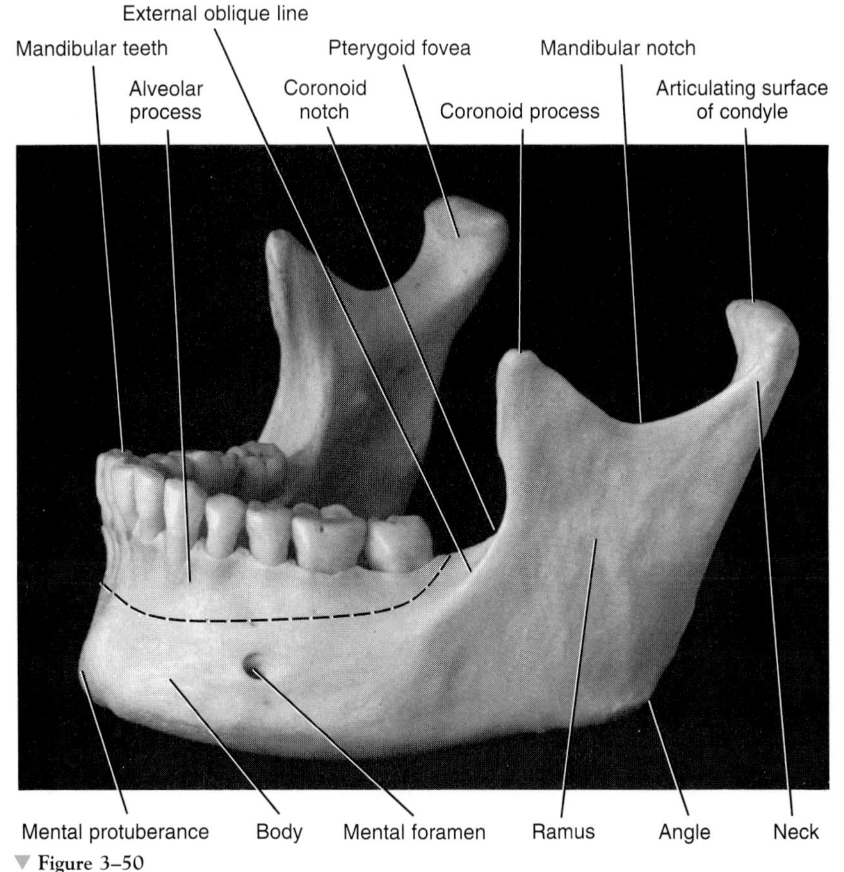

▼ **Figure 3–50**
Slightly oblique lateral view of the mandible and its associated landmarks.

process and the condyle is a depression, the **mandibular notch** (man-**dib**-you-lar).

▼ MEDIAL VIEW OF THE MANDIBLE

Visible on the inner aspect or medial view of the mandible are the **body of the mandible, alveolar process of the mandible** (al-**ve**-o-lar) (Table 3–4), and **ramus** (**ray**-mus) (Figure 3–51). In addition, near the midline of the mandible is a cluster of small projections called the **genial tubercles** (ji-**ni**-il) or mental spines, which is another muscle attachment area.

At the lateral edge of each mandibular alveolar process is a rounded roughened area, the **retromolar triangle** (re-tro-**moh**-lar), just posterior to the most distal molar of the mandibular dentition. The retromolar triangle is a bony landmark that when covered with soft tissue is the retromolar pad in a patient.

Along each medial surface of the body of the mandible is the **mylohyoid line** (my-lo-**hi**-oid) or mylohyoid ridge that extends posteriorly and superiorly, becoming more prominent as it ascends each body. The mylohyoid line is the point of attachment of the mylohyoid muscle that forms the floor of the mouth. The roots of the posterior mandibular teeth often extend internally below the mylohyoid line.

A shallow depression, the **sublingual fossa** (sub-**ling**-gwal), which used to contain the sublingual salivary gland, is located superior to the anterior portion of the mylohyoid line. Below the posterior portion of the mylohyoid line and below the posterior mandibular teeth is a deeper depression, the **submandibular fossa** (sub-man-**dib**-you-lar), which used to contain the submandibular salivary gland.

On the internal surface of the ramus is a central opening, the **mandibular foramen** (man-**dib**-you-lar), which is the opening of the **mandibular canal** (Figure 3–51, Table 3–3). The inferior alveolar nerve and blood vessels exit the mandible through the mandibular foramen after traveling in the mandibular canal. With age and tooth loss, the alveolar process is absorbed so that the mandibular canal is nearer the superior border. Sometimes with excessive alveolar process absorption, the mandibular canal disappears entirely and exposes the inferior alveolar nerve from its bony protection.

Rarely, a patient may have a bifid inferior alveolar nerve, in which case a second mandibular foramen, more inferiorly placed, exists and can be detected by noting a doubled mandibular canal on a radiograph (see Chapter 8 for more information). It is important to keep this anatomical variant concerning the mandibu-

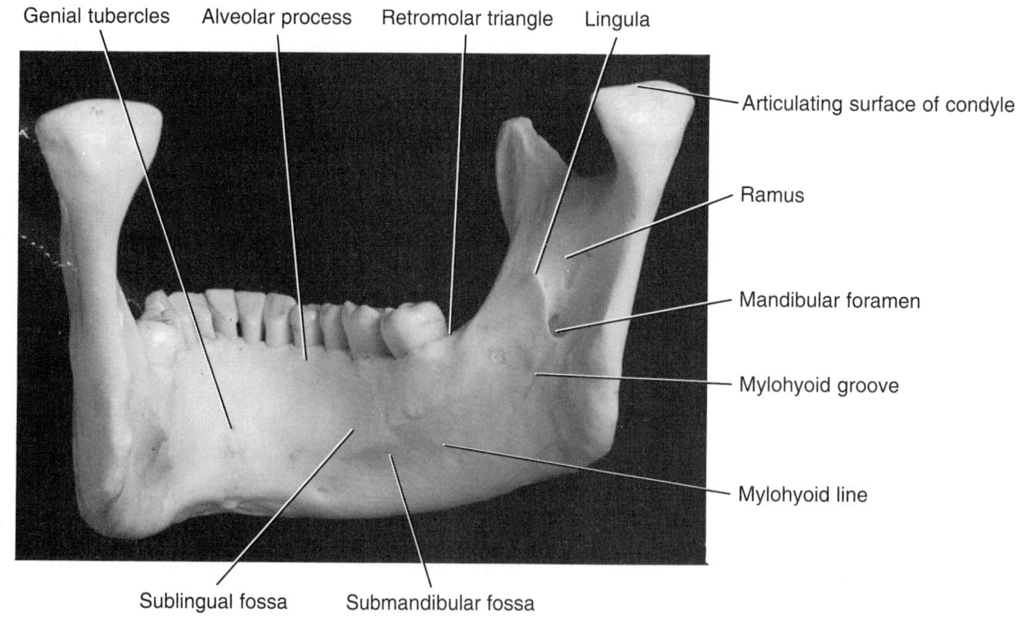

▼ **Figure 3–51**
Internal view of the mandible and its associated landmarks.

lar foramen in mind when administering a local anesthetic for the inferior alveolar block (see Chapter 9).

Overhanging the mandibular foramen is a bony spine, the **lingula** (**ling**-gul-ah), which serves as an attachment for the sphenomandibular ligament associated with the temporomandibular joint (see Chapter 5 for more information). A small groove, the **mylohyoid groove** (my-lo-**hi**-oid), passes forward and downward from the mandibular foramen. The mylohyoid nerve and blood vessels travel in the mylohyoid groove.

The **articulating surface of the condyle** (ar-**tik**-you-late-ing) can be seen in this view. This is where the mandible articulates with the temporal bone at the temporomandibular joint (see Chapter 5 for more information). Below the articular surface of the condyle on the anterior surface is a triangular depression, the **pterygoid fovea** (**teh**-ri-goid fo-**vee**-ah) (Figure 3–50).

PARANASAL SINUSES

The **paranasal sinuses** (pare-ah-**na**-zil **sy**-nuses) are paired air-filled cavities in bone (Figures 3–52 and 3–53). These sinuses are lined with mucous membranes. The paranasal sinuses include the frontal, sphenoid, ethmoid, and maxillary sinuses. The sinuses communicate with the nasal cavity through small ostia or openings in the lateral nasal wall.

The sinuses serve to lighten the skull bones, act as sound resonators, and provide mucus for the nasal cavity. Yet the mucous membranes of the sinuses can become inflamed and congested with mucus as in a ***primary sinusitis*** (sy-nu-**si**-tis). A primary sinusitis can involve allergies or an infection occurring in the sinus. The symptoms of sinusitis are headache, usually near the involved sinus, and foul-smelling nasal or pharyngeal discharge, possibly with some systemic signs of infection such as fever and weakness. The skin over the involved sinus can be tender, hot, and red owing to the inflammatory process in the area.

This congestion of the sinus space with mucus can lead to blockage of the ostia, preventing normal air exchange and drainage into the nasal cavity. In extreme cases of sinusitis, surgery is performed to enlarge the ostia in the lateral walls of the nasal cavity, creating adequate drainage.

An infection in one sinus can travel through the nasal cavity to other sinuses, leading to serious complications for the patient. Since the maxillary posterior teeth are in close proximity to the maxillary sinus, this can also create clinical problems if there are any disease processes, such as an infection, in any of these teeth (discussed later and in Chapter 12). These clinical problems can include a ***secondary sinusitis*** (sy-nu-**si**-tis), inflammation of the sinuses from another source, such as an infection of the adjacent teeth. A ***perforation*** (per-fo-**ray**-shun), an abnormal hole in the wall of the sinus, also can occur with infection.

Frontal Sinuses

The paired **frontal sinuses** (**frunt**-il **sy**-nuses) are located in the frontal bone just superior to the nasal cavity (Figures 3–23 to 3–25). These two paranasal sinuses are asymmetrical (around 2–3 centimeters in diameter), but the left and right sinuses are always separated by a septum. Each frontal sinus communicates with and drains into the nasal cavity by a constricted canal to the middle nasal meatus, the **frontonasal duct** (frunt-il-**na**-zil dukt).

Sphenoid Sinuses

The paired **sphenoid sinuses** (**sfe**-noid **sy**-nuses) are located in the body of the sphenoid bone (Figure 3–33). These two paranasal sinuses are frequently asymmetrical (around 1.5–2.5 centimeters in diameter). The sphenoid sinuses communicate with and drain into the nasal cavity through an opening superior to each superior nasal concha.

Ethmoid Sinuses

The **ethmoid sinuses** (**eth**-moid **sy**-nuses) or ethmoid air cells are a variable number of small cavities in the lateral mass of the ethmoid bone (see Figure 3–37). These paranasal sinuses are roughly divided into the anterior, middle, and posterior ethmoid air cells. The posterior ethmoid air cells open into the superior meatus of the nasal cavity, and the middle and anterior ethmoid air cells open into the middle meatus.

Maxillary Sinuses

The **maxillary sinuses** (**mak**-sil-lare-ee **sy**-nuses) are paired paranasal sinuses, each located in the body of

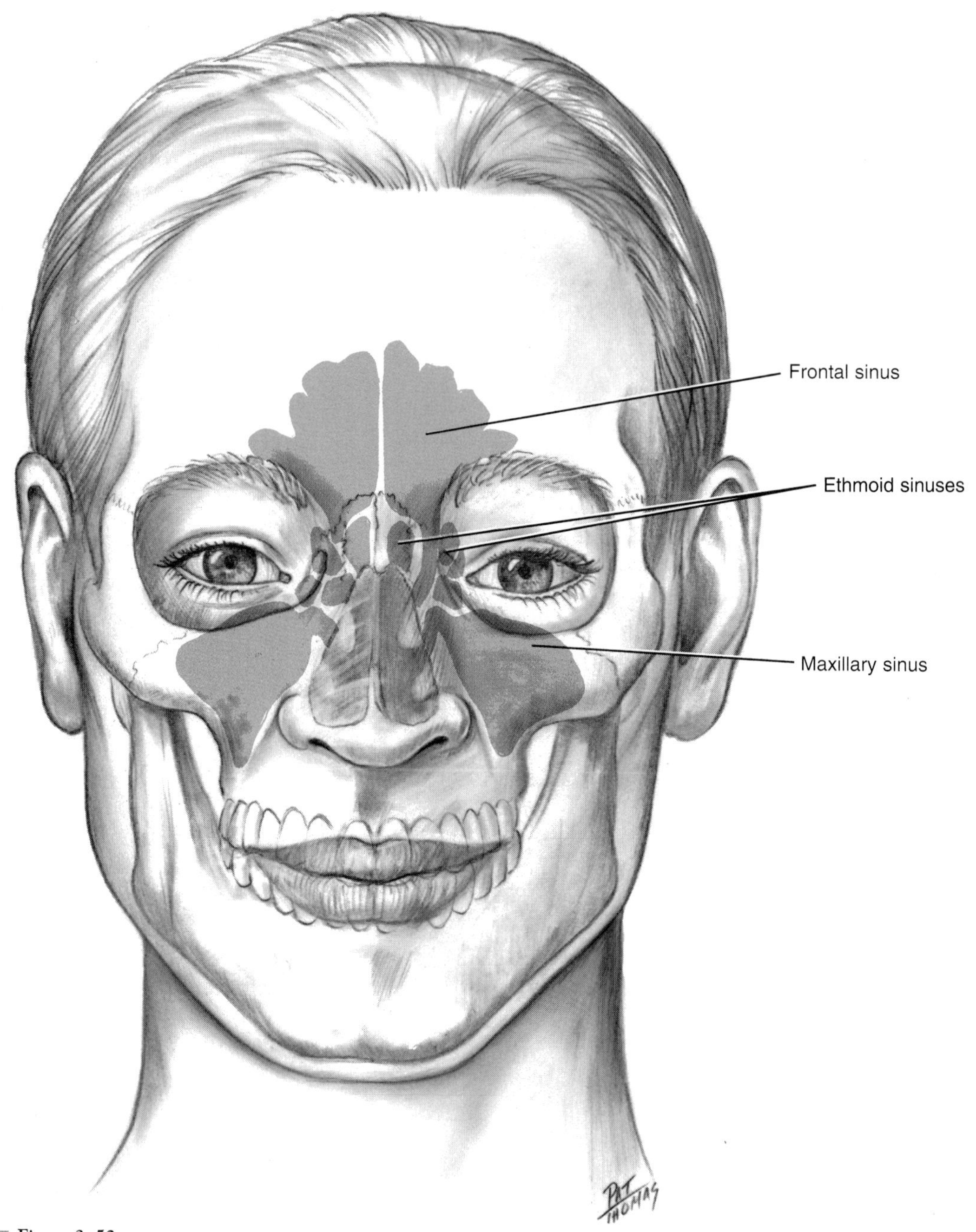

▼ **Figure 3–52**
Anterior view of the skull and the paranasal sinuses.

the maxilla, just posterior to the maxillary canine and premolars (Figure 3–45). The size varies according to the individuals and their age. These pyramid-shaped sinuses are the largest of the paranasal sinuses, and each one has an apex, three walls, a roof, and a floor. A portion of the sinuses can be seen on radiographs of the maxillary posterior teeth.

The apex of the pyramid of the maxillary sinus points into the zygomatic arch, and the medial wall is formed by the lateral wall of the nasal cavity. The anterior wall corresponds to the anterior or facial wall of the maxilla, and the posterior wall is the infratemporal surface of the maxilla, the maxillary tuberosity. The roof of the maxillary sinus is the orbital floor, and

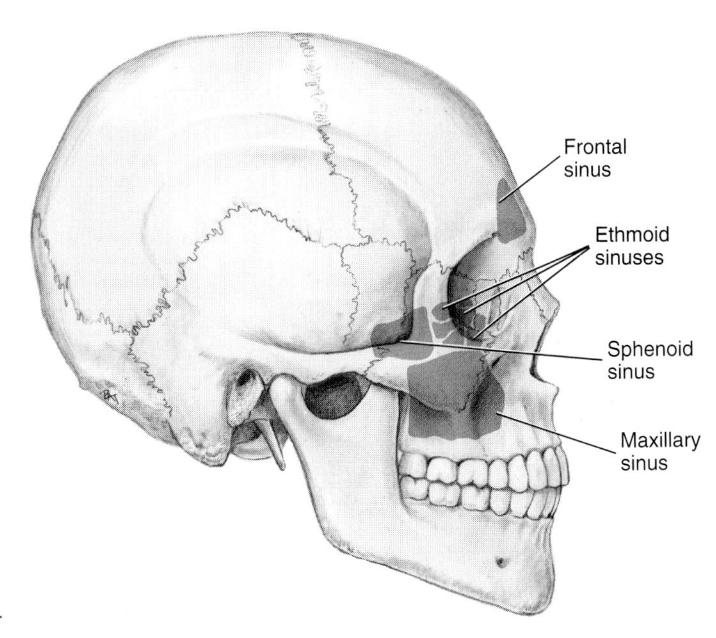

▼ Figure 3–53
Lateral view of the skull and the paranasal sinuses.

the floor is the alveolar process of each maxilla. Each maxillary sinus is divided into communicating compartments by bony walls or septa.

Because of the close proximity of the maxillary sinus to the alveolar process containing the roots of the maxillary posterior teeth, the periodontal tissues of these teeth may be in direct contact with the mucosa of the maxillary sinus. This close proximity can cause serious clinical problems such as secondary sinusitis and perforation during infection (see Chapter 12 for more information), extraction, or trauma related to the maxillary posterior teeth. The discomfort associated with a primary maxillary sinus infection can mimic the same discomfort with an endodontic or periodontal infection of the maxillary posterior teeth.

With age, the enlarging maxillary sinus may surround the roots of the maxillary posterior teeth and extend its margins into the body of the zygoma. If the maxillary posterior teeth are lost, the maxillary sinus may expand even more, thinning the bony floor of the alveolar process, so that only a thin shell of bone is present.

The maxillary sinus drains into the middle meatus on each side. Drainage of the maxillary sinus is complicated and may promote a prolonged or chronic sinusitis because the ostium of each sinus is higher than the floor of the sinus cavity. Surgery may need to be performed in the cases of chronic maxillary sinusitis.

▼ FOSSAE OF THE SKULL

Three deeper depressions or ***fossae*** are present on the external surface of the skull. The bony boundaries for these paired fossae, the temporal, infratemporal, and pterygopalatine fossae, should be located on the skull model and skull diagrams (Table 3–6). These fossae are important landmarks of the skull for locating muscles, blood vessels, and nerves (Table 3–7).

▼ Temporal Fossa

The **temporal fossa** (**tem**-poh-ral) is a flat fan-shaped depression on the lateral surface of the skull (Figures 3–54 and 3–13). The temporal fossa is formed by parts of five bones: the zygomatic, frontal, greater wing of the sphenoid, temporal, and parietal bones.

The boundaries of the temporal fossa are, superiorly and posteriorly, the inferior temporal line; anteriorly, the frontal process of the zygomatic bone; medially, the surface of the temporal bone; and laterally, the zygomatic arch. Inferiorly, the boundary between the temporal fossa and the infratemporal fossa is the infratemporal crest on the greater wing of the sphenoid bone.

The temporal fossa includes a narrow strip of the parietal bone, the squamous portion of the temporal bone, the temporal surface of the frontal bone, and the

Table 3–6

BOUNDARIES OF FOSSAE OF THE SKULL

Boundaries of Fossae	Temporal Fossa	Infratemporal Fossa	Pterygopalatine Fossa
Superior	Inferior temporal line	Greater wing of sphenoid bone	Inferior surface of sphenoid bone body
Anterior	Frontal process of zygomatic bone	Maxillary tuberosity	Maxillary tuberosity
Medial	Surface of temporal bone	Lateral pterygoid plate	Vertical plate of palatine bone
Lateral	Zygomatic arch	Mandibular ramus and zygomatic arch	Pterygomaxillary fissure
Inferior	Infratemporal crest of sphenoid bone	No bony border	Pterygopalatine canal
Posterior	Inferior temporal line	No bony border	Pterygoid process of sphenoid bone

temporal surface of the greater wing of the sphenoid bone. The temporal fossa contains the body of the temporalis muscle and area blood vessels and nerves.

Infratemporal Fossa

The **infratemporal fossa** (in-frah-**tem**-poh-ral) is a paired depression that is inferior to the anterior part of the temporal fossa (Figure 3–54). The temporal fossa and infratemporal fossa are divided by the infratemporal crest on the greater wing of the sphenoid bone. The infratemporal fossa can also be viewed from the inferior aspect of the skull model after temporarily removing the mandible (Figure 3–55).

The boundaries of the infratemporal fossa include superiorly, the greater wing of the sphenoid bone; anteriorly, the maxillary tuberosity; medially, the lateral pterygoid plate; and laterally, the ramus of the mandible and zygomatic arch. There is no bony inferior or posterior boundary.

Many structures pass from the infratemporal fossa into the orbit through the inferior orbital fissure, which is located at the anterior, superior end of the fossa. Other structures pass into the infratemporal fossa from the cranial cavity.

The infratemporal fossa contains the mandibular division of the fifth cranial or trigeminal nerve (including the inferior alveolar and lingual nerves), which enters by way of the foramen ovale (Table 3–3). The fossa also contains the pterygoid plexus and the pterygoid muscles.

The maxillary artery and its branches are also located here, including the middle meningeal artery, which goes into the cranial cavity through the foramen spinosum, the inferior alveolar artery which enters the mandible through the mandibular foramen, and the posterior alveolar artery which enters the maxilla through the posterior superior alveolar foramina (Table 3–3).

Pterygopalatine Fossa

The **pterygopalatine fossa** (**teh**-ri-go-**pal**-ah-tine) is a cone-shaped depression, deep to the infratemporal

Table 3–7

MUSCLES, BLOOD VESSELS, AND NERVES OF FOSSAE OF THE SKULL

	Temporal Fossa	Infratemporal Fossa	Pterygopalatine Fossa
Muscles	Temporalis muscle	Pterygoid muscles	
Blood vessels	Area blood vessels	Pterygoid plexus and maxillary artery (second part) and branches, including middle meningeal artery, inferior alveolar artery, and posterior superior alveolar artery	Maxillary artery (third part) and branches, including infraorbital and sphenopalatine arteries
Nerves	Area nerves	Mandibular nerve, including inferior alveolar and lingual nerves	Pterygopalatine ganglion and maxillary nerve

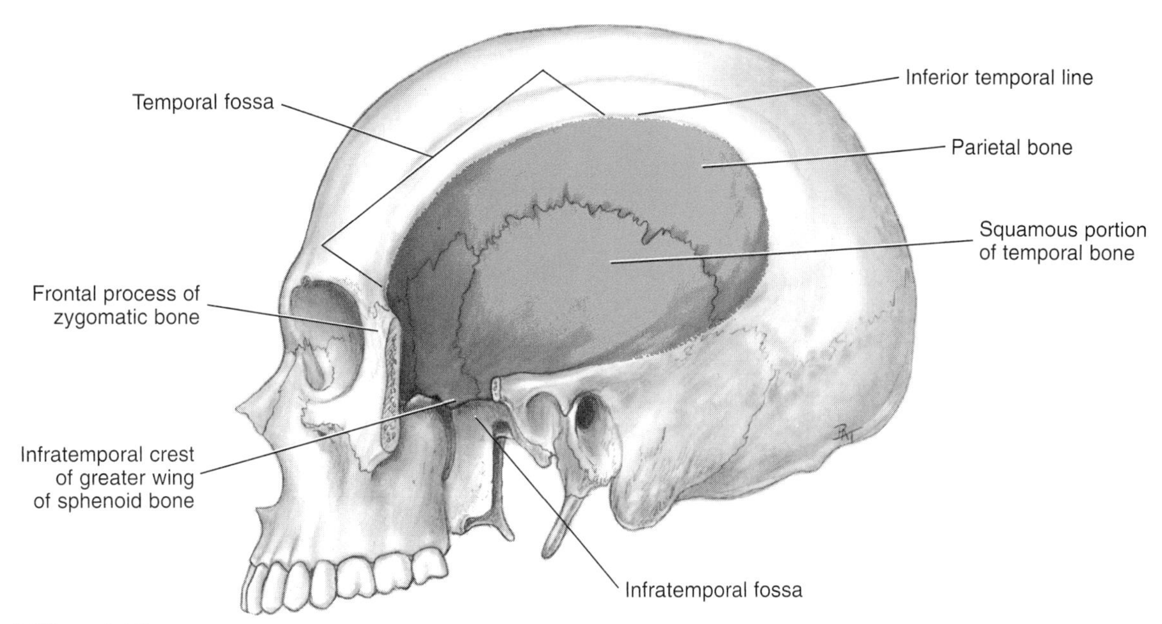

▼ **Figure 3–54**
Lateral view of the skull and the temporal fossa and its boundaries (with portions of the zygomatic and temporal bones removed).

fossa (Figure 3–56). This small but important fossa is located between the pterygoid process and the maxillary tuberosity, close to the apex of the orbit.

The boundaries of the pterygopalatine fossa are superiorly, the inferior surface of the body of the sphenoid bone; anteriorly, the maxillary tuberosity; medially, the vertical plate of the palatine bone; laterally, the pterygomaxillary fissure; inferiorly, the pterygopalatine canal; and posteriorly, the pterygoid process of the sphenoid bone.

The pterygopalatine fossa contains the maxillary artery and nerve and their branches arising here, including the infraorbital and sphenopalatine arteries, the maxillary division of the trigeminal or fifth cranial nerve and branches, and the pterygopalatine ganglion (Table 3–3). The foramen rotundum is the entrance route for the maxillary nerve. A second foramen, the pterygoid canal, in the pterygoid process, transmits autonomic fibers to the ganglion. The pterygopalatine canal connects with the greater and lesser palatine foramina of the palatine bones of the hard palate.

▼ BONES OF THE NECK

▼ Cervical Vertebrae

The **cervical vertebrae** (**ser**-vi-kal **ver**-teh-bray) are located in the vertebral column between the skull and the thoracic vertebrae. There are seven cervical vertebrae. All vertebrae have a central **vertebral foramen** (**ver**-teh-brahl) for the spinal cord and associated tissues. In contrast to other vertebrae, the cervical vertebrae are characterized by the presence of a **transverse foramen** in the **transverse process** on each side of the vertebral foramen. The vertebral artery runs through these transverse foramina.

Damage to any of the vertebrae can affect dental treatment as the patient may experience a range of problems from difficulty in movement to paralysis. Only the first two cervical vertebrae will be described because their anatomy is unusual and they are located near the skull.

▼ FIRST CERVICAL VERTEBRA

The first cervical vertebra or **atlas** (**at**-lis) articulates with the skull at the occipital condyles of the occipital bone (Figure 3–57). The atlas has the form of an irregular ring consisting of two **lateral masses** (masz) connected by a short **anterior arch** and a longer **posterior arch** (Figure 3–58). This cervical bone lacks a body and a spine.

The lateral masses can be effectively palpated by placing fingers between the two mastoid processes and the angles of the mandible. More medially, the lateral masses present large concave **superior articular**

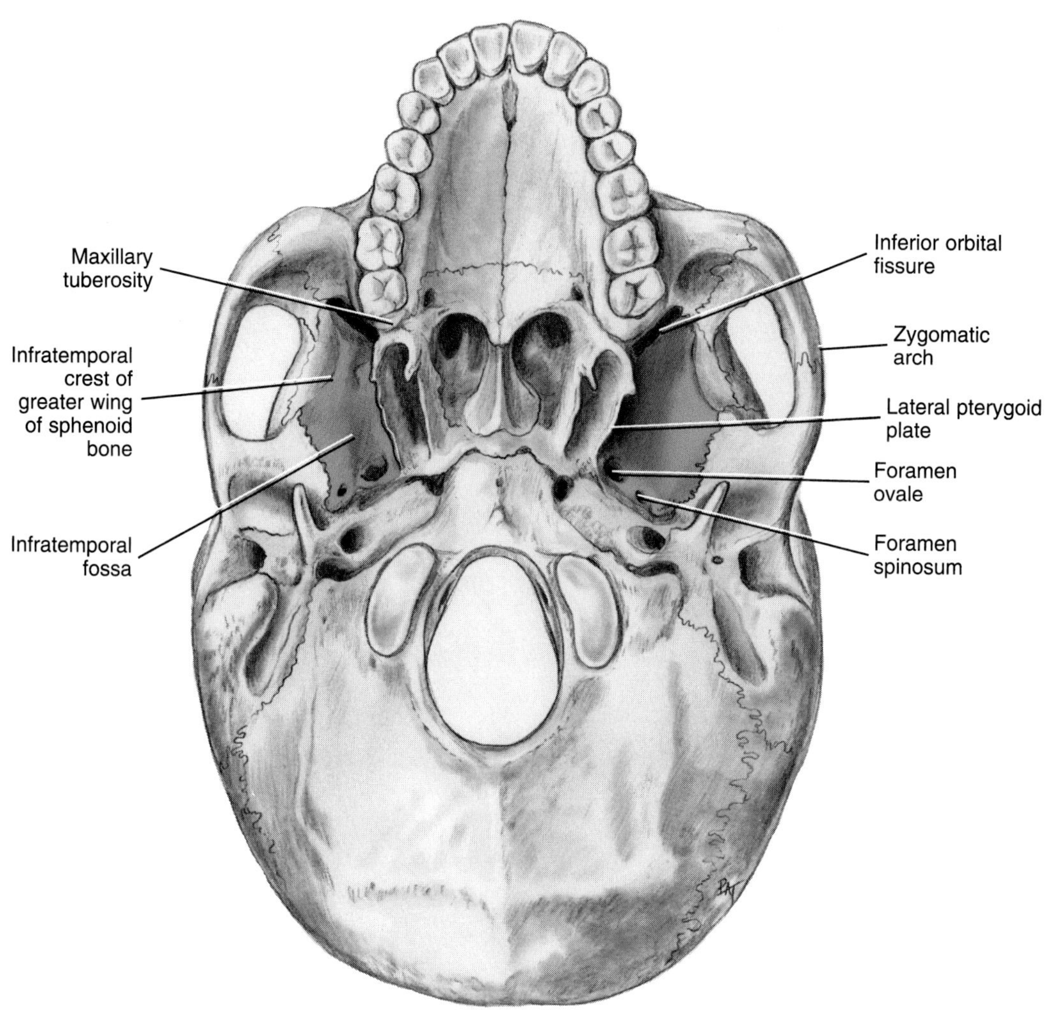

▼ **Figure 3–55**
Inferior view of the skull (with the mandible removed) and the infratemporal fossae and boundaries.

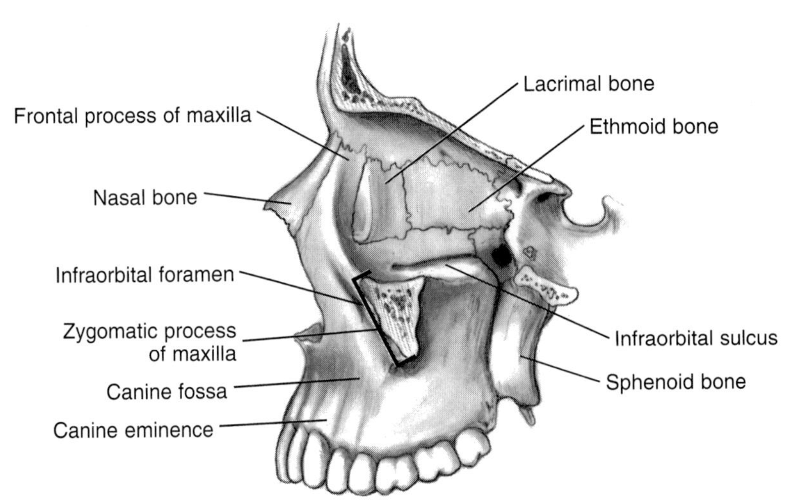

▼ **Figure 3–56**
Oblique lateral view of the base of the skull and the roof of the pterygopalatine fossa and its boundaries.

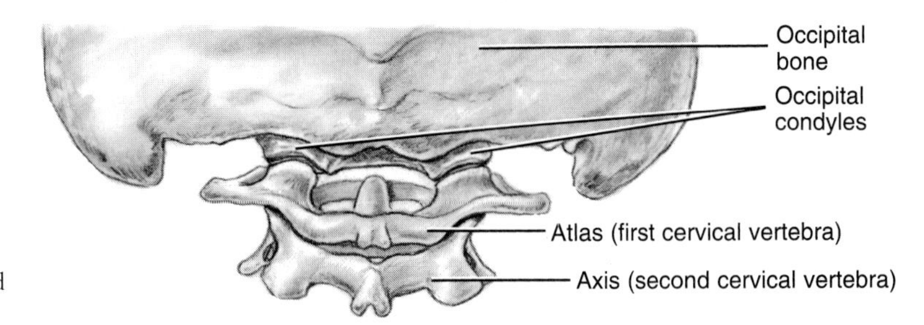

▼ **Figure 3–57**
Posterior view of the skull and the first and second cervical vertebrae.

processes (ar-**tik**-you-lar) for the corresponding occipital condyles of the skull. The lateral masses also have circular **inferior articular processes** for articulation with the second cervical vertebra.

▼ SECOND CERVICAL VERTEBRA

The second cervical vertebra or **axis** (**ak**-sis) is characterized by the **dens** (denz) or odontoid process (Figure 3–59). The dens articulates anteriorly with the anterior arch of the first cervical vertebra (Figure 3–57). The body of the axis is inferior to the dens. The spine of the axis is located posterior to the body. The body and the adjoining transverse process present **superior articular processes** (ar-**tik**-you-lar) for an additional articulation with the inferior articulating surfaces of the atlas. The inferior aspect of the axis presents **inferior articular processes** for articulating with the articular processes of the third cervical vertebra.

▼ Hyoid Bone

The **hyoid bone** (**hi**-oid) is suspended in the neck. It forms the base of the tongue and larynx. Many muscles attach to the hyoid bone (see Chapter 4 for more information). The hyoid bone is superior and anterior to the thyroid cartilage of the larynx. This bone is typically at the level of the third cervical vertebra but is raised during swallowing and other activities. The hyoid bone can be effectively palpated by feeling below and medial to the angles of the mandible. Do not confuse the hyoid bone with the thyroid cartilage (the Adam's apple).

The hyoid bone does not articulate with any other bones, giving it its characteristic mobility needed for mastication, swallowing, and speech. This bone is horizontally suspended from the end of the styloid process by the stylohyoid ligament and connected by the broad thyrohyoid membrane with the thyroid cartilage (see Chapter 4 for more information).

The U-shaped hyoid bone consists of five parts as seen from an anterior view (Figure 3–60). The anterior portion is the midline **body of the hyoid bone.** There is also a pair of projections on each side of the hyoid bone, the **greater cornu** and **lesser cornu.** These horns serve as attachments for muscles and ligaments.

Abnormalities of Bone

Abnormalities of bone can include bony enlargements and fractures that may heal with abnormal contours.

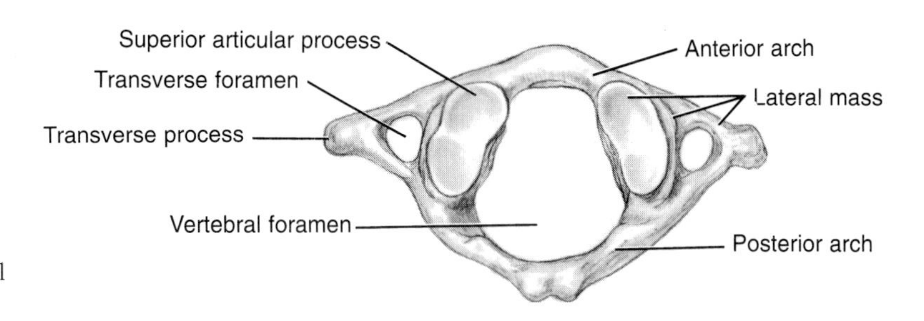

▼ **Figure 3–58**
Superior view of the first cervical vertebra, the atlas.

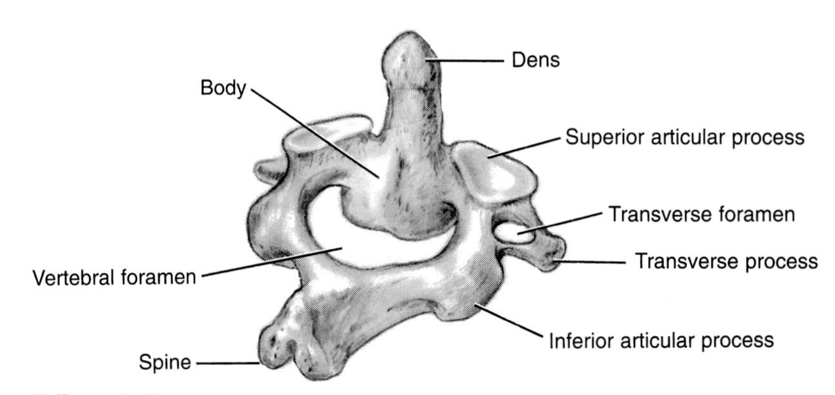

Figure 3–59
Posterosuperior view of the second cervical vertebra, the axis.

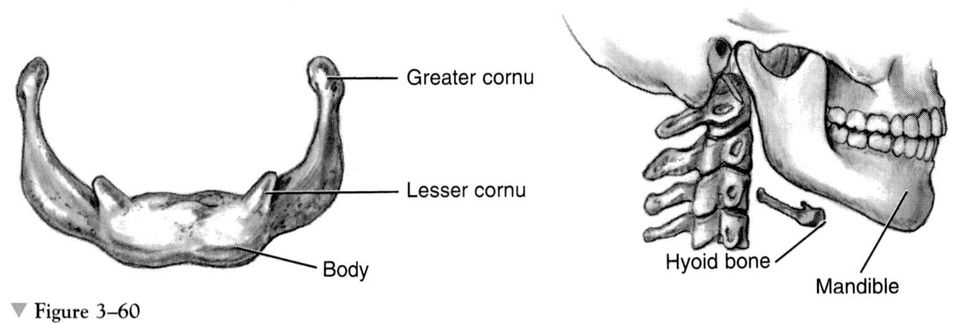

Figure 3–60
Anterior view of the hyoid bone with its location demonstrated on a lateral view of the lower skull and the upper vertebral bones.

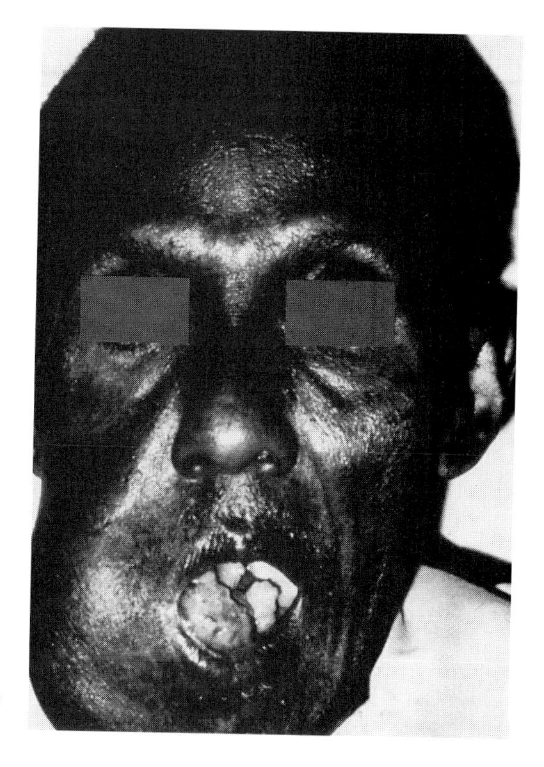

▼ **Figure 3–61**
Bony enlargements may lead to facial asymmetry, such as in this patient with an ameloblastoma, a tumor of dental tissues.

The dental professional needs to record any abnormal areas of bone and make any appropriate referrals. It is important to remember that since many bones of the face are shared by two or more soft tissue components of the face, an abnormality of one facial bone often involves many soft tissue components. A fracture of the frontal bone may clinically involve both the forehead and the eyes.

Bony enlargements can lead to facial asymmetry (Figure 3–61) or nodular intraoral areas (Figure 3–62). Some, such as palatal or mandibular tori, are normal variations, but others can be due to endocrine diseases, causing abnormal bone growth. Bone can also enlarge with tumorous growth of bone or other tissues, such as dental tissues, as in the case of an ameloblastoma.

Bone may also fracture with severe blows to the face. Fractures of the facial skeleton tend to occur at its points of buttress with the cranium. These buttress points include the medial aspect of the orbit, articulation of the zygoma with the frontal and temporal bones, articulation of the pterygoid plates, and the palatine bones and maxillae. If the fracture is bilateral, the entire facial skeleton can be pushed posteriorly, resulting in upper respiratory tract obstruction. These fractures may heal poorly and result in abnormal bony contours.

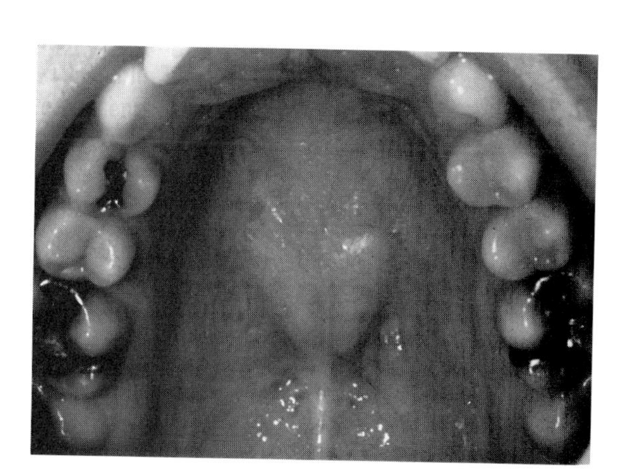

▼ **Figure 3–62**
Nodular bony enlargements may occur in the oral cavity, such as in this case of a palatal torus, a benign growth of bone.

Identification Exercise *Continued*

5. (Figure 3–32)

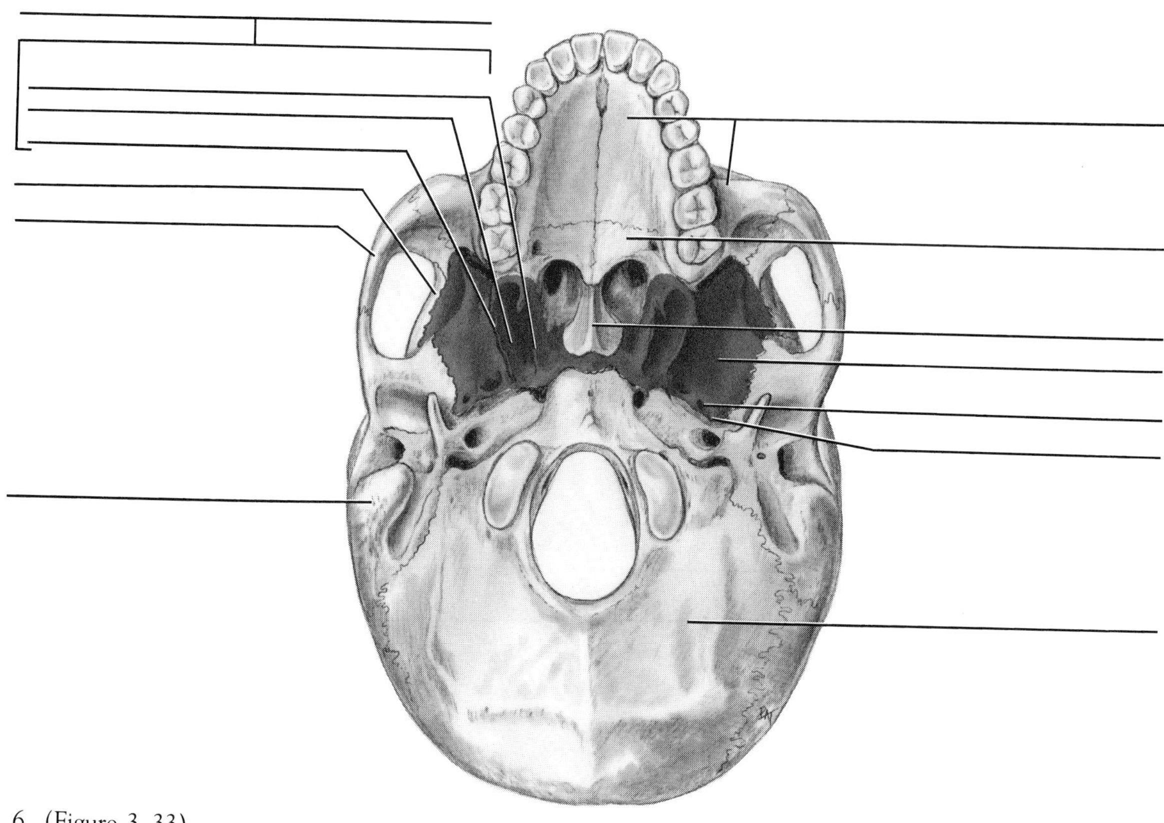

6. (Figure 3–33)

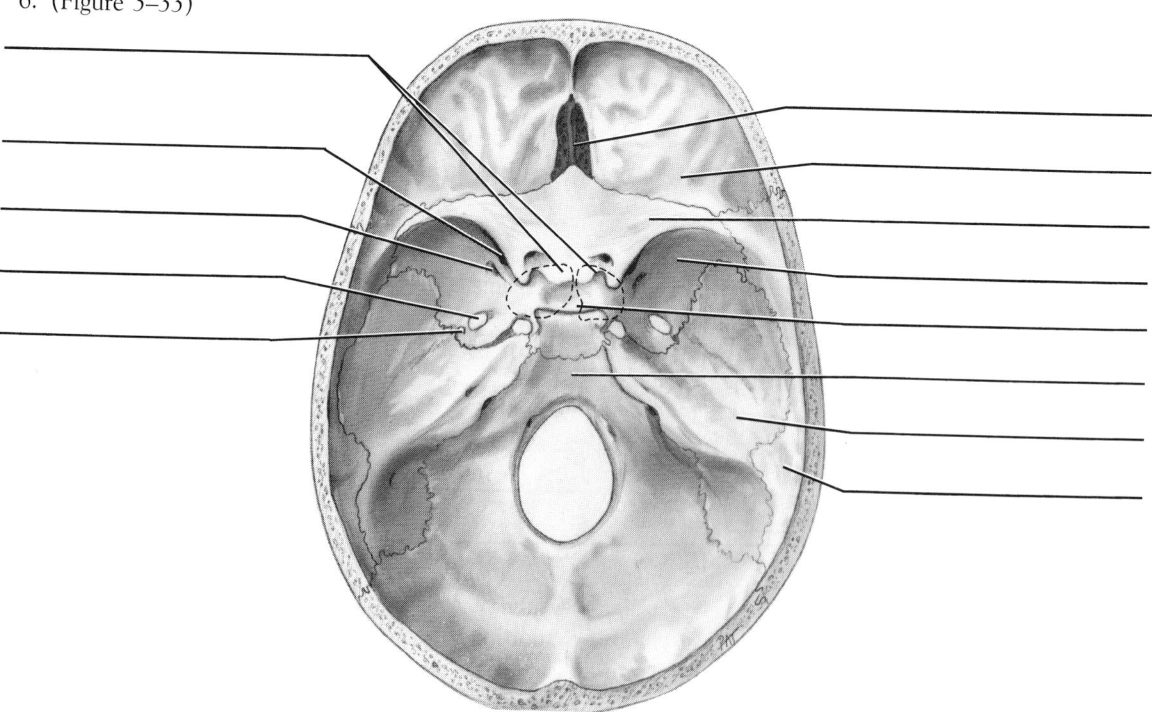

Identification Exercise *Continued*

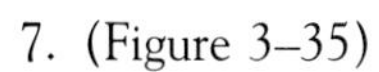

7. (Figure 3–35)

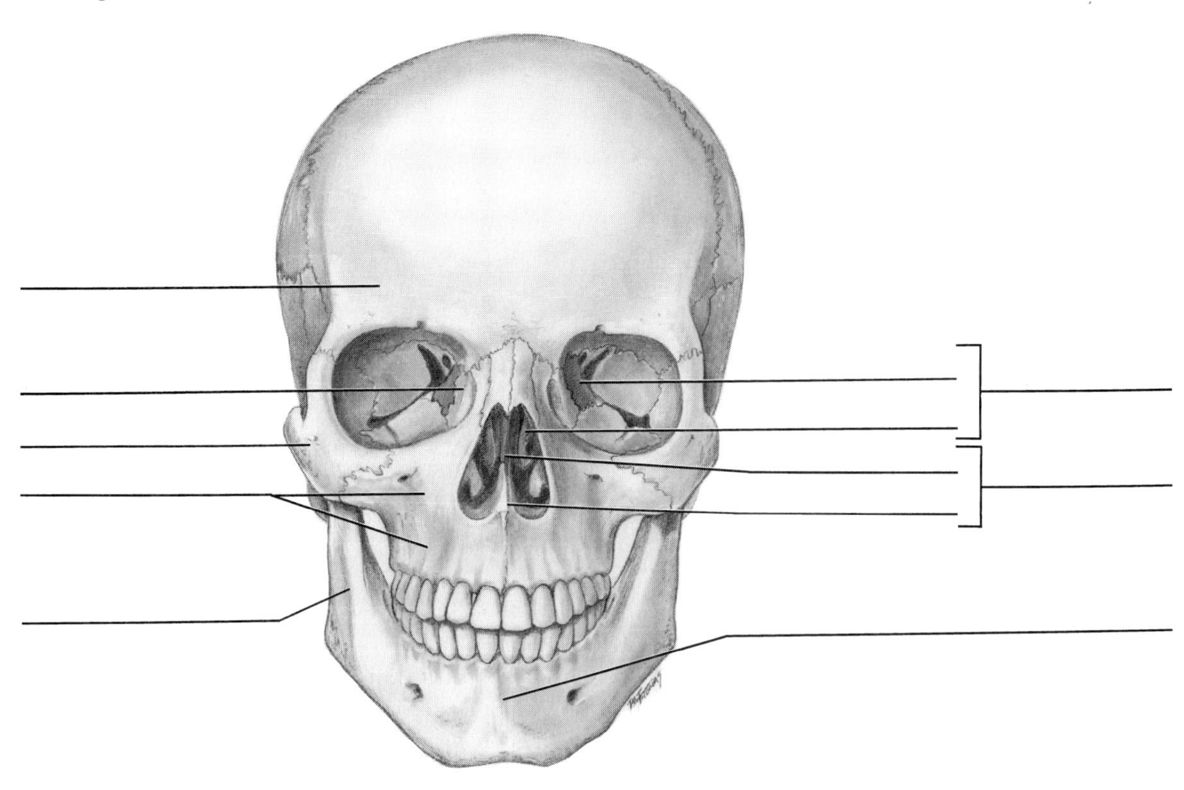

8. (Figure 3–38)

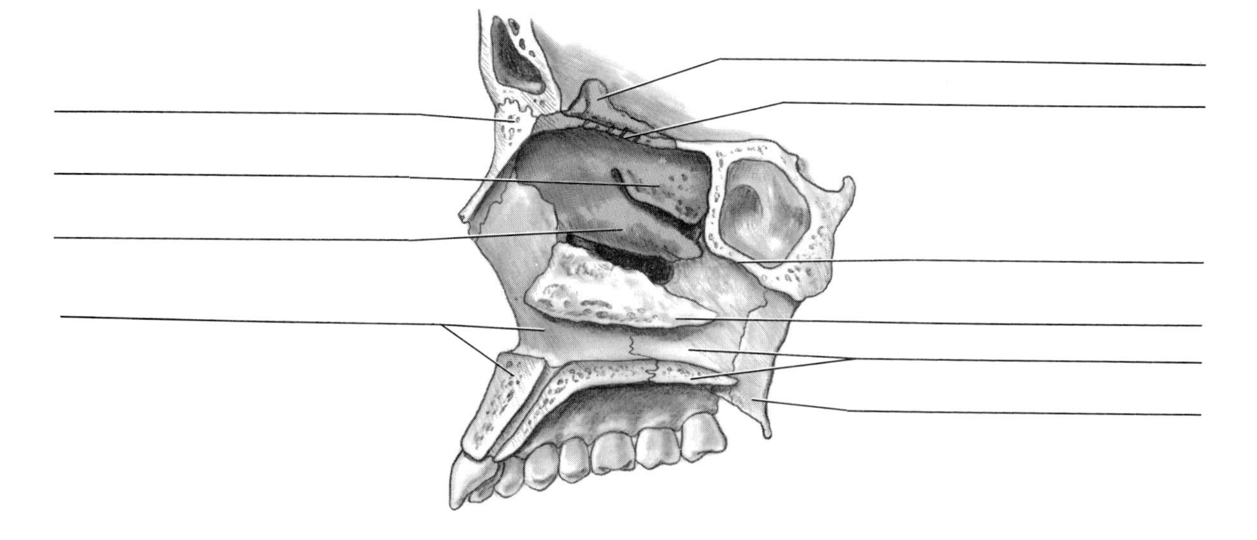

Identification Exercise *Continued*

9. (Figure 3–40)

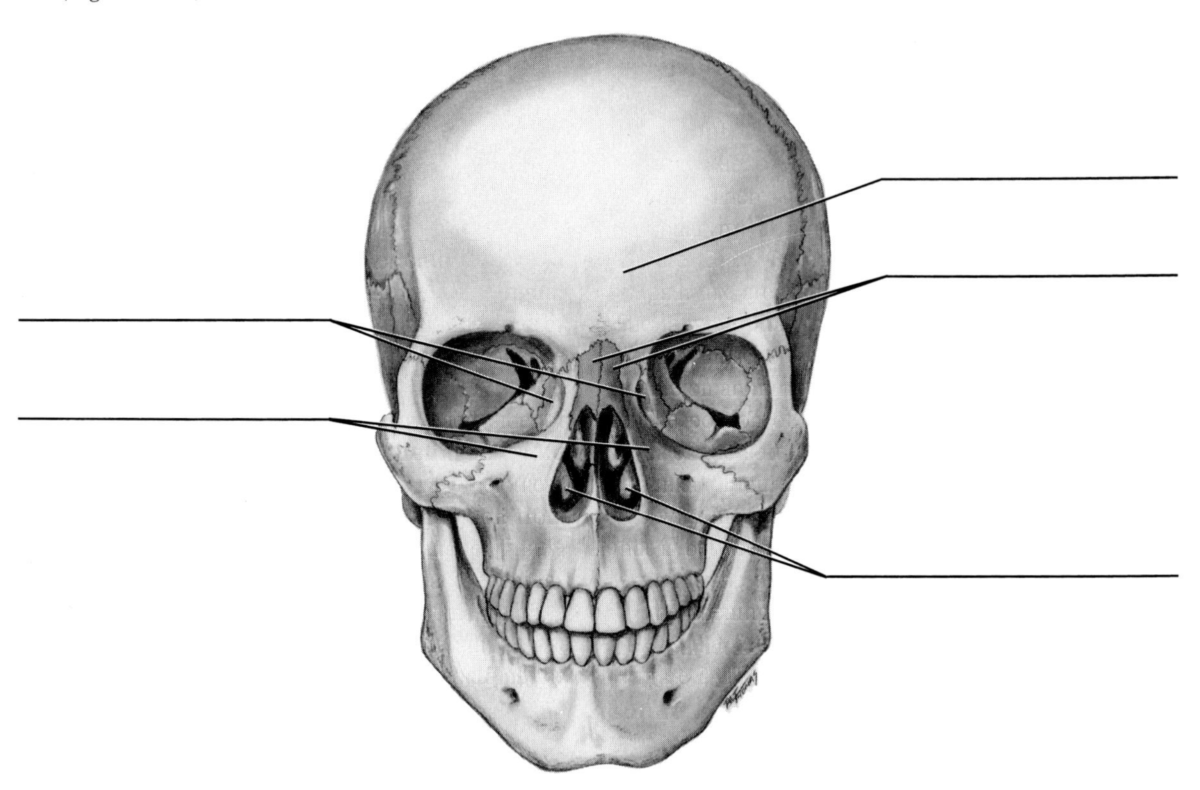

10. (Figure 3–42)

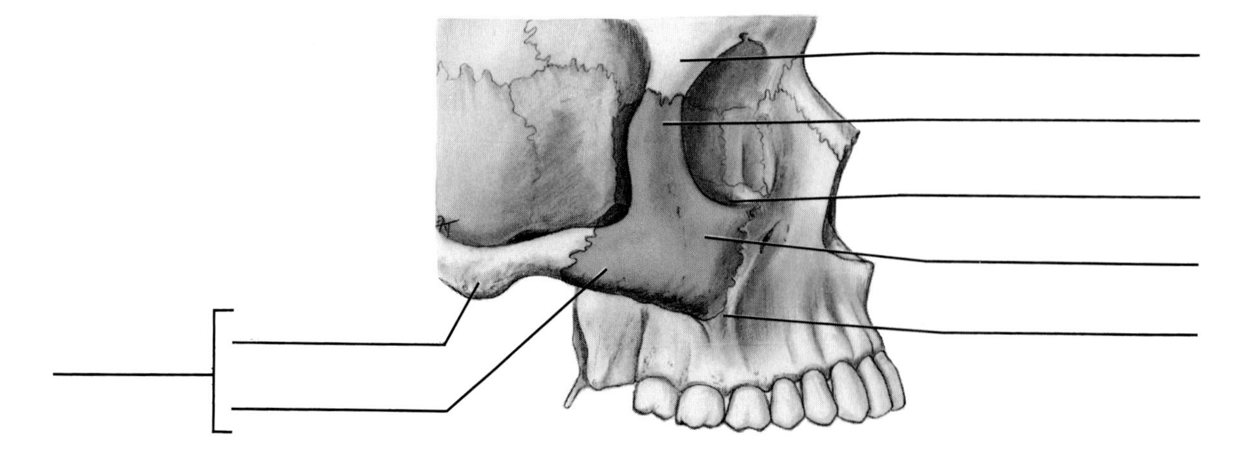

Identification Exercise *Continued*

11. (Figure 3–44)

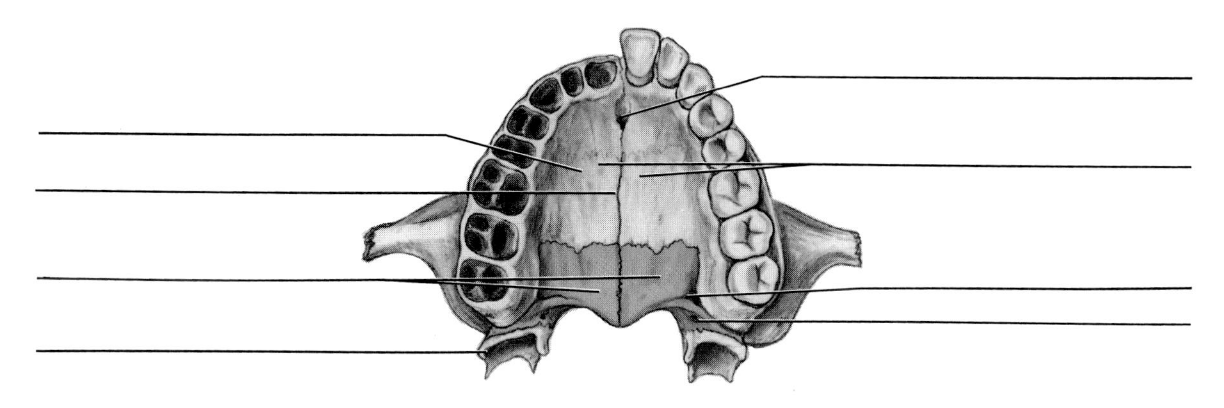

12. (Figure 3–45)

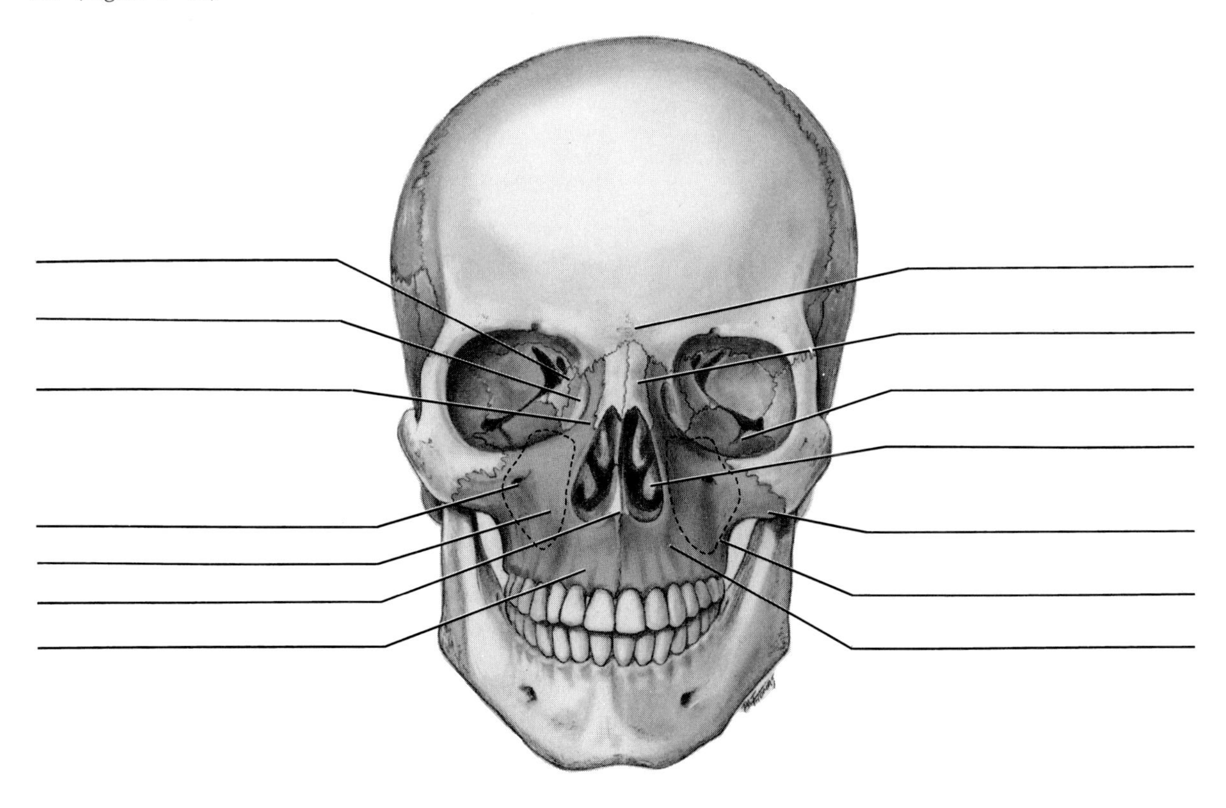

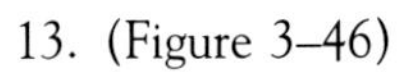

Identification Exercise *Continued*

13. (Figure 3–46)

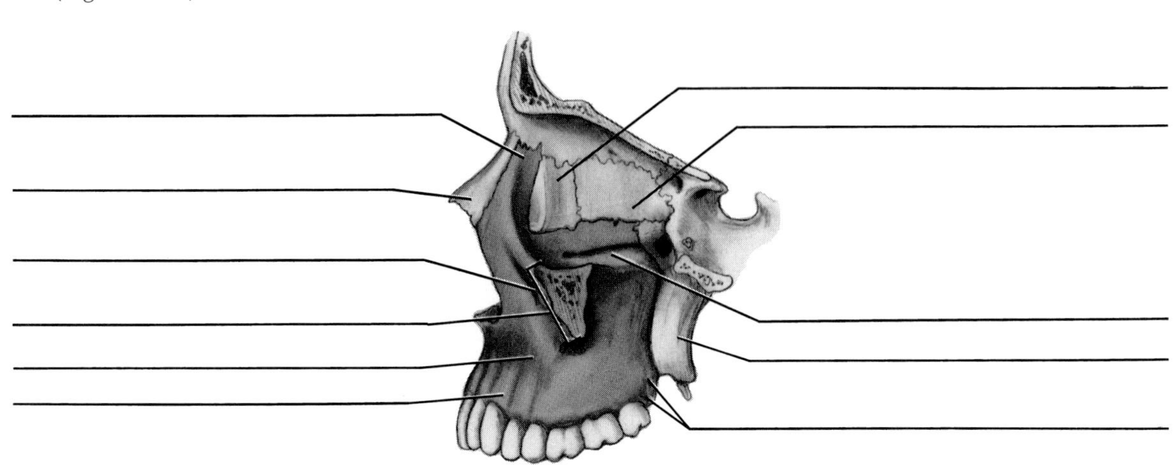

14. (Figure 3–50) (Figure 3–53)

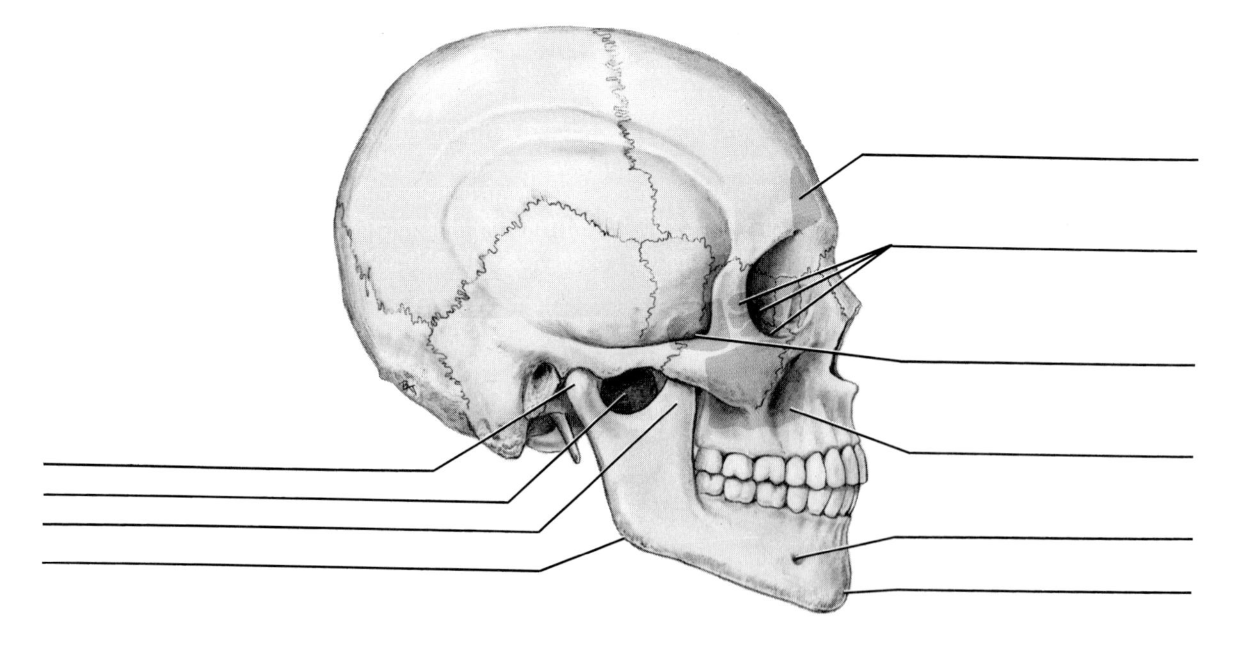

Identification Exercise *Continued*

15. (Figure 3–55)

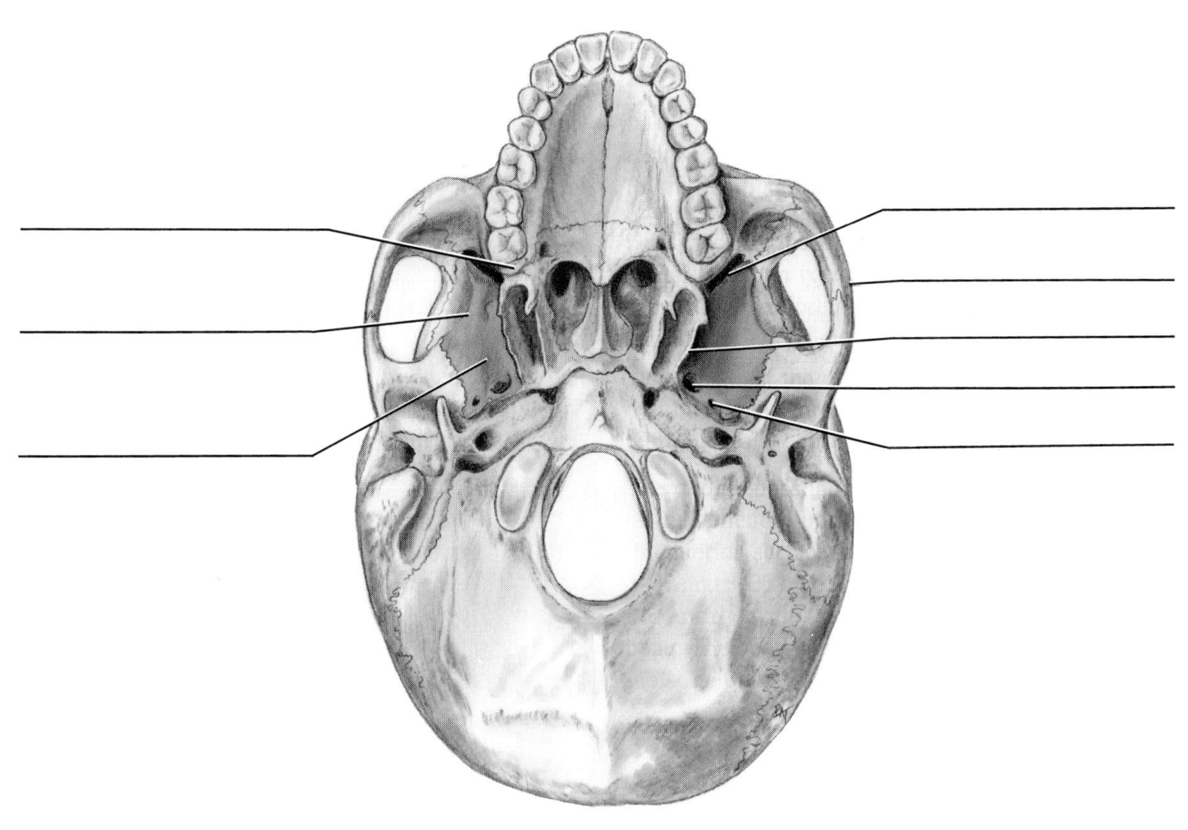

16. (Figure 3–56)

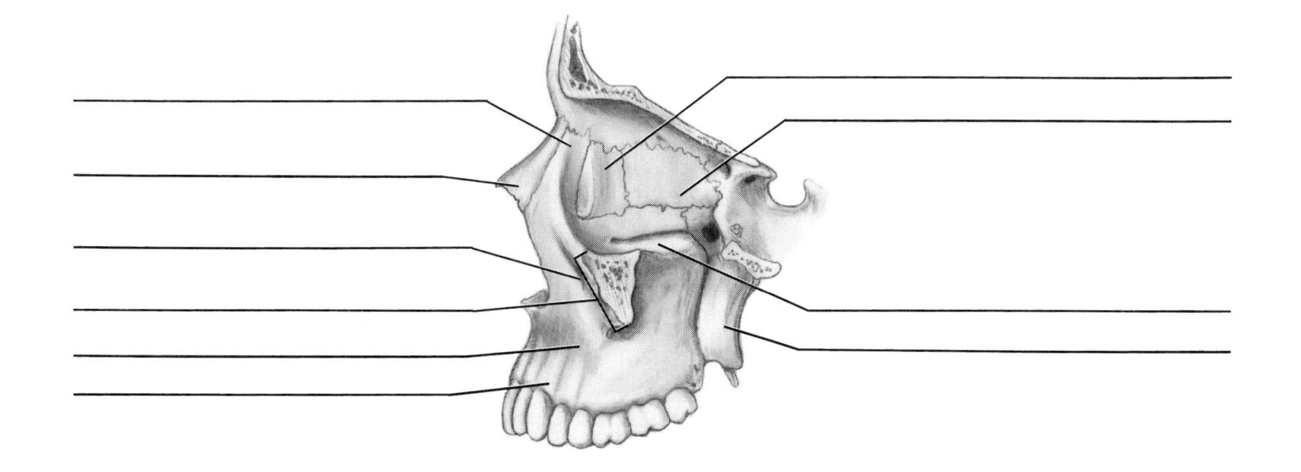

Identification Exercise *Continued*

17. (Figure 3–57)

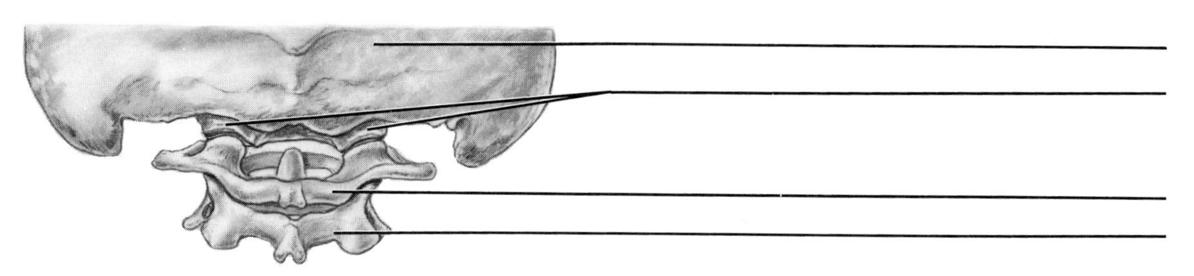

18. (Figure 3–60)

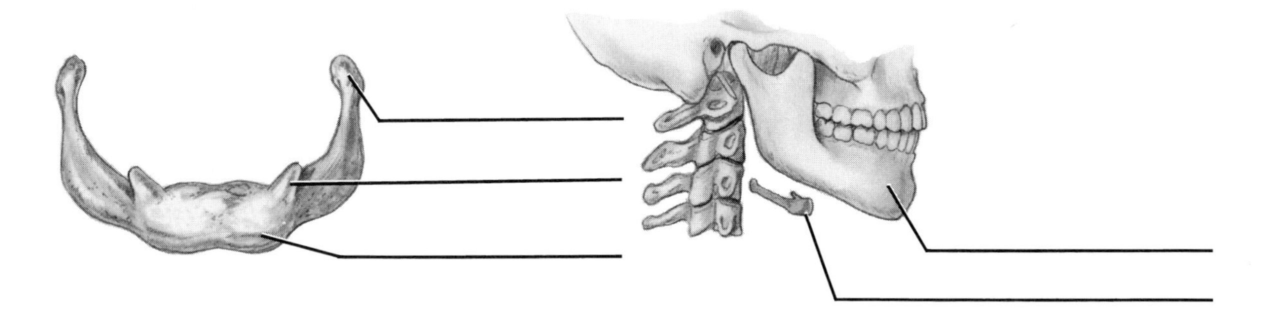

REVIEW QUESTIONS

1. Which of the following features is located on the temporal bone?
   A. Superior temporal line
   B. Foramen rotundum
   C. External acoustic meatus
   D. Cribriform plate
   E. Lamina orbitalis

2. The area immediately posterior to the most distal tooth in the upper arch is called the:
   A. Retromolar triangle.
   B. Postglenoid process.
   C. Cribriform plate.
   D. Maxillary tuberosity.
   E. Hamular process.

3. In addition to the zygomatic bone, which of the following bones has a process that forms the zygomatic arch?
   A. Temporal bone
   B. Maxilla
   C. Sphenoid bone
   D. Palatine bone

4. Which of the following is the name of the articulation of the parietal bones and the occipital bone?
   A. Coronal suture
   B. Squamosal suture
   C. Sagittal suture
   D. Lambdoidal suture

5. Which of the following bony landmarks form an articulation?
   A. Occipital condyles with the atlas
   B. Occipital condyles with the axis
   C. Mandibular fossa with the coronoid notch
   D. Mandibular fossa with the coronoid process

6. Which of the following features is located on the lateral surface of the mandible?
   A. Lingula
   B. Submandibular fossa
   C. Genial tubercles
   D. External oblique line
   E. Mandibular foramen

7. The orbital apex is composed of the lesser wing of the sphenoid bone and the:
   A. Ethmoid bone.
   B. Frontal bone.
   C. Maxilla.
   D. Palatine bone.
   E. Lacrimal bone.

8. Which of the following landmarks is formed by the maxillae?
   A. Mental spine
   B. Median palatine suture
   C. Retromolar triangle
   D. Hamulus
   E. Inferior orbital fissure

9. Which of the following is located in the infratemporal fossa?
   A. Masseter muscle
   B. Pterygopalatine ganglion
   C. Posterior superior alveolar artery
   D. Maxillary division of the fifth cranial nerve

10. The concavity on the anterior border of the coronoid process of the ramus is called the:
    A. Mandibular notch.
    B. Coronoid notch.
    C. Temporal fossa.
    D. Infratemporal fossa.

11. Which of the following landmarks serves to help locate the hyoid bone?
    A. Level of the first cervical vertebra
    B. Superior and anterior to the thyroid cartilage
    C. Articulation with the cartilage of the larynx
    D. Inferior and posterior to the Adam's apple

12. Which of the following forms the floor of each maxillary sinus?
    A. Alveolar process of the maxilla
    B. Facial wall of the maxilla
    C. Infratemporal surface of the maxilla
    D. Lateral wall of the nasal cavity

REVIEW QUESTIONS *Continued*

13. Which of the following processes is located just inferior and medial to the external acoustic meatus?
    A. Pterygoid process
    B. Styloid process
    C. Mastoid process
    D. Hamulus

14. The spaces under the three conchae of the lateral walls of the nasal cavity are called:
    A. Ostia.
    B. Ducts.
    C. Meatus.
    D. Inferior nasal conchae.
    E. Vestibules.

15. Which of the following bones and their processes form the hard palate?
    A. Maxillary processes of the maxillae and the horizontal plates of the palatine bones
    B. Palatal processes of the maxillae and the maxillary plates of the palatine bones
    C. Horizontal plates of the palatine bones and the palatine processes of the maxillae
    D. Maxillary plates of the palatine bones and the horizontal processes of the maxillae

16. Which of the following nerves is associated with the stylomastoid foramen?
    A. Fifth cranial nerve
    B. Seventh cranial nerve
    C. Ninth cranial nerve
    D. Tenth cranial nerve
    E. Eleventh cranial nerve

17. Which of the following bones of the skull is paired?
    A. Sphenoid bone
    B. Ethmoid bone
    C. Occipital bone
    D. Vomer
    E. Parietal bone

18. Which of the following plates is perforated to allow the passage of the olfactory nerves for the sense of smell?
    A. Medial plate of the sphenoid bone
    B. Lateral plate of the sphenoid bone
    C. Perpendicular plate of the ethmoid bone
    D. Cribriform plate of the ethmoid bone

19. Which of the following bones of the skull is considered a cranial bone?
    A. Vomer
    B. Maxilla
    C. Sphenoid bone
    D. Zygomatic bone
    E. Mandible

20. In which portion of the temporal bone is the temporomandibular joint located?
    A. Squamous portion
    B. Tympanic portion
    C. Petrous portion
    D. Mastoid portion

Chapter 4

Muscles

▼ *This chapter discusses the following topics:*

I. Muscular System

II. Muscles of the Head and Neck
   A. Cervical Muscles
   B. Muscles of Facial Expression
   C. Muscles of Mastication
   D. Hyoid Muscles
   E. Muscles of the Tongue
   F. Muscles of the Pharynx

▼ *Key words*

Action (**ak**-shun) Movement accomplished by the muscle when the muscle fibers contract.

Facial paralysis (pah-**ral**-i-sis) Loss of action of the facial muscles.

Insertion (in-**sir**-shun) End of the muscle that is attached to the more movable structure.

Muscle (**mus**-il) Type of body tissue that shortens under neural control, causing soft tissue and bony structures to move.

Origin (**or**-i-jin) End of the muscle that is attached to the least movable structure.

Muscles

▼ ***After studying this chapter, the reader should be able to:***

1. Define and pronounce all the key words and anatomical terms in this chapter.
2. Locate and identify the muscles of the head and neck on a diagram, a skull, and a patient.
3. Describe the origin, insertion, and action of the muscles of the head and neck.
4. State the nerve(s) innervating each of the muscles of the head and neck.
5. Discuss the processes of mastication, speech, and swallowing in regard to anatomical considerations.
6. Correctly complete the identification exercise and review questions for this chapter.
7. Integrate the knowledge about the muscles of the head and neck into the clinical practice of patient examination and related muscular diseases.

Muscular System

A ***muscle*** (**mus**-il) in the muscular system shortens under neural control, causing soft tissue and bony structures of the body to move. Each muscle has two ends attached to these structures, and they are categorized according to their role in movement. The ***origin*** (**or**-i-jin) is the end of the muscle that is attached to the least movable structure. The ***insertion*** (in-**sir**-shun) is the other end of the muscle and is attached to the more movable structure.

Generally, the insertion of the muscle moves toward the origin when the muscle is contracted. The movement that is accomplished when the muscle fibers contract is the ***action*** (**ak**-shun) of the muscle. The muscles have specific innervation that is discussed in this chapter, but a more thorough explanation of the nervous system can be found in Chapter 8. The blood supply to the muscular area is discussed in Chapter 6.

Muscles of the Head and Neck

The dental professional needs to determine the location and action of many muscles of the head and neck in order to perform a thorough patient examination. This information is important since the placement of many other structures, such as bones, blood vessels, nerves, and lymph nodes, is related to the location of the muscles. The muscles may also malfunction and be involved in temporomandibular joint disorders, occlusal dysfunction, and certain nervous system diseases. Muscles of the head and neck and their attachments also are a consideration in the spread of dental infections (see Chapter 12 for more information).

The muscles of the head and neck are divided into seven main groups: the cervical muscles, muscles of facial expression, muscles of mastication, hyoid muscles, muscles of the tongue, muscles of the soft palate, and muscles of the pharynx. There are also muscle groups to the ears, eyes, and nose that are not included in this chapter. As the muscles of the head and neck are studied, refer frequently to a skull model and a mirror image, using your own muscle movement and palpation. Palpating the muscles during an examination of a patient may also be helpful.

▼ CERVICAL MUSCLES

The two **cervical muscles** (**ser**-vi-kal) considered in this text are both superficial and easily palpated on the

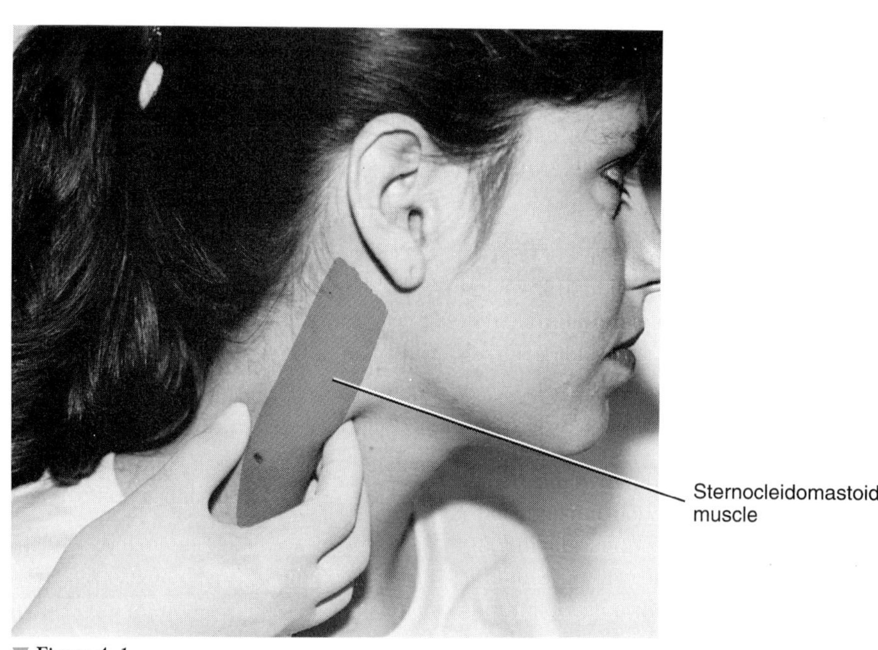

▼ **Figure 4–1**
Palpating the sternocleidomastoid muscle on a patient by having the patient turn the head to the opposite side (muscle highlighted).

neck. These cervical muscles are the sternocleidomastoid and trapezius muscles.

▼ Sternocleidomastoid Muscle

One of the largest and most superficial cervical muscles is the paired **sternocleidomastoid muscle (SCM)** (stir-no-klii-do-**mass**-toid), or sternomastoid muscle. The SCM is thick and serves as a primary muscular landmark of the neck during an extraoral examination of a patient. The SCM is effectively palpated when the patient moves the head to the side (Figure 4–1). The SCM divides the neck region into anterior and posterior cervical triangles (see Chapter 2).

Origin and Insertion. The SCM originates from the medial portion of the clavicle and the sternum's superior and lateral surfaces and passes posteriorly and superiorly to insert on the mastoid process of the temporal bone (Figure 4–2). This insertion is just posterior and inferior to the external acoustic meatus of each ear.

Action. If one SCM contracts, the head and neck bend to the same side, and the face and front of the neck rotate to the opposite side. If both muscles contract, the head will flex at the neck and extend at the junction between the neck and skull.

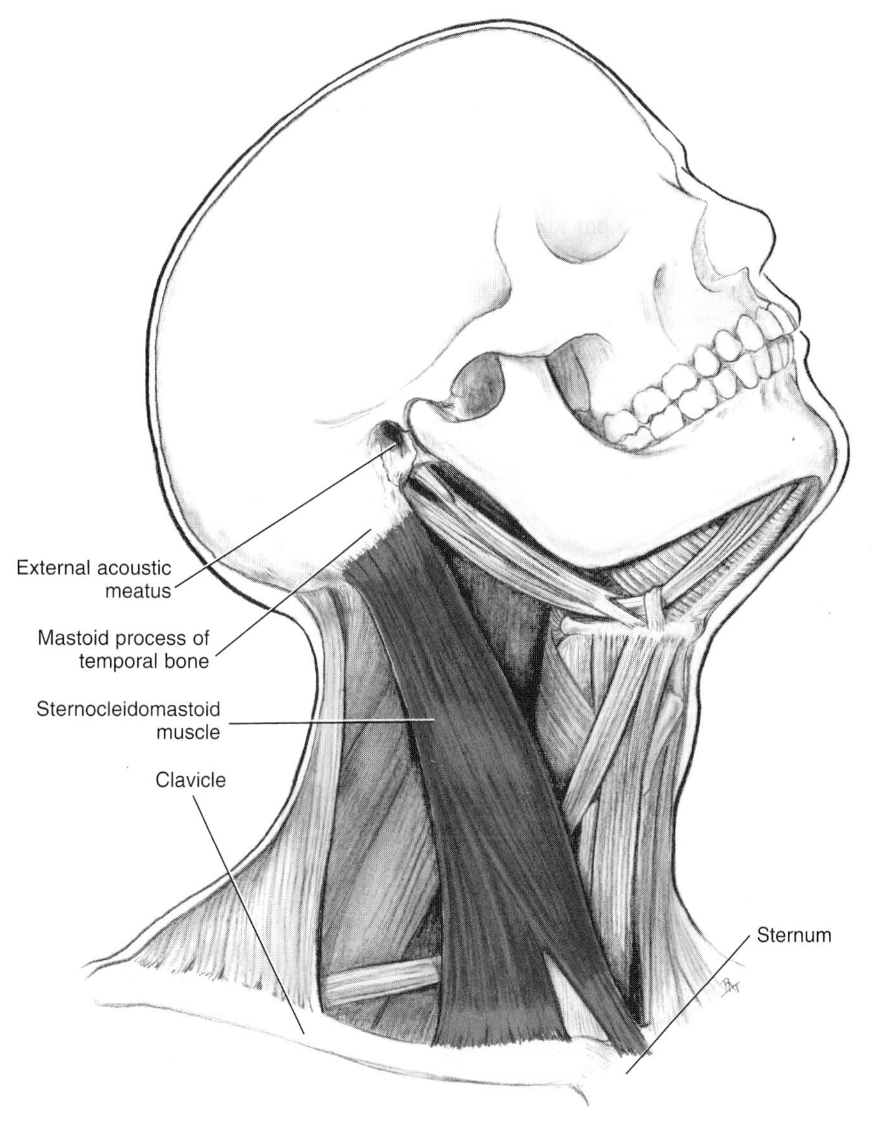

Figure 4–2
Origin and insertion of the sternocleidomastoid muscle.

Innervation. The SCM is innervated by the eleventh cranial or accessory nerve.

Trapezius Muscle

The other important superficial cervical muscle is the paired **trapezius muscle** (trah-**pee**-zee-us), which covers the lateral and posterior surfaces of the neck. The trapezius muscle is a broad, flat triangular muscle.

Origin and Insertion. The trapezius muscle originates from the external surface of the occipital bone and the posterior midline of the cervical and thoracic regions. This muscle inserts on the lateral third of the clavicle and parts of the scapula (Figure 4–3).

Action. The cervical fibers of the trapezius muscle act to lift the clavicle and scapula, as when the shoulders are shrugged.

Innervation. The trapezius muscle is innervated by the eleventh cranial or accessory nerve and third and fourth cervical nerves.

MUSCLES OF FACIAL EXPRESSION

The **muscles of facial expression** are paired muscles in the superficial fascia of the facial tissues (Figures

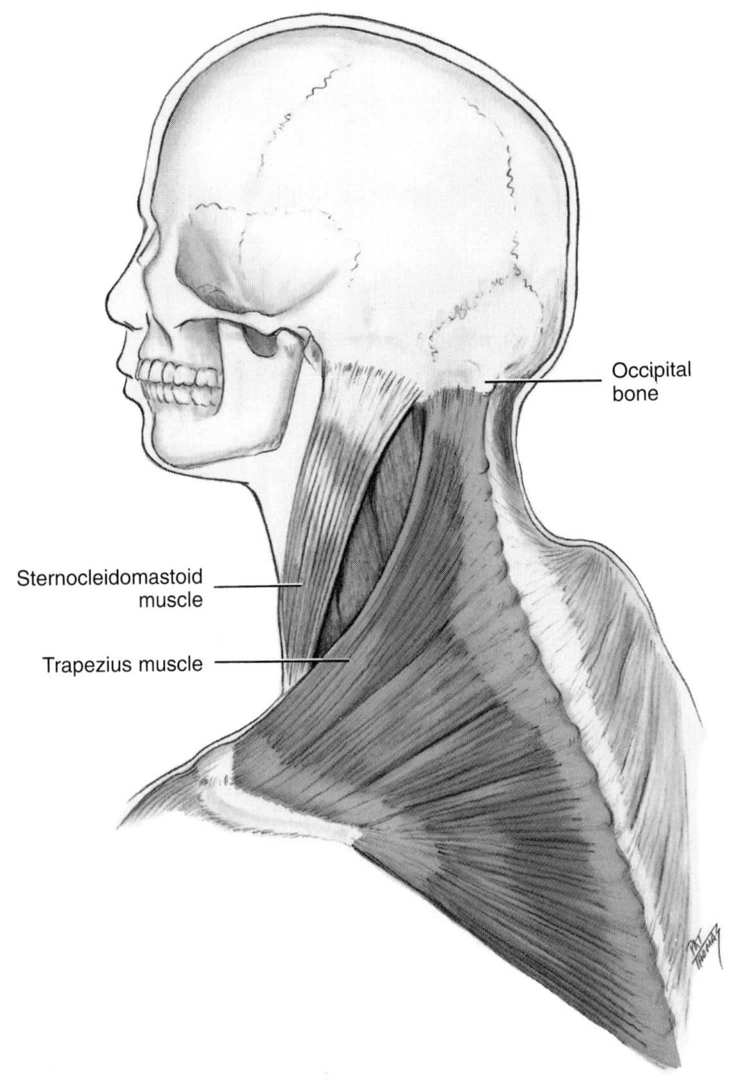

Figure 4–3
Origin and insertion of the trapezius muscle.

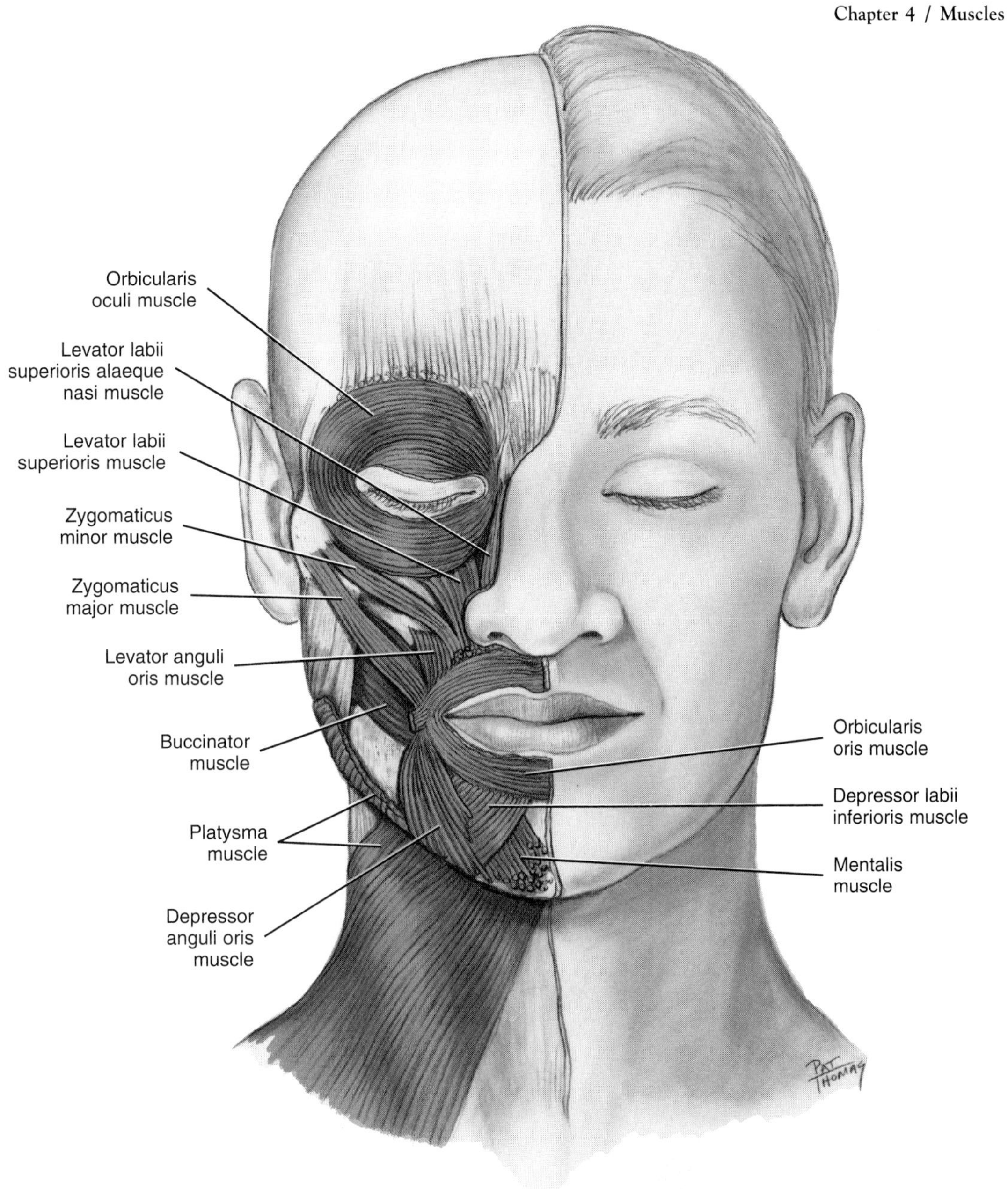

Figure 4–4
Frontal view of most of the muscles of facial expression.

4–4 and 4–5). All the muscles of facial expression originate from bone (rarely the fascia) and insert on skin tissue. These muscles also cause wrinkles at right angles to the muscle's action line. Again, the use of your mirror image as various facial expressions are made is very helpful in learning about these muscles.

Origin and Insertion. The locations of the muscles of facial expression are varied. These muscles may be further grouped according to whether they are situated in the scalp, eye, or mouth region. The specific origin and insertion of each muscle of facial expression are discussed (Table 4–1).

Action. During facial expression, the muscles of facial expression all act in various combinations, similar to the muscles of mastication, to vary the appearance of the face (Table 4–2). An inability to form facial expressions on one side of the face may be the first sign of damage to the nerve of these muscles.

Innervation. All the muscles of facial expression are innervated by the seventh cranial or facial nerve,

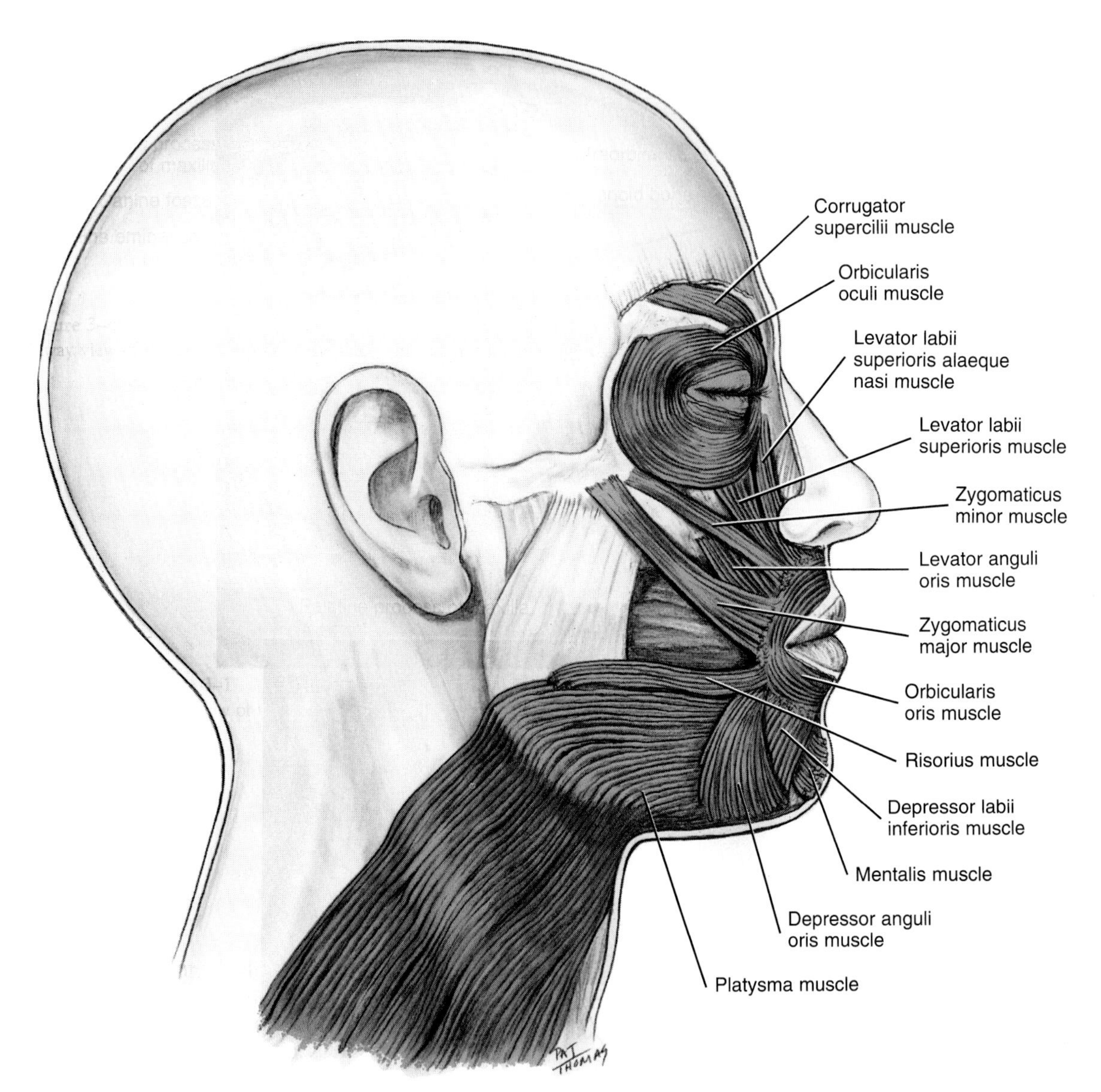

Figure 4–5
Lateral view of most of the muscles of facial expression.

Table 4–1

ORIGIN AND INSERTION OF THE MUSCLES OF FACIAL EXPRESSION

Muscles	Origin	Insertion
Epicranial	Frontal belly: epicranial aponeurosis	Frontal belly: eyebrow and root of nose
	Occipital belly: occipital and temporal bone	Occipital belly: epicranial aponeurosis
Orbicularis oculi	Orbital rim, frontal and maxillary bones	Lateral region of eye, some encircle eye
Corrugator supercilii	Frontal bone	Eyebrow
Orbicularis oris	Encircle mouth	Angle of mouth
Buccinator	Maxilla, mandible, and pterygomandibular raphe	Angle of mouth
Risorius	Fascia superficial to masseter muscle	Angle of mouth
Levator labii superioris	Maxilla	Upper lip
Levator labii superioris alaeque nasi	Maxilla	Ala of nose and upper lip
Zygomaticus major	Zygomatic bone	Angle of mouth
Zygomaticus minor	Zygomatic bone	Upper lip
Levator anguli oris	Maxilla	Angle of mouth
Depressor anguli oris	Mandible	Angle of mouth
Depressor labii inferioris	Mandible	Lower lip
Mentalis	Mandible	Chin
Platysma	Clavicle and shoulder	Mandible and muscles of mouth

Table 4–2

MUSCLES OF FACIAL EXPRESSION AND THEIR ASSOCIATED FACIAL EXPRESSIONS

Muscles	Facial Expression
Epicranial	Surprise
Orbicularis oculi	Closing eyelid
Corrugator supercilii	Frowning
Orbicularis oris	Closing or pursing lips
Buccinator	Chewing
Risorius	Smiling widely
Levator labii superioris	Raising upper lip
Levator labii superioris alaeque nasi	Raising upper lip and dilating nostrils in a sneer
Zygomaticus major	Smiling
Zygomaticus minor	Raising upper lip
Levator anguli oris	Smiling
Depressor anguli oris	Frowning
Depressor labii inferioris	Lowering lower lip
Mentalis	Raising chin and protruding lower lip
Platysma	Grimacing

each nerve serving one side of the face. Damage to the facial nerve results in ***facial paralysis*** (pah-**ral**-i-sis) of facial expression on the involved side (Figure 4–6). Paralysis is the loss of voluntary muscle action. The facial nerve can become damaged permanently or temporarily (see Chapter 8 for more information). This damage can occur with a stroke (cerebrovascular accident), Bell's palsy, and parotid salivary gland cancer (malignant neoplasm) since the facial nerve travels through the gland. The gland can also be damaged by surgery or trauma, as with an incorrectly given inferior alveolar local anesthetic block (see Chapter 9). These disease processes not only damage facial expression but also seriously impair the patient's ability to masticate and speak.

Muscles of Facial Expression in the Scalp Region

EPICRANIAL MUSCLE

The **epicranial muscle** (ep-ee-**kray**-nee-al) or epicranius is a muscle of facial expression located in the

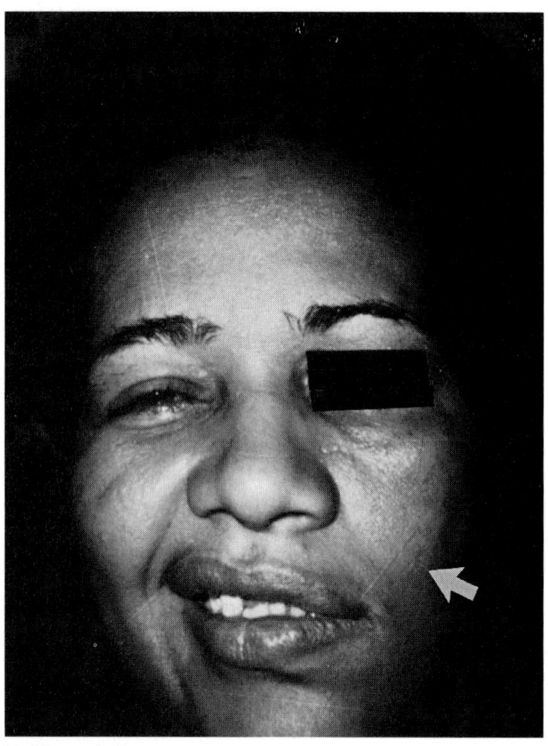

Figure 4–6
Patient with unilateral paralysis of the facial muscles due to muscle damage. Patient is unable to show facial expression on that side (arrow).

scalp region. This muscle has two bellies, the frontal and occipital bellies. The bellies are separated by a large, spread-out scalpal tendon. This muscle and its tendon are one of the layers that form the scalp.

Origin and Insertion. The frontal belly arises from the **epicranial aponeurosis** (ap-o-new-**row**-sis) or galea aponeurotica, a scalpal tendon (Figure 4–7). The epicranial aponeurosis is located over the area where the parietal and occipital bones meet, the most superior portion of the skull. The frontal belly then inserts into the skin tissue of the eyebrow and root of the nose. The occipital belly originates from the occipital bone and mastoid process of the temporal bone and then inserts in the epicranial aponeurosis.

Action. The epicranial muscle raises the eyebrows and scalp, as when a person shows surprise (Figure 4–8).

Muscles of Facial Expression in the Eye Region

The muscles of facial expression in the eye region include the orbicularis oculi and corrugator supercilii muscles (Figure 4–9).

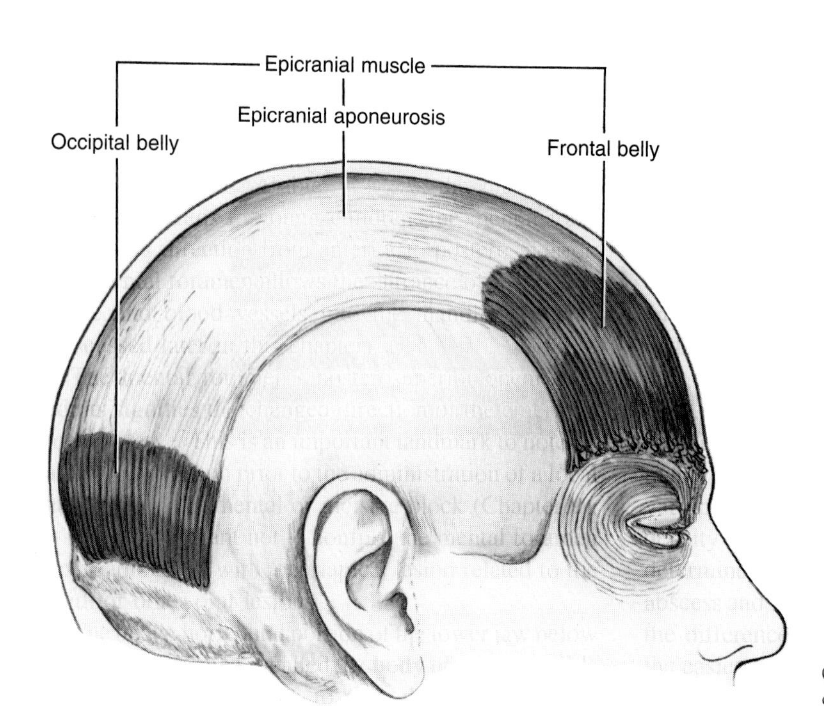

Figure 4–7
Origin and insertion of the frontal belly and occipital belly of the epicranial muscle.

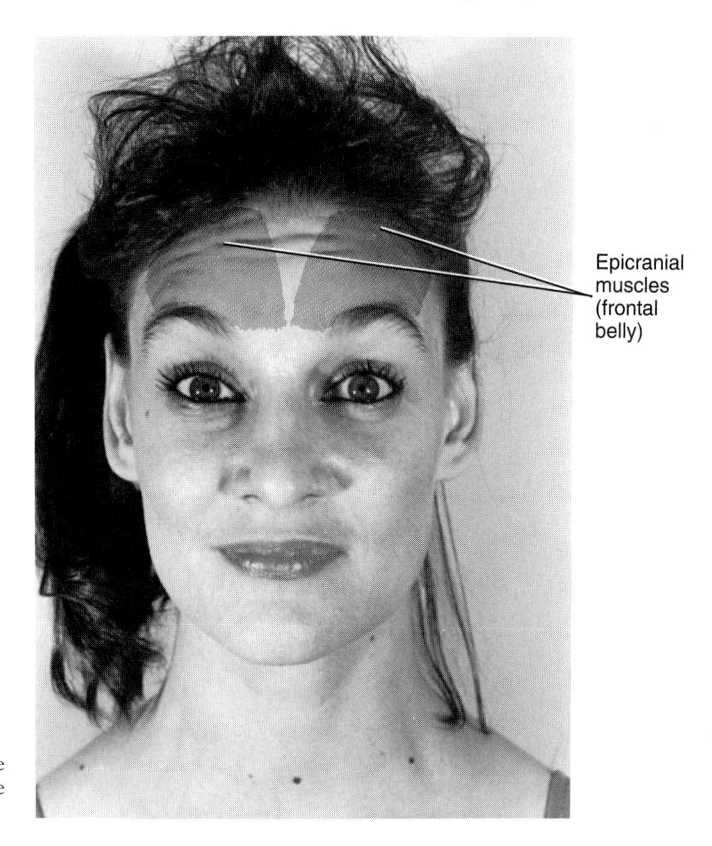

Figure 4–8
Facial expression of surprise on a patient using the epicranial muscle, raising the eyebrows and scalp (muscle highlighted).

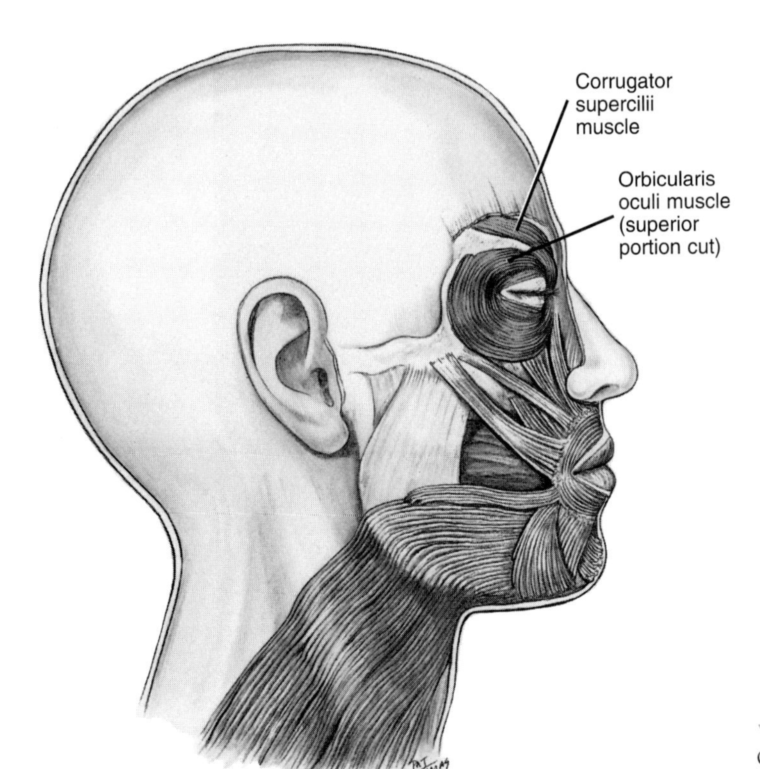

Figure 4–9
Orbicularis oculi muscle and corrugator supercilii muscle.

▼ ORBICULARIS OCULI MUSCLE

The **orbicularis oculi muscle** (or-bik-you-**laa**-ris **oc**-yule-eye) is a muscle of facial expression that encircles the eye.

Origin and Insertion. The orbicularis oculi muscle originates on the orbital rim, nasal process of the frontal bone, and frontal process of the maxilla. Most of the fibers insert in the skin tissue at the lateral region of the eye, although some of the inner fibers completely encircle the eye.

Action. The orbicularis oculi muscle closes the eyelid. If all the fibers are active, the eye can be squinted, and wrinkles form in the lateral portions of the eye (crow's feet).

▼ CORRUGATOR SUPERCILII MUSCLE

The **corrugator supercilii muscle** (cor-rew-**gay**-tor soo-per-**sili**-eye) is a muscle of facial expression in the eye region, deep to the superior portion of the orbicularis oculi muscle.

Origin and Insertion. The corrugator supercilii muscle originates on the frontal bone in the supraorbital region. The muscle then passes superiorly and laterally to insert in the skin tissue of the eyebrow.

Action. The corrugator supercilii muscle draws the skin tissue of the eyebrow medially and inferiorly toward the nose. This movement causes vertical wrinkles in the forehead, as when a person frowns (see Figure 4–14B).

▼ Muscles of Facial Expression in the Mouth Region

The muscles of facial expression in the mouth region include the orbicularis oris, buccinator, risorius, levator labii superioris, levator labii superioris alaeque nasi, zygomaticus major, zygomaticus minor, levator anguli oris, depressor anguli oris, depressor labii inferioris, mentalis, and platysma muscles.

▼ ORBICULARIS ORIS MUSCLE

The **orbicularis oris muscle** (or-bik-you-**laa**-ris **or**-is) is an important muscle of facial expression in the mouth region.

Origin and Insertion. The orbicularis oris muscle encircles the mouth and inserts in the skin tissue at the angle of the mouth (Figure 4–10). In the upper lip, fibers also insert on the ridges of the philtrum.

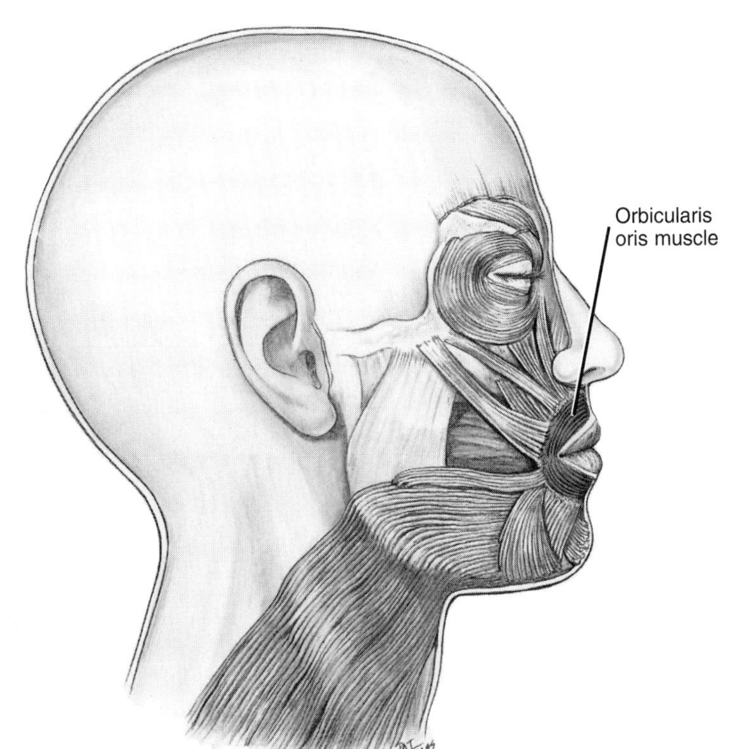

▼ **Figure 4–10**
Orbicularis oris muscle.

Action. The orbicularis oris muscle closes the lips. When the fibers are very active, the lips are pursed.

▼ BUCCINATOR MUSCLE

The **buccinator muscle** (buck-**sin**-nay-tor) is a muscle of facial expression that forms the anterior portion of the cheek.

Origin and Insertion. The buccinator muscle originates from three areas: the alveolar processes of the maxilla and mandible and a fibrous structure, the **pterygomandibular raphe** (**teh**-ri-go-man-**dib**-yule-lar **ra**-fe) (Figure 4–11). The pterygomandibular raphe extends from the hamulus and passes inferiorly to attach to the posterior end of the mandible's mylohyoid line. The buccinator and superior pharyngeal constrictor muscles of the pharynx are attached to each other at the raphe. The pterygomandibular raphe is noted in the oral cavity as the **pterygomandibular fold.** The buccinator runs horizontally to insert into the skin tissue at the angle of the mouth. Thus the buccinator muscle has different fiber groups, the deep vertical fibers between the alveolar processes and the superficial horizontal fibers from the raphe to the corner of the mouth.

Action. The buccinator muscle pulls the angle of the mouth laterally and shortens the cheek both vertically and horizontally. This action causes the muscle to keep food pushed back on the occlusal surface of teeth, as when a person chews. By keeping the food in the correct position when chewing, the buccinator muscle assists the muscles of mastication.

▼ RISORIUS MUSCLE

The **risorius muscle** (ri-**soh**-ree-us) is a thin muscle of facial expression in the mouth region.

Origin and Insertion. The risorius muscle originates from fascia superficial to the masseter muscle and passes anteriorly to insert in the skin tissue at the angle of the mouth (Figure 4–12).

Action. The risorius muscle widens the mouth, as when a person smiles widely.

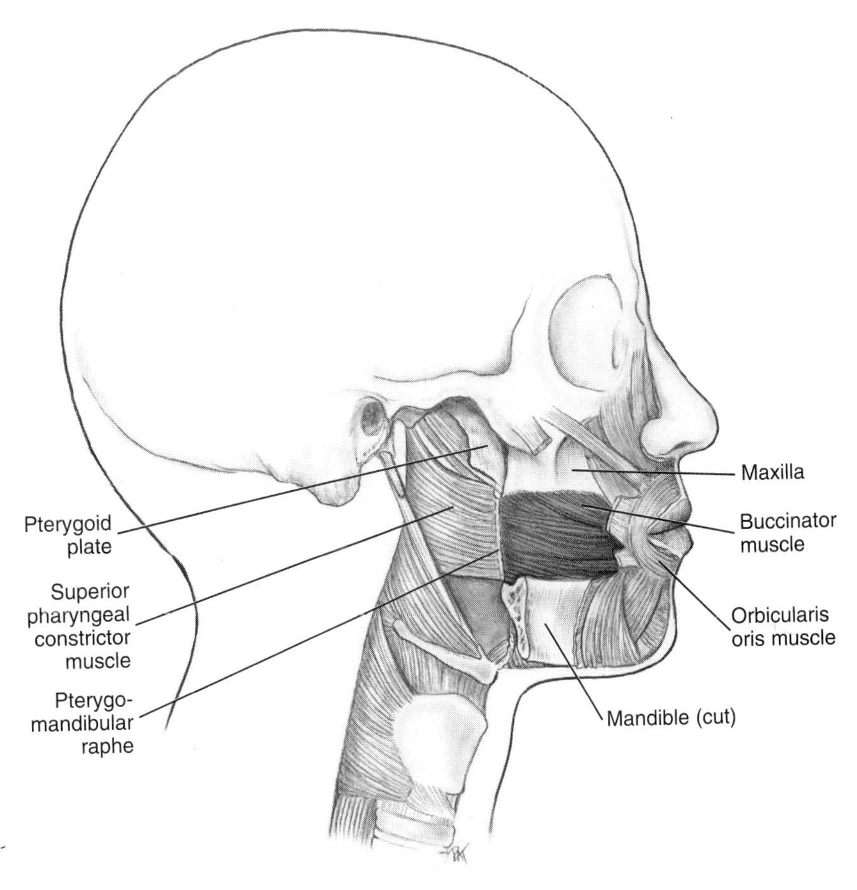

▼ **Figure 4–11**
Origin and insertion of the buccinator muscle.

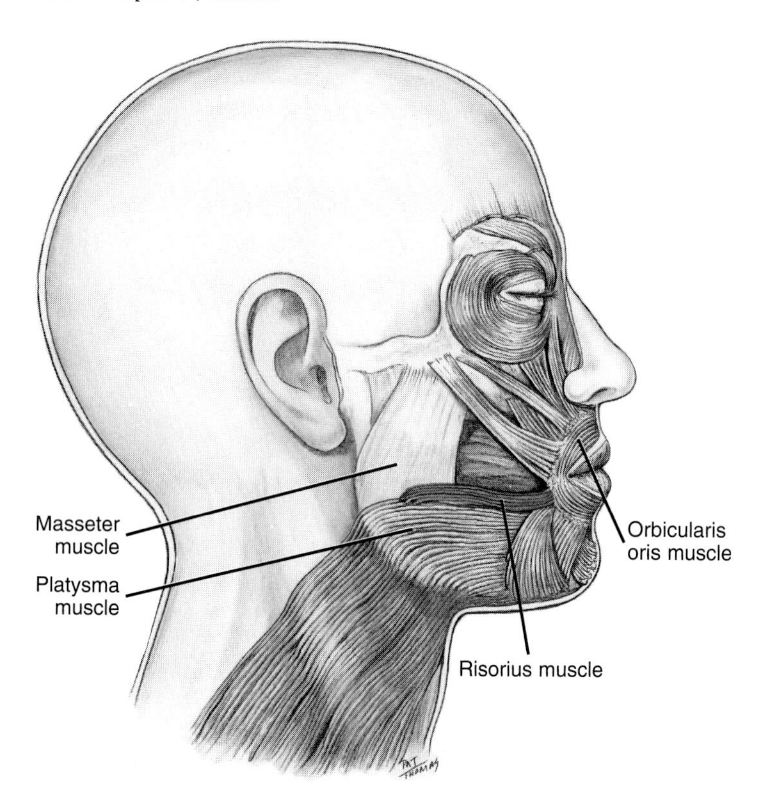

▼ **Figure 4–12**
Origin and insertion of the risorius muscle.

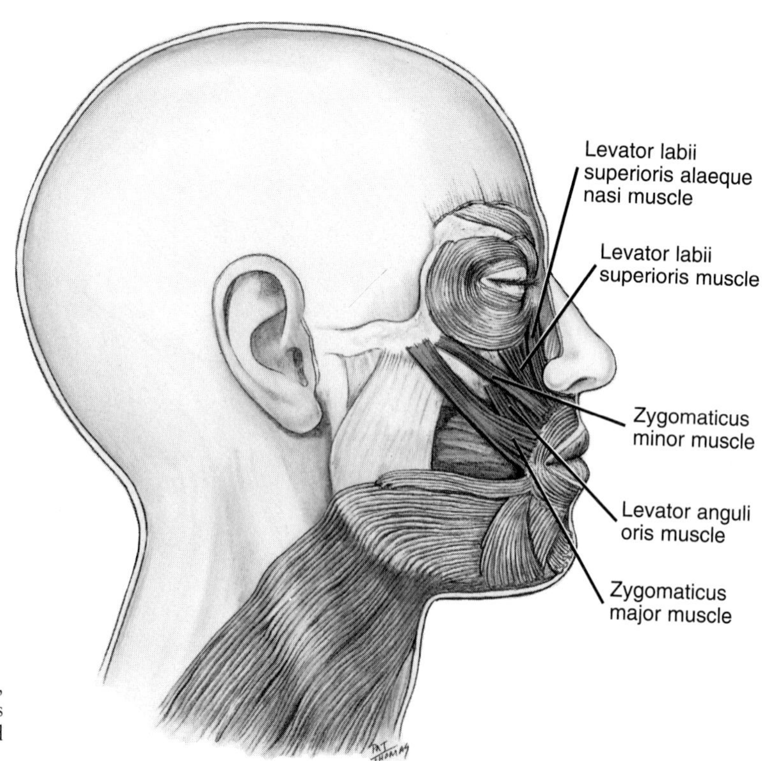

▼ **Figure 4–13**
Levator labii superioris alaeque nasi muscle, levator labii superioris muscle, zygomaticus minor muscle, levator anguli oris muscle, and zygomaticus major muscle.

▼ LEVATOR LABII SUPERIORIS MUSCLE

A broad, flat muscle of facial expression in the mouth region is the **levator labii superioris muscle** (le-**vate**-er **lay**-be-eye soo-per-ee-**or**-is).

Origin and Insertion. The levator labii superioris muscle originates from the infraorbital rim of the maxilla. The muscle then passes inferiorly to insert in the skin tissue of the upper lip (Figure 4–13).

Action. The levator labii superioris muscle elevates the upper lip (Figure 4–14A).

▼ LEVATOR LABII SUPERIORIS ALAEQUE NASI MUSCLE

The **levator labii superioris alaeque nasi muscle** (le-**vate**-er **lay**-be-eye soo-per-ee-**or**-is **a**-lah-cue **naz**-eye) is also a muscle of facial expression in the mouth region.

Origin and Insertion. The levator labii superioris alaeque nasi muscle originates from the frontal process of the maxilla. The muscle then passes inferiorly to insert into two areas: the skin tissue of the ala (or wing) of the nose and the upper lip (see Figure 4–13).

Action. The levator labii superioris alaeque nasi muscle elevates the upper lip and ala of the nose, thus also dilating the nostrils, as in a sneering expression (Figure 4–14B).

▼ ZYGOMATICUS MAJOR MUSCLE

Another muscle of facial expression in the mouth region is the **zygomaticus major muscle** (zy-go-**mat**-i-kus **may**-jer).

Origin and Insertion. The zygomaticus major muscle originates from the zygomatic bone, lateral to the zygomaticus minor muscle. The muscle then passes anteriorly and inferiorly to insert in the skin tissue at the angle of the mouth (see Figure 4–13).

Action. The zygomaticus major muscle elevates the angle of the upper lip and pulls it laterally, as when a person smiles (see Figure 4–14A).

▼ ZYGOMATICUS MINOR MUSCLE

The **zygomaticus minor muscle** (zy-go-**mat**-i-kus **my**-ner) is a small muscle of facial expression in the mouth region, medial to the zygomaticus major muscle.

Origin and Insertion. The zygomaticus minor muscle originates on the body of the zygomatic bone. The muscle then inserts in the skin tissue of the upper lip adjacent to the insertion of the levator labii superioris muscle (see Figure 4–13).

Action. The zygomaticus minor elevates the upper lip (see Figure 4–14A).

▼ LEVATOR ANGULI ORIS MUSCLE

Deep to both the zygomaticus major and zygomaticus minor muscles is the **levator anguli oris muscle** (le-**vate**-er **an**-gu-lie **or**-is), another muscle of facial expression in the mouth region.

Origin and Insertion. The levator anguli oris muscle originates on the canine fossa of the maxilla, usually superior to the root of the maxillary canine. The muscle then passes inferiorly to insert in skin tissues at the angle of the mouth (see Figure 4–13).

Action. The levator anguli oris muscle elevates the angle of the mouth, as when a person smiles (see Figure 4–14A).

▼ DEPRESSOR ANGULI ORIS MUSCLE

The **depressor anguli oris muscle** (de-**pres**-er **an**-gu-lie **or**-is) is a triangular muscle of facial expression in the lower mouth region.

Origin and Insertion. The depressor anguli oris muscle originates on the lower border of the mandible and passes superiorly to insert in the skin tissue at the angle of the mouth (Figure 4–15).

Action. The depressor anguli oris muscle depresses the angle of the mouth, as when a person frowns (see Figure 4–14B).

▼ DEPRESSOR LABII INFERIORIS MUSCLE

Deep to the depressor anguli oris muscle is the **depressor labii inferioris muscle** (de-**pres**-er **lay**-be-eye in-**fere**-ee-o-ris), another muscle of facial expression in the mouth region.

Origin and Insertion. The depressor labii inferioris muscle also originates from the lower border of the mandible and passes superiorly to insert in the skin tissue of the lower lip (see Figure 4–15).

Action. The depressor labii inferioris muscle depresses the lower lip, exposing the mandibular incisor teeth.

▼ MENTALIS MUSCLE

The **mentalis muscle** (men-**ta**-lis) is a short, thick muscle of facial expression superior and medial to the mental nerve in the mouth region.

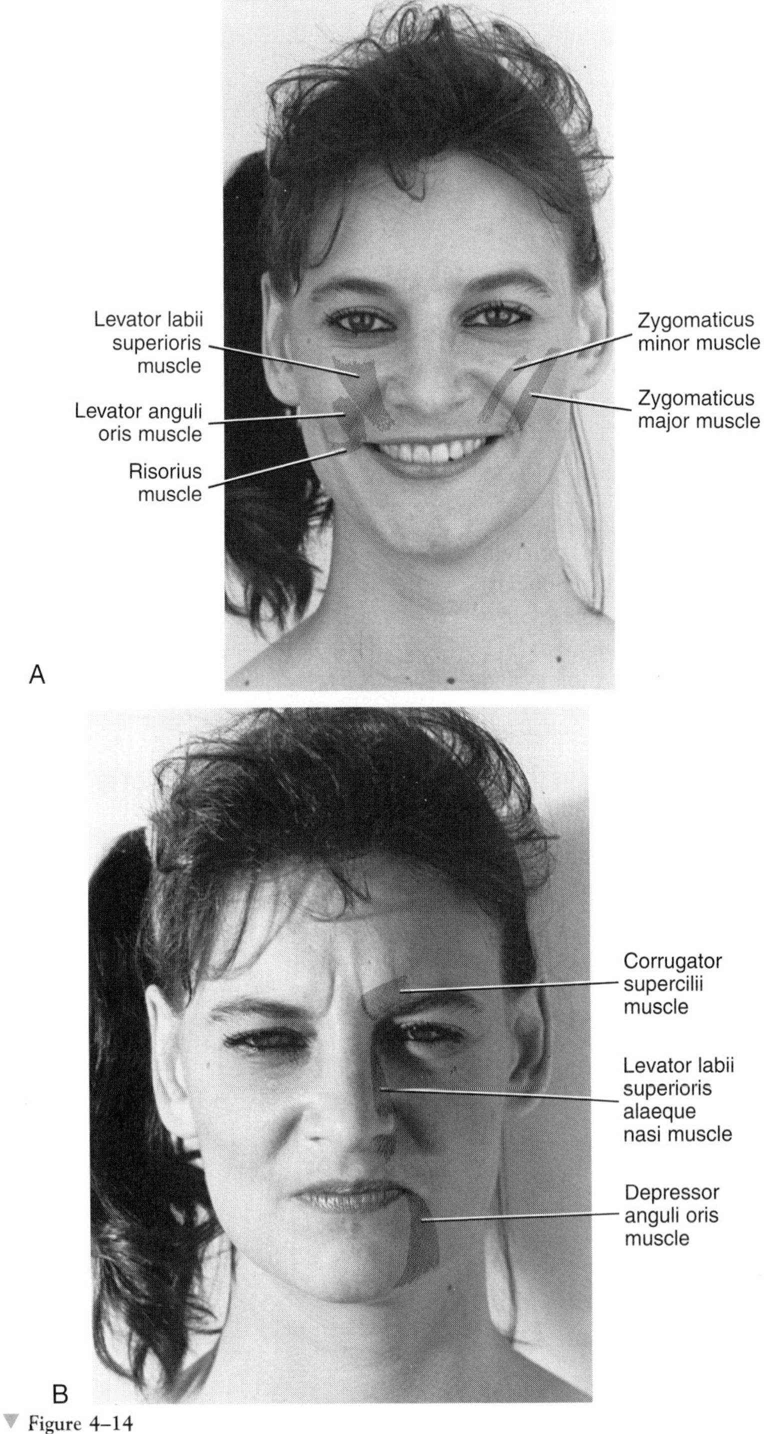

Figure 4–14

A: Patient using the risorius muscle, levator labii superioris muscle, levator anguli oris muscle, zygomaticus minor muscle, and zygomaticus major muscle when smiling (muscles highlighted and labeled). **B:** Patient using the corrugator supercilii muscle, levator labii superioris alaeque nasi muscle, and depressor anguli oris muscle while looking disgusted (muscles highlighted and labeled).

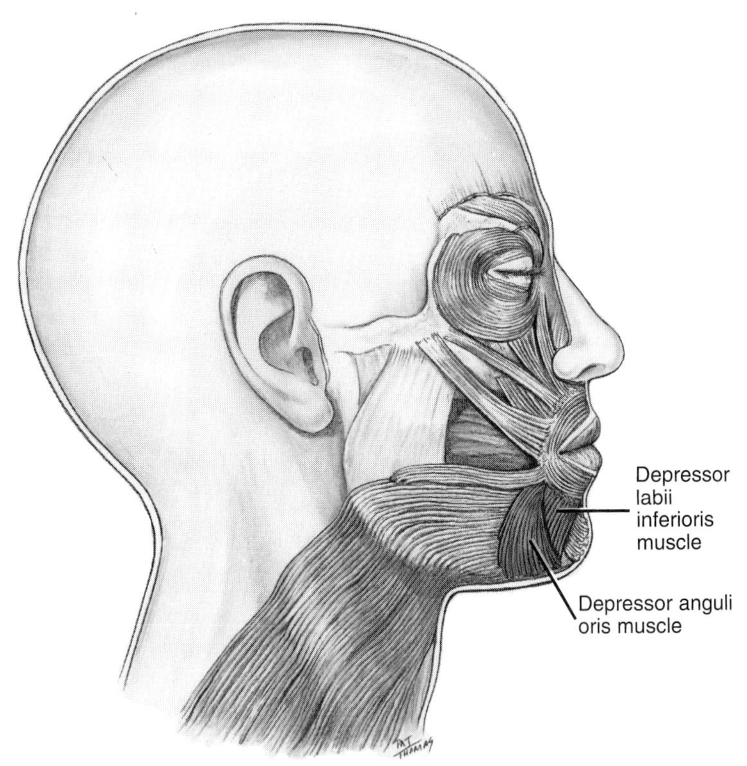

▼ **Figure 4–15**
Depressor labii inferioris muscle and depressor anguli oris muscle.

Origin and Insertion. The mentalis muscle originates on the mandible near the midline and inserts in the skin tissue of the chin (Figure 4–16).

Action. The mentalis muscle raises the chin, causing the displaced lower lip to protrude and narrowing the oral vestibule. These fibers when active may dislodge a complete denture in an edentulous patient who has lost alveolar ridge height.

▼ PLATYSMA MUSCLE

The **platysma muscle** (plah-**tiz**-mah) is a muscle of facial expression that runs from the neck all the way to the mouth, covering the anterior cervical triangle.

Origin and Insertion. The platysma muscle originates in the skin tissue superficial to the clavicle and shoulder. The muscle then passes anteriorly to insert on the lower border of the mandible and the muscles surrounding the mouth (Figure 4–17).

Action. The platysma muscle raises the skin of the neck and also pulls the corner of the mouth down, as when a person grimaces (Figure 4–18).

▼ MUSCLES OF MASTICATION

The **muscles of mastication** (mass-ti-**kay**-shun) are four pairs of muscles attached to the mandible: the masseter, temporalis, medial pterygoid, and lateral pterygoid muscles.

Origin and Insertion. The origin and insertion of each muscle are discussed (Table 4–3).

Action. The muscles of mastication are responsible for closing the jaws, bringing the lower jaw forward or backward, and shifting the lower jaw to one side. These jaw movements involve the movement of the mandible, while the rest of the skull remains relatively stable. These muscles of mastication work with the temporomandibular joint to accomplish these move-

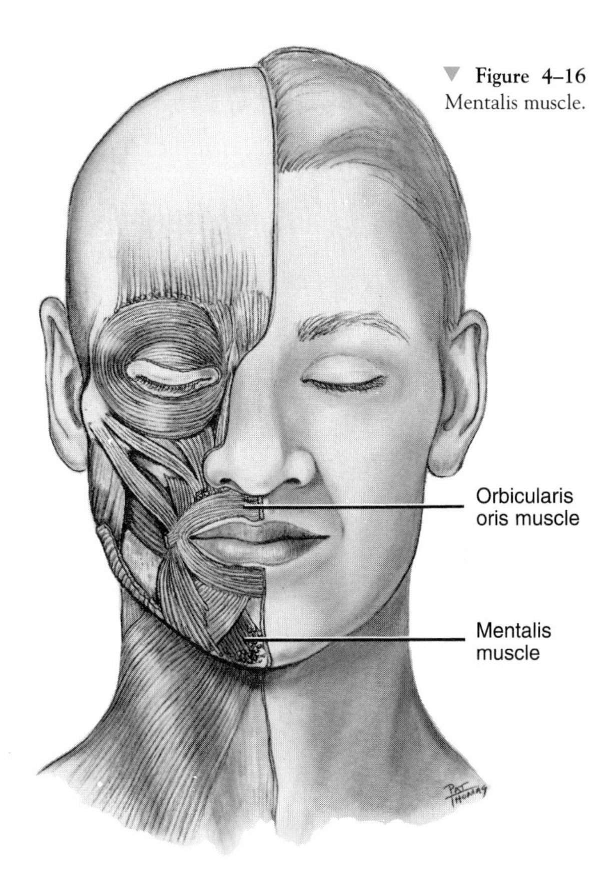

Figure 4–16
Mentalis muscle.

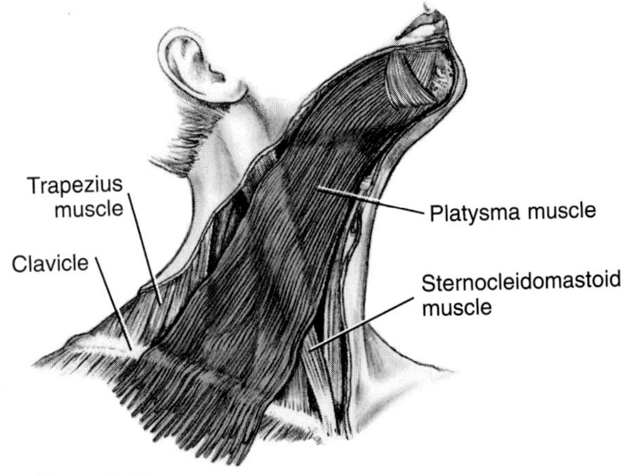

Figure 4–17
Origin and insertion of the platysma muscle.

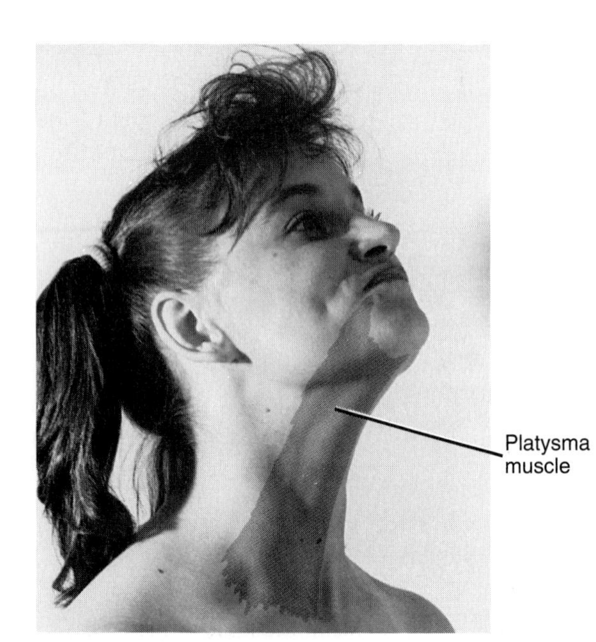

Figure 4–18
Patient using the platysma muscle to raise the skin of the neck and pull the corner of the mouth down, as when a person grimaces (muscle highlighted).

Table 4–3

ORIGIN AND INSERTION OF THE MUSCLES OF MASTICATION

Muscles	Origin	Insertion
Masseter	Superficial head: anterior two thirds of lower border of zygomatic arch Deep head: posterior one third and medial surface of zygomatic arch	Superficial head: angle of mandible Deep head: ramus of mandible
Temporalis	Temporal fossa	Coronoid process of mandible
Medial pterygoid	Pterygoid fossa of sphenoid bone	Angle of mandible
Lateral pterygoid	Superior head: greater wing of sphenoid bone Inferior head: Lateral pterygoid plate from sphenoid bone	Both heads: pterygoid fovea of mandible

ments of the mandible (see Chapter 5 for more information). The dental professional needs to understand the association of the muscles of mastication with the movements of the mandible: depression, elevation, protrusion, retraction, and lateral deviation (Table 4–4).

Innervation. All the muscles of mastication are innervated by the mandibular division of the fifth cranial or trigeminal nerve.

Masseter Muscle

The most obvious muscle of mastication is the **masseter muscle** (**mass**-et-er) since it is the most superficially located and one of the strongest. To make the masseter more prominent for examination, have patients clench their teeth (see Figure 2–6). This muscle can become enlarged in patients who habitually clenches or grinds their teeth. The masseter muscle is a broad, thick rectangular muscle on each side of the face, anterior to the parotid salivary gland. This muscle has two heads, a superficial head and a deep head.

Origin and Insertion. The superficial head of the masseter muscle originates from the anterior two thirds of the lower border of the zygomatic arch. The deep head originates from the posterior third and the entire medial surface of the zygomatic arch. Both these heads of the masseter muscle pass inferiorly to insert on the mandible. The superficial head inserts on the lateral surface of the angle, and the deep head inserts on the ramus above the angle (Figure 4–19).

Action. The action of the masseter muscle when there is bilateral contraction of the entire muscle is to elevate the mandible, raising the lower jaw. Elevation of the mandible occurs during the closing of the jaws.

Temporalis Muscle

The **temporalis muscle** (tem-poh-**ral**-is) is a broad fan-shaped muscle of mastication on each side of the head that fills the temporal fossa, superior to the zygomatic arch.

Table 4–4

MUSCLES OF MASTICATION WITH ASSOCIATED MOVEMENTS OF MANDIBLE

Muscles	Mandibular Movements
Masseter	Elevation of mandible (during jaw closing)
Temporalis	Elevation of mandible (during jaw closing) Retraction of mandible (lower jaw backward)
Medial pterygoid	Elevation of mandible (during jaw closing)
Lateral pterygoid	Inferior heads: slight depression of mandible (during jaw opening) One muscle: lateral deviation of mandible (shift lower jaw to opposite side) Both muscles: protrusion of mandible (lower jaw forward)

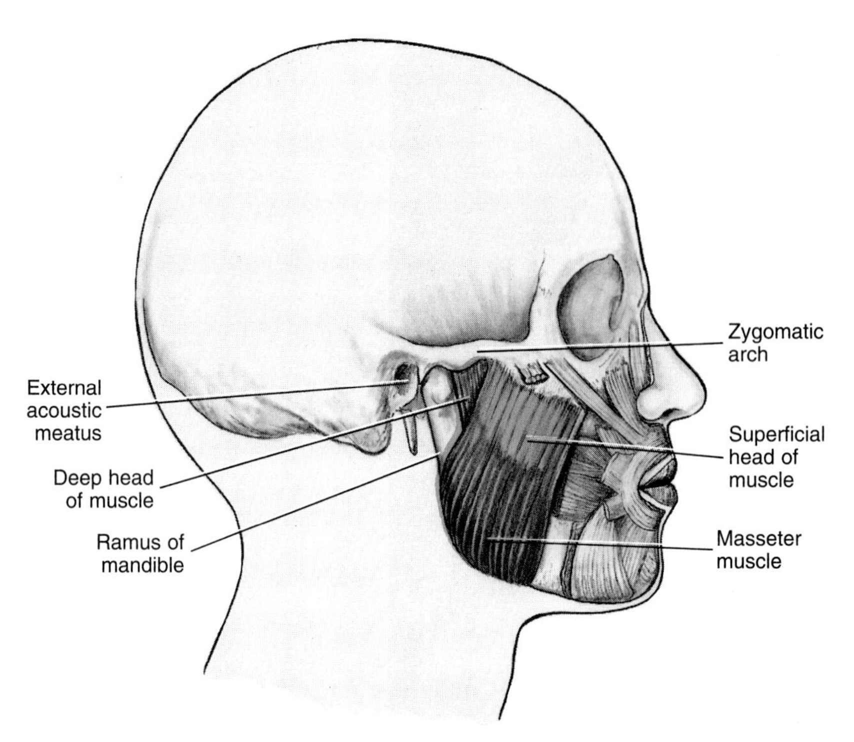

▼ **Figure 4–19**
Masseter muscle with its superficial head and deep head.

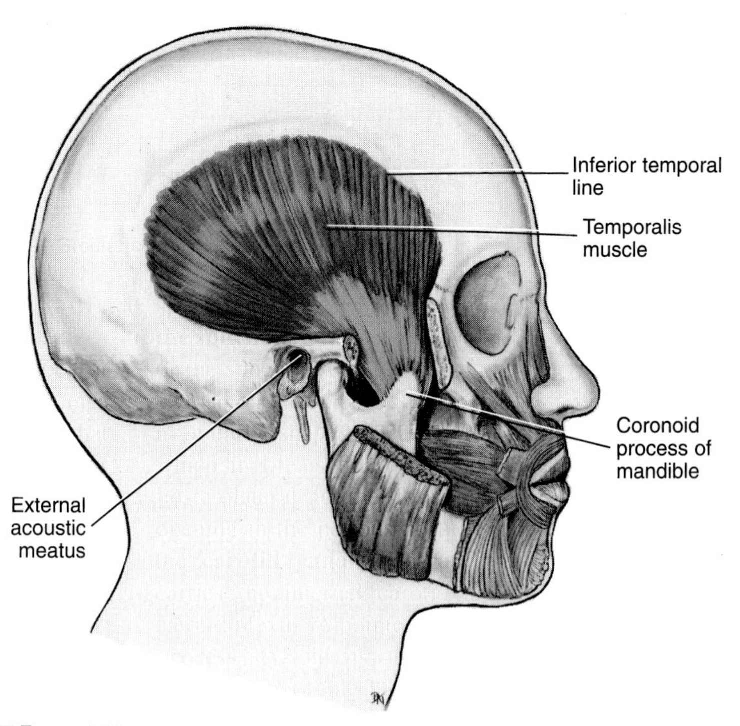

▼ **Figure 4–20**
Origin and insertion of the temporalis muscle (zygomatic arch and superior part of the masseter muscle have been removed).

Origin and Insertion. The temporalis muscle originates from the entire temporal fossa that is bound at the top by the inferior temporal line and at the bottom by the infratemporal crest. The muscle then passes inferiorly to insert on the coronoid process of the mandible (Figure 4–20).

Action. If the entire temporalis muscle contracts, the main action of the temporalis muscle is to elevate the mandible, raising the lower jaw. Elevation of the mandible occurs during the closing of the jaws. If only the posterior portion contracts, the muscle moves the lower jaw backward. Moving the lower jaw backward causes retraction of the mandible. Retraction of the jaw often accompanies the closing of the jaws.

Medial Pterygoid Muscle

Deeper, yet similar in form to the superficial masseter muscle, is another muscle of mastication, the **medial pterygoid muscle** (**teh**-ri-goid) or internal pterygoid muscle.

Origin and Insertion. The medial pterygoid muscle originates from the pterygoid fossa on the medial surface of the lateral pterygoid plate of the sphenoid bone. The muscle then passes inferiorly, posteriorly, and laterally to insert on the medial surface of the angle of the mandible (Figure 4–21).

Action. The action of the medial pterygoid muscle is to elevate the mandible, raising the lower jaw. Elevation of the mandible occurs during the closing of the jaws. This muscle is weaker than the masseter muscle in this action.

Lateral Pterygoid Muscle

The final muscle of mastication to consider is the **lateral pterygoid muscle** (**teh**-ri-goid) or external pterygoid muscle. This muscle has two separate heads of origin, the superior head and inferior head. The two heads of the muscle are separated by a slight interval anteriorly but fuse posteriorly. The entire muscle lies within the infratemporal fossa, deep to the temporalis muscle (see Chapter 3).

Origin and Insertion. The superior head of the lateral pterygoid muscle originates from the inferior surface of the greater wing of the sphenoid bone (the

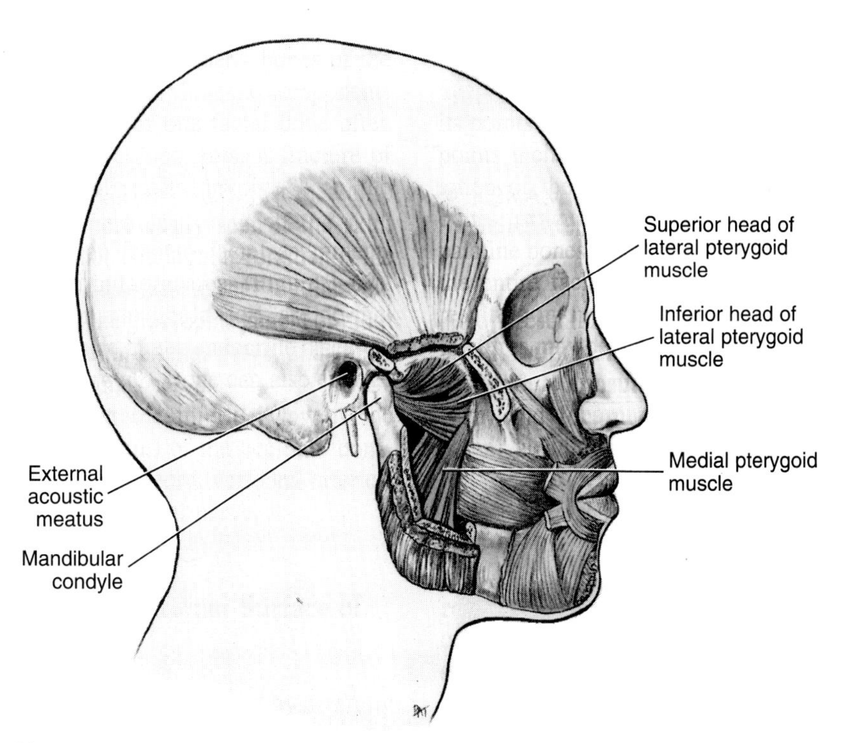

Figure 4–21
Medial pterygoid muscle and lateral pterygoid muscle with its superior head and inferior head (lower part of the temporalis muscle, zygomatic arch, and most of the mandibular ramus have been removed).

roof of the infratemporal fossa). The inferior head of the muscle originates from the lateral surface of the lateral pterygoid plate of the sphenoid bone. Both heads of the muscle then unite, passing posteriorly to insert on the anterior surface of the neck of mandibular condyle at the pterygoid fovea (see Figure 4–21). A few of the most superior fibers insert on the capsule of the temporomandibular joint.

Action. The inferior head has a slight tendency to depress the mandible, lowering the lower jaw. Depression of the mandible occurs during the opening of the jaws. The main action when both muscles contract is to bring the lower jaw forward, thus causing the protrusion of the mandible. Protrusion of the mandible often occurs during opening of the jaws. If only one lateral pterygoid muscle is contracted, the lower jaw shifts to the opposite side, causing lateral deviation of the mandible.

HYOID MUSCLES

The **hyoid muscles** (**hi**-oid) assist in the actions of mastication and swallowing. Most of the hyoid muscles are in a superficial position in the neck tissues. The hyoid muscles can be grouped according to whether they are suprahyoid or infrahyoid muscles (Table 4–5).

Origin and Insertion. Both groups of the hyoid muscles are attached in some way to the **hyoid bone** (Figure 4–22). The hyoid bone is a horseshoe-shaped bone suspended inferiorly to the mandible (see Chapter 3 for more information). The hyoid bone does not articulate with any other bone and has only muscular and ligamentary attachments. The designation for each muscle group, suprahyoid or infrahyoid muscles, is based on its vertical position in relationship to the hyoid bone. The specific origin and insertion of each hyoid muscle also are discussed (Table 4–6).

Table 4–5

HYOID MUSCLES AND THEIR GROUP, SUPRAHYOID OR INFRAHYOID, BASED ON THEIR RELATIONSHIP TO THE HYOID BONE

Hyoid Muscles
Suprahyoid
Digastric
Mylohyoid
Geniohyoid
Stylohyoid
Infrahyoid
Sternothyroid
Sternohyoid
Thyrohyoid

Suprahyoid Muscles

The **suprahyoid muscles** (soo-prah-**hi**-oid) are located superior to the hyoid bone (Figures 4–22, 4–23, and 4–24). These muscles may further be divided according to their anterior or posterior position to the hyoid bone. The **anterior suprahyoid muscle group** includes the anterior belly of the digastric, mylohyoid, and geniohyoid muscles. The **posterior suprahyoid muscle group** includes the posterior belly of the digastric and stylohyoid muscles.

Action. There are two actions associated with mastication that result from the contraction of the suprahyoid muscles. One action of both the anterior and posterior suprahyoid muscles is to cause the elevation of the hyoid bone and larynx if the mandible is stabilized by contraction of the muscles of mastication. This action occurs during swallowing.

The other action associated with mastication is from only the contraction of the anterior suprahyoid muscles, which causes the mandible to depress and the jaws to open when the hyoid bone is stabilized by the contraction of the posterior suprahyoid muscles and infrahyoid muscles, the other hyoid muscle group. Thus normal jaw opening involves the lateral pterygoid muscles, which protrude the mandible, and the anterior suprahyoid muscles, which lower the mandible. Some of the suprahyoid muscles have additional specific actions that are also discussed.

DIGASTRIC MUSCLE

The **digastric muscle** (di-**gas**-trik) is a suprahyoid muscle that has two separate bellies, the anterior and posterior bellies. The anterior belly is part of the anterior suprahyoid muscle group, and the posterior belly is part of the posterior suprahyoid muscle group. Each digastric muscle demarcates the superior portion of the anterior cervical triangle, forming (with the mandible) a submandibular triangle on each side of the neck. The right and left anterior bellies of the muscle form a midline submental triangle.

Origin and Insertion. The anterior belly of the digastric muscle originates on a tendon loosely at-

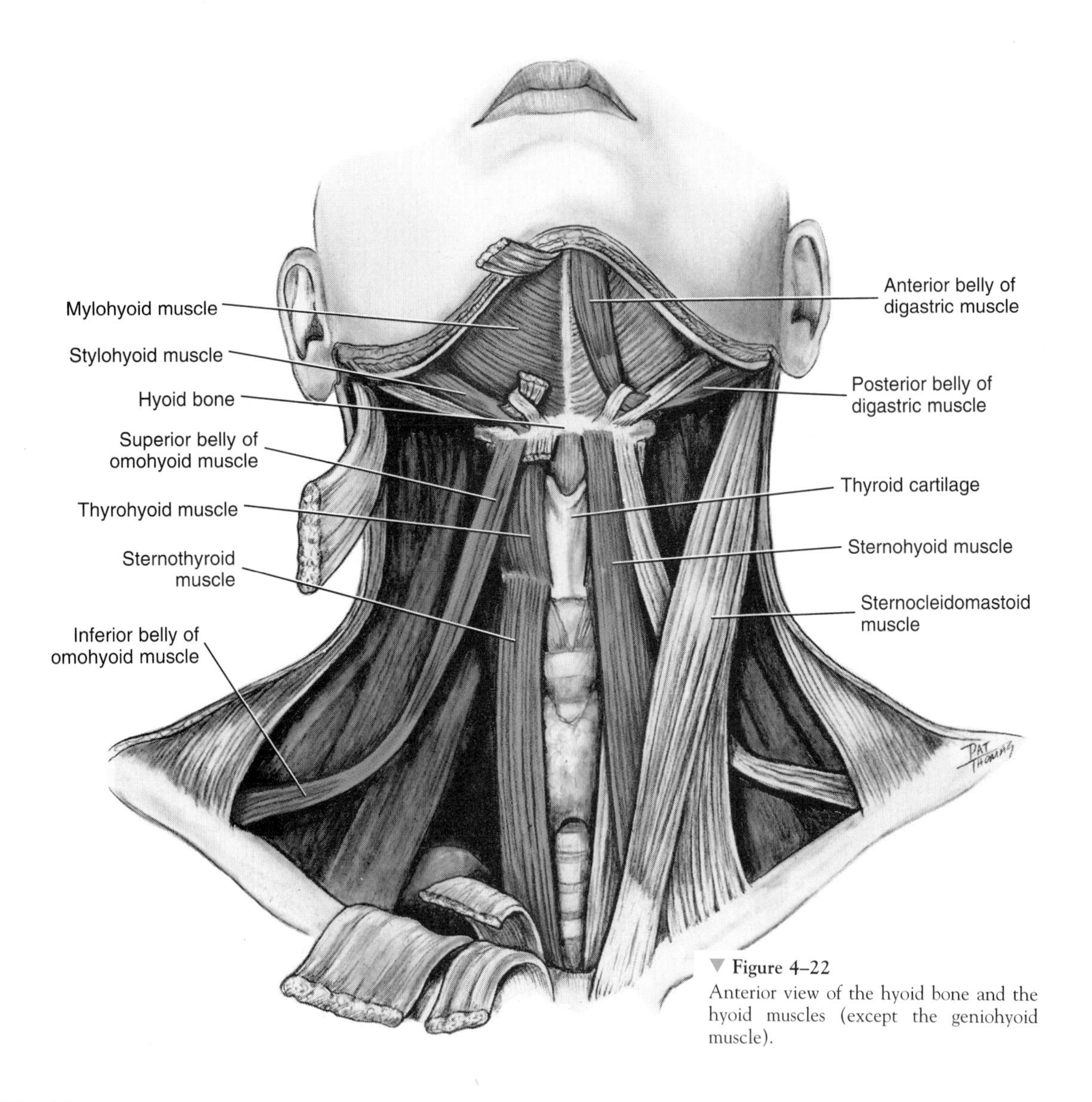

▼ Figure 4–22
Anterior view of the hyoid bone and the hyoid muscles (except the geniohyoid muscle).

▼ Table 4–6

ORIGIN AND INSERTION OF THE HYOID MUSCLES

Muscles	Origin	Insertion
Suprahyoid		
Digastric	Anterior belly: intermediate tendon	Anterior belly: medial surface of mandible
	Posterior belly: mastoid notch of temporal bone	Posterior belly: intermediate tendon
Mylohyoid	Mylohyoid line of mandible	Body of hyoid bone
Geniohyoid	Genial tubercles of mandible	Body of hyoid bone
Stylohyoid	Styloid process of temporal bone	Body of hyoid bone
Infrahyoid		
Sternothyroid	Posterior surface of sternum	Thyroid cartilage
Sternohyoid	Posterior and superior surfaces of sternum	Body of hyoid bone
Omohyoid	Inferior belly: scapula	Inferior belly: superior belly
	Superior belly: inferior belly	Superior belly: body of hyoid bone
Thyrohyoid	Thyroid cartilage	Body and greater cornu of hyoid bone

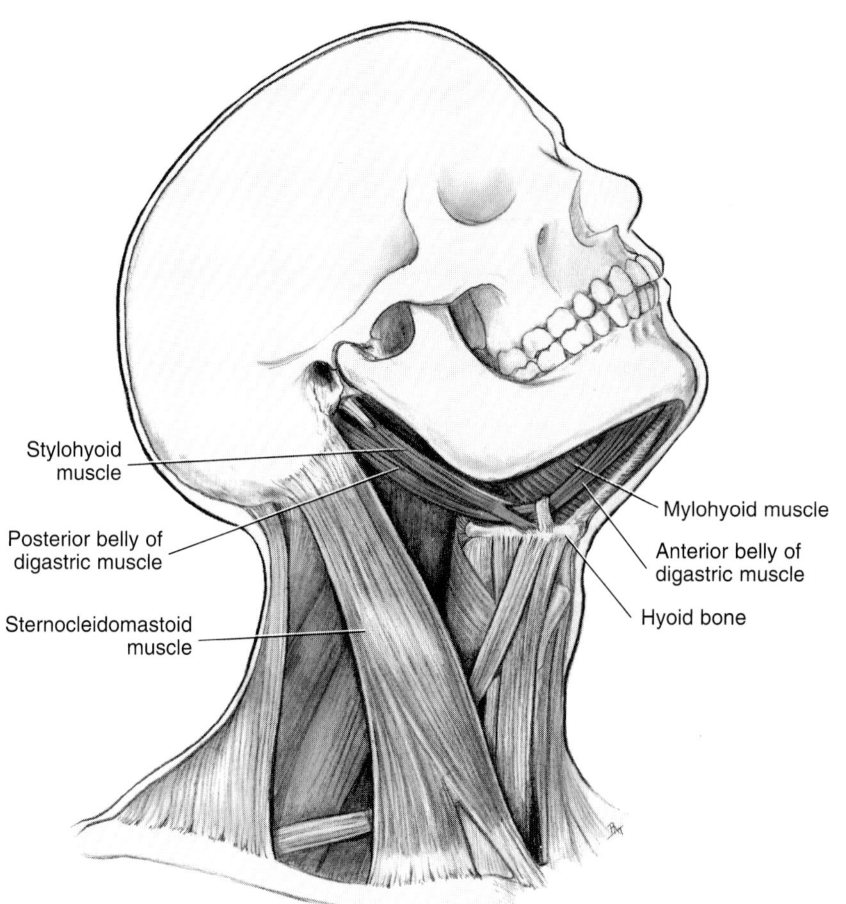

▼ **Figure 4–23**
Lateral view of the hyoid bone and the suprahyoid muscles (except the geniohyoid muscle).

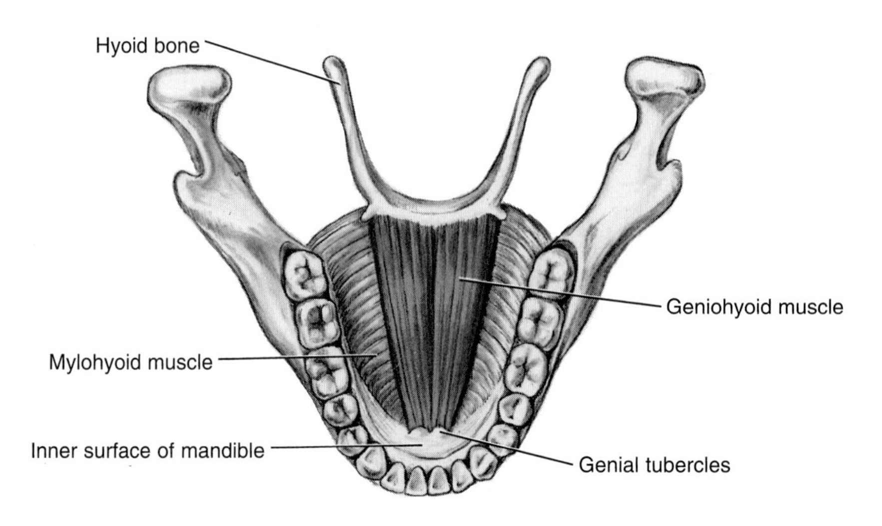

▼ **Figure 4–24**
View from above of the floor of the oral cavity showing the origin and insertion of the geniohyoid muscle.

tached to the body and greater cornu of the hyoid bone called the **intermediate tendon** (in-ter-**me**-dee-it-**ten**-don) and passes superiorly and anteriorly to insert close to the symphysis on the inner surface of the mandible. The posterior belly arises from the mastoid notch, medial to the mastoid process of the temporal bone, and passes anteriorly and inferiorly to insert on the intermediate tendon.

Innervation. The digastric muscle's anterior belly is innervated by the mylohyoid nerve, a branch of the mandibular division of the fifth cranial or trigeminal nerve. The posterior belly of the muscle is innervated by the posterior digastric nerve, a branch of the seventh cranial or facial nerve.

▼ MYLOHYOID MUSCLE

The **mylohyoid muscle** (my-lo-**hi**-oid) is an anterior suprahyoid muscle that is deep to the digastric muscle, with fibers running transversely between the two sides of the mandible.

Origin and Insertion. The mylohyoid muscle originates from the mylohyoid line on the inner surface of the mandible. The right and left muscles pass inferiorly to unite medially, forming the floor of the mouth. The most posterior fibers of the muscle insert on the body of the hyoid bone.

Action. In addition to either elevating the hyoid bone or depressing the mandible, the mylohyoid muscle also forms the floor of the mouth and helps elevate the tongue.

Innervation. The mylohyoid muscle is innervated by the mylohyoid nerve, a branch of the mandibular division of the fifth cranial or trigeminal nerve.

▼ STYLOHYOID MUSCLE

The **stylohyoid muscle** (sty-lo-**hi**-oid) is a thin posterior suprahyoid muscle, anterior and superficial to the posterior belly of the digastric muscle.

Origin and Insertion. The stylohyoid muscle originates from the styloid process of the temporal bone and passes anteriorly and inferiorly to insert on the body of the hyoid bone.

Innervation. The stylohyoid muscle is innervated by the stylohyoid nerve, a branch of the seventh cranial or facial nerve.

▼ GENIOHYOID MUSCLE

The **geniohyoid muscle** (ji-nee-o-**hi**-oid) is an anterior suprahyoid muscle that is deep to the mylohyoid muscle.

Origin and Insertion. The geniohyoid muscle originates from the medial surface of the mandible. On this surface of the mandible, the muscle is attached near the symphysis at the genial tubercles (see Figure 4–24). The muscle then passes posteriorly and inferiorly to insert on the body of the hyoid bone.

Innervation. The geniohyoid muscle is innervated by the first cervical nerve, conducted by way of the twelfth cranial or hypoglossal nerve.

▼ Infrahyoid Muscles

The **infrahyoid muscles** (in-frah-**hi**-oid) are four pairs of hyoid muscles inferior to the hyoid bone (Figure 4–25). The infrahyoid muscles include the sternohyoid, sternothyroid, thyrohyoid, and omohyoid muscles.

Action. Most of the infrahyoid muscles depress the hyoid bone. Some of the infrahyoid muscles have additional specific actions that are also discussed.

Innervation. All the infrahyoid muscles are innervated by the second and third cervical nerves.

▼ STERNOTHYROID MUSCLE

The **sternothyroid muscle** (ster-no-**thy**-roid) is an infrahyoid muscle located superficial to the thyroid gland.

Origin and Insertion. The sternothyroid muscle originates from the posterior surface of the sternum, deep and medial to the sternothyroid muscle, at the level of the first rib. The muscle then passes superiorly to insert on the thyroid cartilage.

Action. The sternothyroid muscle depresses the thyroid cartilage and larynx, yet does not directly depress the hyoid bone.

▼ STERNOHYOID MUSCLE

The **sternohyoid muscle** (ster-no-**hi**-oid) is an infrahyoid muscle superficial to the sternothyroid muscle, as well as the thyroid gland and cartilage.

Origin and Insertion. The sternohyoid muscle originates from the posterior and superior surfaces of the sternum, close to where the sternum joins each clavicle. The muscle then passes superiorly to insert on the body of the hyoid bone.

▼ OMOHYOID MUSCLE

The **omohyoid muscle** (o-mo-**hi**-oid) is an infrahyoid muscle lateral to both the sternothyroid and

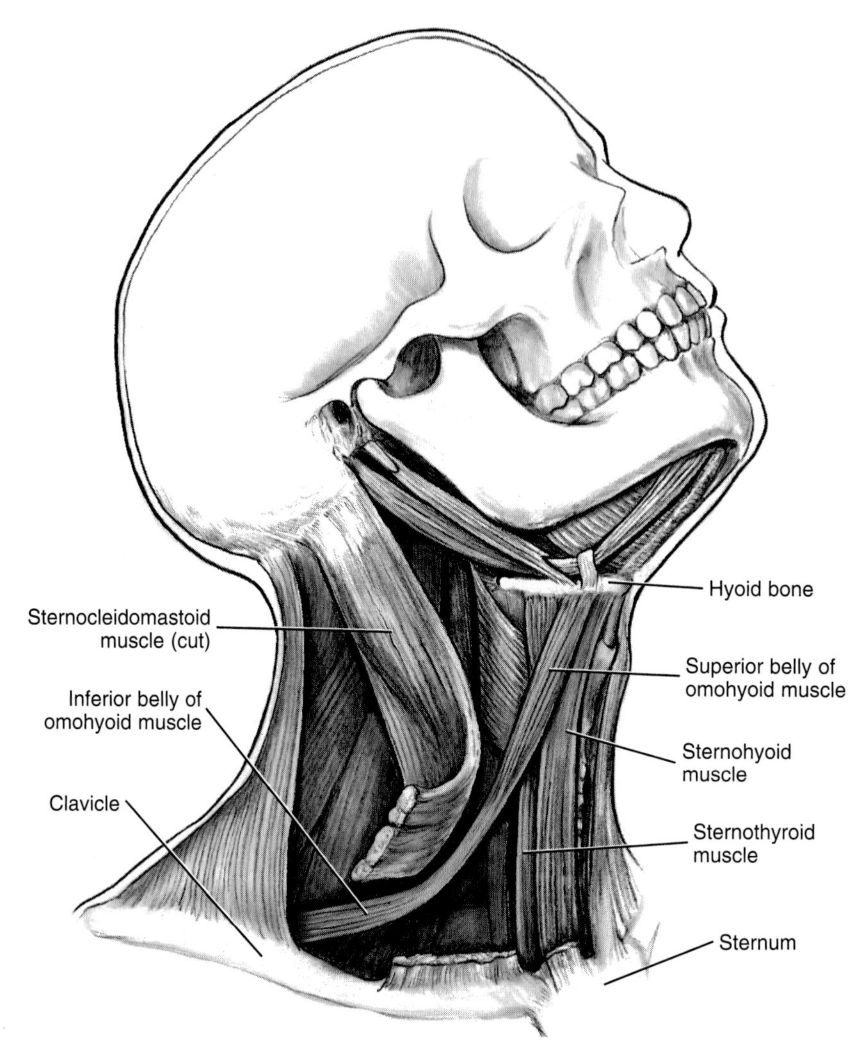

▼ **Figure 4–25**
Lateral view of the hyoid bone and the infrahyoid muscles.

thyrohyoid muscles. This muscle has a superior belly and inferior belly. The superior belly of each muscle divides the inferior portion of the anterior cervical triangle into the carotid and muscular triangles. In the posterior cervical triangle, the inferior belly of each muscle serves to demarcate the subclavian triangle (inferiorly) from the occipital triangle (superiorly).

Origin and Insertion. The inferior belly of the omohyoid muscle originates from the scapula. The inferior belly then passes anteriorly and superiorly, crossing the internal jugular vein, beneath the SCM, where it attaches by a short tendon to the superior belly. The muscle's superior belly originates from the short tendon attached to the inferior belly and then inserts on the lateral border of the body of the hyoid bone.

▼ THYROHYOID MUSCLE

The **thyrohyoid muscle** (thy-ro-**hi**-oid), the final infrahyoid muscle to be discussed, is covered by the omohyoid and sternohyoid muscles.

Origin and Insertion. The thyrohyoid muscle originates on the thyroid cartilage and inserts on the body and greater cornu of the hyoid bone. This muscle appears as a continuation of the sternothyroid muscle.

Action. In addition to depressing the hyoid bone, the thyrohyoid muscle raises the thyroid cartilage and larynx.

MUSCLES OF THE TONGUE

The **muscles of the tongue** can be grouped according to whether they are intrinsic or extrinsic tongue muscles (Figure 4–26). The tongue consists of symmetrical halves divided from each other by the **median septum** (**sep**-tum), a deep fibrous structure in the midline of the tongue. The median septum corresponds to the midline depression on the tongue's dorsal surface called the **median lingual sulcus** (see Chapter 1 for more information). Each half of the tongue has muscular groups arranged in various directions, with the intrinsic and extrinsic tongue muscles intertwining with each other. The tongue is further divided into a base and body with an apex.

Origin and Insertion. The intrinsic tongue muscles are all located inside the tongue. The extrinsic tongue muscles all have their origin outside the tongue yet have their insertion inside the tongue.

Action. The tongue has complex movements during mastication, speaking, and swallowing. These movements are a result of the combined action of muscles of the tongue. The intrinsic tongue muscles change the shape of the tongue. The extrinsic tongue muscles also move the tongue, while suspending and anchoring the tongue to the mandible, styloid process, and hyoid bone. When studying the muscles of the tongue, try to visualize the movements produced by these muscles. The specific actions of each muscle of the tongue will be given.

Intrinsic Tongue Muscles

There are four sets of **intrinsic tongue muscles** (in-**trin**-sik) that are all located entirely inside the tongue (see Figure 4–26). These muscles of the tongue are named by their orientation. The intrinsic tongue muscles include the superior longitudinal, transverse, vertical, and inferior longitudinal muscles. Anatomists consider the intrinsic tongue muscles to be inseparable, so they are not usually treated as separate muscles.

Origin and Insertion. The superior longitudinal muscle is the most superficial of the intrinsic tongue muscles. This muscle runs in an oblique and longitudinal direction in the dorsal surface from the base to the apex. Deep to the superior longitudinal muscle is the **transverse muscle.** The transverse muscle runs in a transverse direction from the median septum to pass

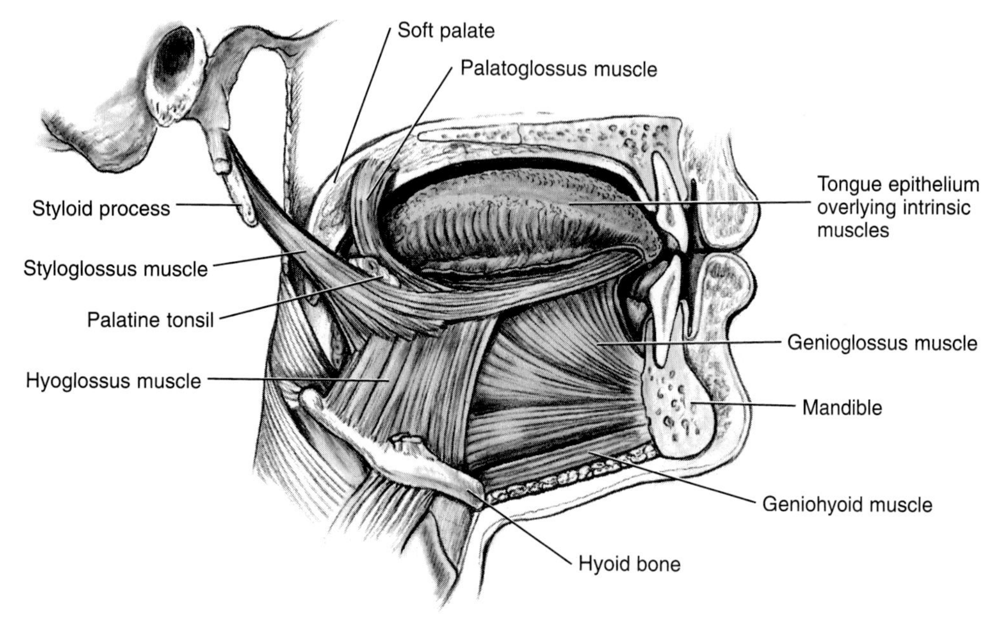

Figure 4–26
The tongue with its intrinsic and extrinsic muscles.

outward toward the lateral surface of the tongue. The **vertical muscle** (**ver**-ti-kel) runs in a vertical direction from the dorsal surface inward to the ventral surface in the body of the tongue. The inferior longitudinal muscle is located in the ventral surface of the tongue. This muscle runs in a longitudinal direction from the base to the apex of the tongue.

Action. The superior and inferior longitudinal muscles act together to change the shape of the tongue by shortening and thickening it and act singly to help it curl in various directions. The transverse and vertical muscles act together to make the tongue long and narrow.

Innervation. All the intrinsic tongue muscles are innervated by the twelfth cranial or hypoglossal nerve.

Extrinsic Tongue Muscles

There are three pairs of **extrinsic tongue muscles** (eks-**trin**-sik) with different origins outside the tongue and all their insertions inside the tongue (see Figure 4–26, Table 4–7). The extrinsic tongue muscles include the genioglossus, styloglossus, and hyoglossus muscles. Some anatomists also include the palatoglossus muscle in this category since it is involved in tongue movement; the palatoglossus muscle is discussed with the muscles of the soft palate in this chapter.

Innervation. All the extrinsic tongue muscles are innervated by the twelfth cranial or hypoglossal nerve.

▼ GENIOGLOSSUS MUSCLE

The **genioglossus muscle** (ji-nee-o-**gloss**-us) is a fan-shaped extrinsic tongue muscle superior to the geniohyoid muscle.

Origin and Insertion. The genioglossus muscle arises from the genial tubercles (or mental spines) on the internal surface of the mandible. A few of the most inferior fibers insert on the hyoid bone, but most of the fibers insert in the tongue from its base almost to the apex. The right and left muscles are separated by the tongue's median septum.

Action. Different parts of the genioglossus muscle can protrude the tongue out of the oral cavity or depress parts of the tongue surface. The protrusive activity of the genioglossus muscle helps to prevent the tongue from sinking back and obstructing respiration; therefore during general anesthesia, the mandible is sometimes pulled forward to achieve the same effect.

▼ STYLOGLOSSUS MUSCLE

The **styloglossus muscle** (sty-lo-**gloss**-us) is another extrinsic tongue muscle.

Origin and Insertion. The styloglossus muscle originates from the styloid process of the temporal bone. The muscle then passes inferiorly and anteriorly to insert into two parts of the lateral surface of the tongue, its apex and at the border of the body and base.

Action. The styloglossus muscle retracts the tongue, moving it superiorly and posteriorly.

▼ HYOGLOSSUS MUSCLE

The **hyoglossus muscle** (hi-o-**gloss**-us) is the final extrinsic tongue muscle to be considered.

Origin and Insertion. The hyoglossus muscle originates on the greater cornu and part of the body of the hyoid bone. The muscle then inserts into the lateral surface of the body of the tongue.

Action. The hyoglossus muscle depresses the tongue.

MUSCLES OF THE PHARYNX

The **pharynx** (**far**-inks) is part of both the respiratory and digestive tract, connected to both the nasal and oral cavity. The pharynx consists of three portions, the

Table 4–7

EXTRINSIC TONGUE MUSCLES WITH THEIR ORIGINS, INSERTIONS, AND ACTIONS

Muscles	Origin	Insertion	Action
Genioglossus	Genial tubercles on mandible	Hyoid bone and tongue	Protrudes tongue and depresses portions
Styloglossus	Styloid process of temporal bone	Tongue	Retracts tongue
Hyoglossus	Greater cornu and body of hyoid bone	Tongue	Depresses tongue

nasopharynx, oropharynx, and laryngopharynx (see Chapter 1 for more information). The **muscles of the pharynx** are involved in speaking, swallowing, and middle ear function. The muscles of the pharynx include the stylopharyngeus, pharyngeal constrictor, and soft palate muscles.

Stylopharyngeus Muscle

The **stylopharyngeus muscle** (sty-lo-fah-**rin**-je-us) is a paired longitudinal muscle of the pharynx.

Origin and Insertion. The stylopharyngeus originates from the styloid process of the temporal bone. The muscle then inserts into the lateral and posterior pharyngeal walls (Figure 4–27A and B).

Action. The stylopharyngeus muscle elevates the pharynx, simultaneously widening the pharynx.

Innervation. The stylopharyngeus muscle is innervated by the ninth cranial or glossopharyngeal nerve.

Pharyngeal Constrictor Muscles

The **pharyngeal constrictor muscles** (fah-**rin**-je-il kon-**strik**-tor) consist of three paired muscles, the superior, middle, and inferior pharyngeal constrictor muscles, which form the lateral and posterior walls of the pharynx.

Origin and Insertion. The origin for each of the three pharyngeal constrictor muscles is different, although the muscles overlap each other and all

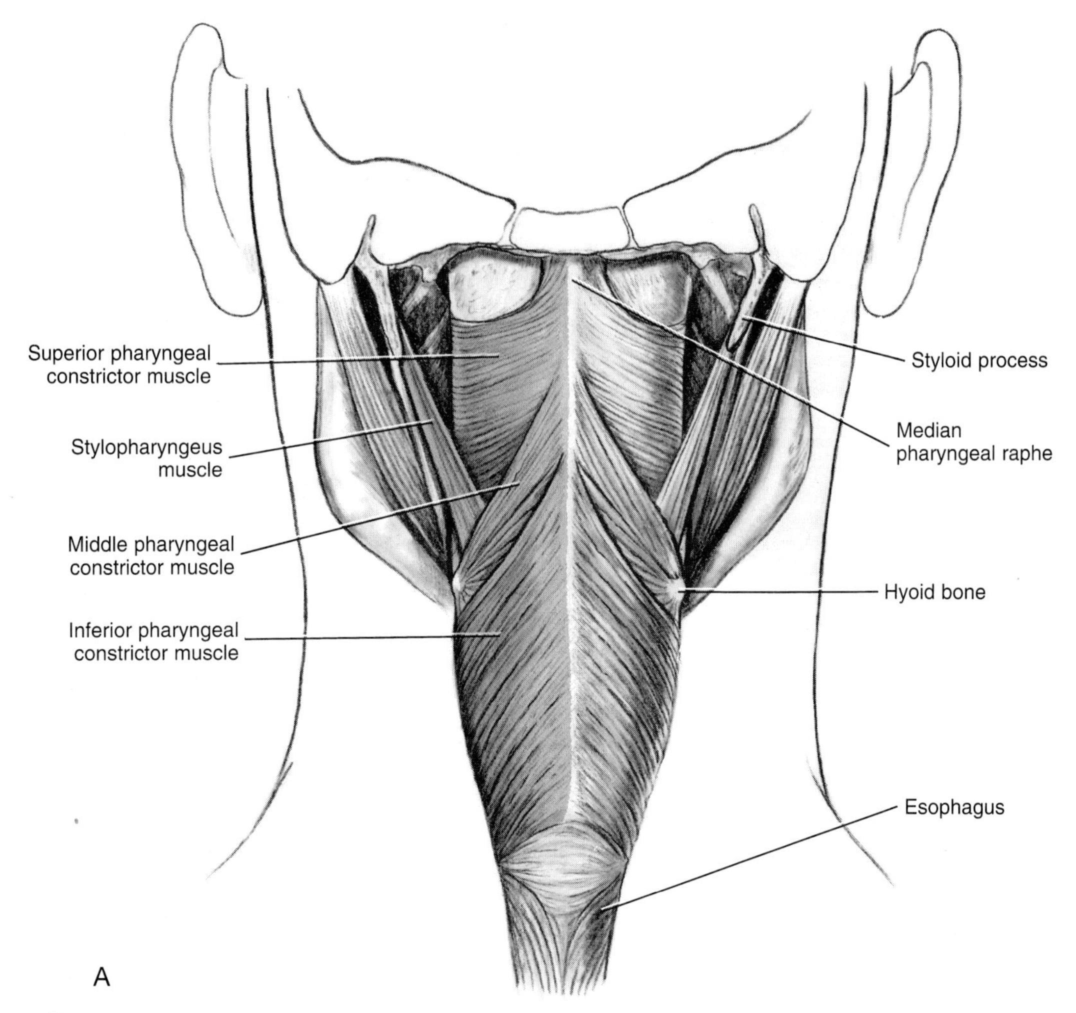

Figure 4–27
A: Posterior view of the muscles of the pharynx.

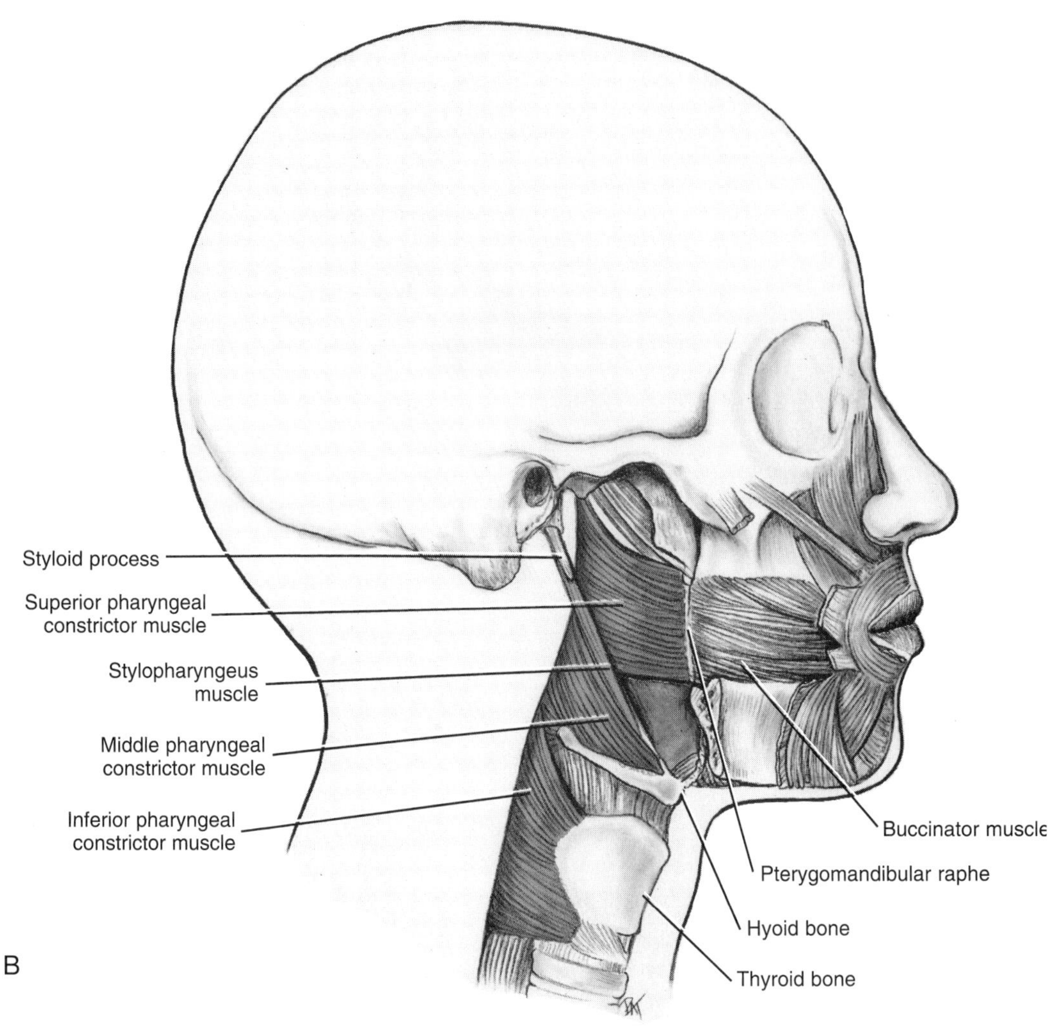

Figure 4-27–Continued
B: Lateral view of the muscles of the pharynx.

have similar insertions (see Figure 4–27A and B). The superior pharyngeal constrictor muscle originates from the pterygoid hamulus, mandible, and pterygomandibular raphe. The middle pharyngeal constrictor muscle originates on the hyoid bone and stylohyoid ligament. The inferior pharyngeal constrictor muscle originates from the thyroid and cricoid cartilages of the larynx. All three pharyngeal constrictor muscles overlap each other, the inferior being most superficial. These muscles then insert into the **median pharyngeal raphe** (fah-**rin**-je-al **ra**-fe), a midline fibrous band of the posterior wall of the pharynx that is itself attached to the base of the skull.

Action. The pharyngeal constrictor muscles raise the pharynx and larynx and help drive food inferiorly into the esophagus during swallowing.

Innervation. All three pharyngeal constrictor muscles are innervated by the pharyngeal plexus.

Muscles of the Soft Palate

The **soft palate** (**pal**-it) forms the nonbony posterior portion of the roof of the mouth or the oropharynx and connects laterally with the tongue (see Chapter 1 for more information). The muscles of the soft palate are involved in speaking and swallowing. The **muscles of the soft palate** include the palatoglossus, palatopharyngeus, levator veli palatini, and tensor veli palatini muscles and the muscle of the uvula (Figures 4–28 and 4–29, Table 4–8). Some anatomists consider the palatoglossus muscle to be an extrinsic muscle of the tongue since it is involved in tongue movement, but it will only be considered a muscle of the soft palate in this chapter.

Action. When the muscles of the soft palate are relaxed, the soft palate extends posteriorly over the anterior oropharynx. The combined actions of several muscles of the soft palate move the soft palate superiorly and posteriorly to contact the posterior pharyngeal wall that is being moved anteriorly. This movement of both the soft palate and pharyngeal wall brings a separation between the nasopharynx and oral cavity during swallowing to prevent food from entering the nasal cavity while eating. Specific actions of each muscle of the soft palate will also be discussed.

Innervation. All the muscles of the soft palate are innervated by the pharyngeal plexus, except the tensor veli palatini muscle. This muscle is supplied by the mandibular division of the fifth cranial or trigeminal nerve.

▼ PALATOGLOSSUS MUSCLE

The **palatoglossus muscle** (pal-ah-to-**gloss**-us) forms the **anterior tonsillar pillar** (**ton**-sil-ar **pil**-er) in the oral cavity, a vertical fold anterior to each palatine tonsil (see Figures 4–26 and 4–29).

Origin and Insertion. The palatoglossus muscle originates from the **median palatine raphe**

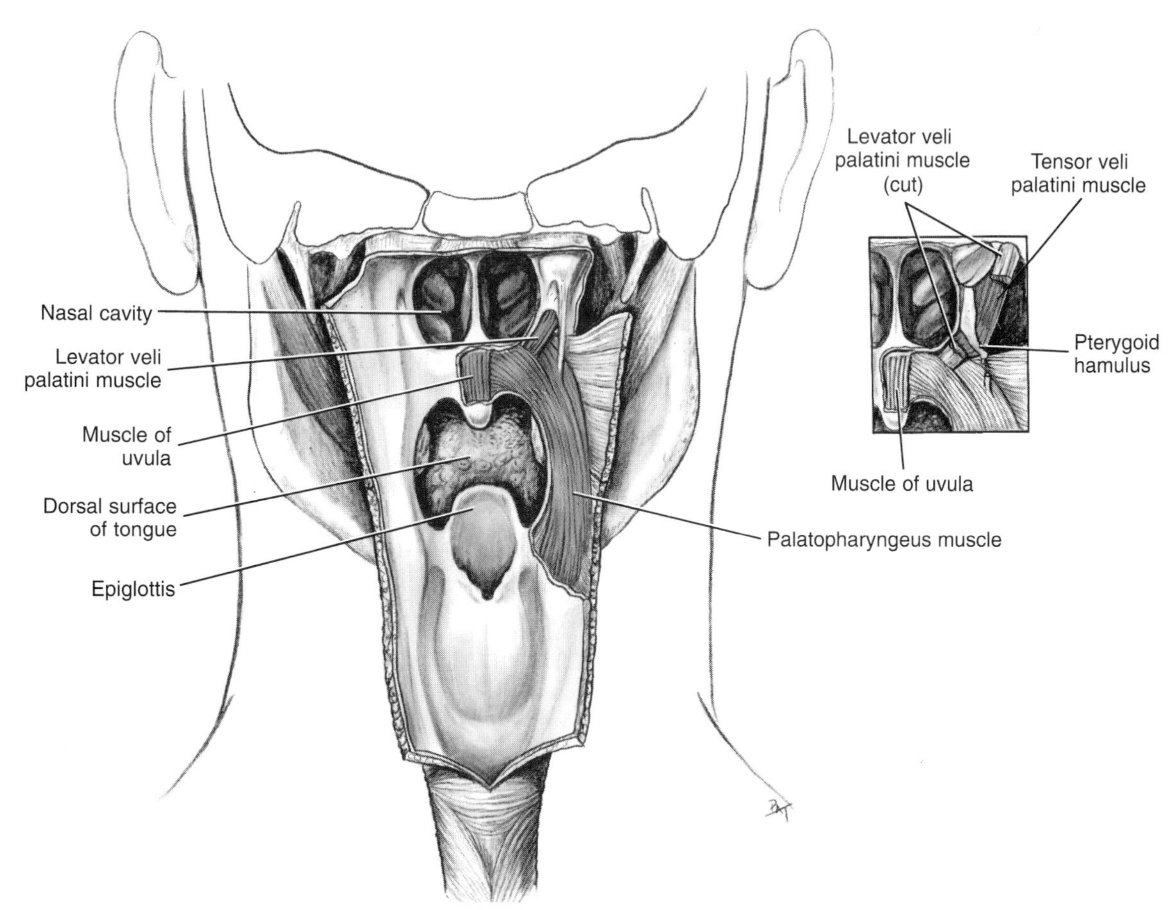

▼ **Figure 4–28**
Posterior view of the muscles of the soft palate (pharyngeal constrictor muscles have been cut and mucous membranes partially removed).

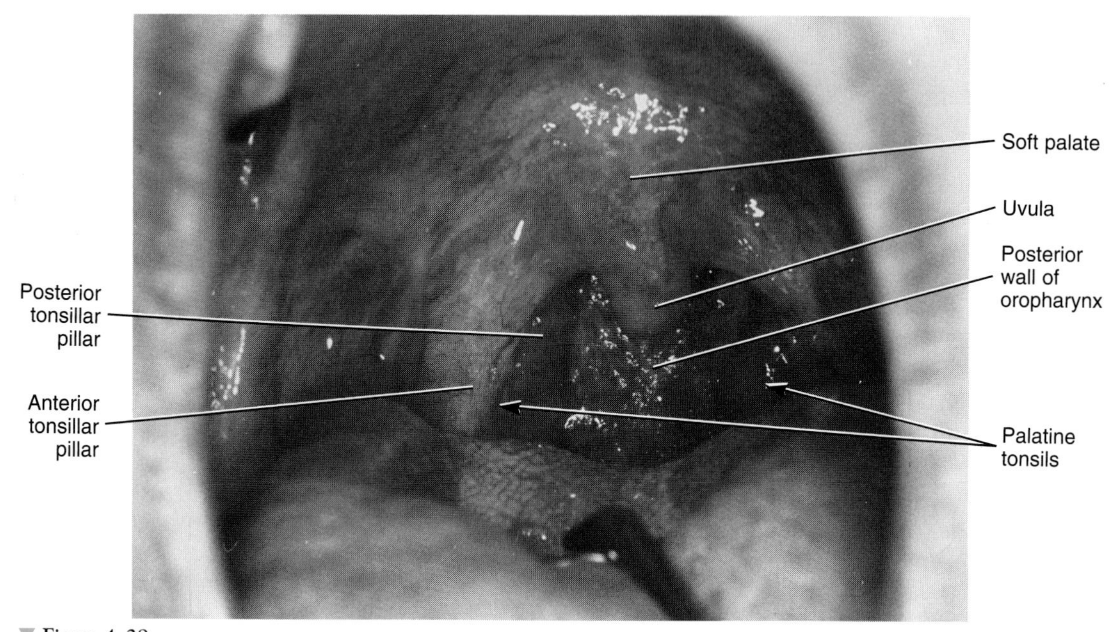

Figure 4–29
Intraoral view of the soft palate, uvula, and anterior and posterior tonsillar pillars with palatine tonsils. These oral landmarks are related to muscular tissue.

Table 4–8
MUSCLES OF THE SOFT PALATE WITH THEIR ORIGINS, INSERTIONS, AND ACTIONS

Muscles	Origin	Insertion	Action
Palatoglossus (anterior tonsillar pillar)	Median palatine raphe	Tongue	Elevates and arches tongue, depressing soft palate toward tongue
Palatopharyngeus (posterior tonsillar pillar)	Soft palate	Laryngopharynx and thyroid cartilage	Moves palate posteroinferiorly and posterior pharyngeal wall anterosuperiorly
Levator veli palatini	Temporal bone	Median palatine raphe	Raises soft palate to contact the posterior pharyngeal wall
Tensor veli palatini	Auditory tube and sphenoid bone	Median palatine raphe	Tenses and slightly lowers soft palate
Muscle of the uvula	Tissue projection that hangs inferiorly from posterior soft palate		Soft palate closely adapts to posterior pharyngeal wall

(**pal**-ah-tine **ra**-fe), a midline fibrous band of the palate. The muscle then inserts into the lateral surface of the tongue.

Action. The palatoglossus muscle elevates the base of the tongue, arching the tongue against the soft palate, and depresses the soft palate toward the tongue. The muscles on both sides form a sphincter, separating the oral cavity from the pharynx.

▼ PALATOPHARYNGEUS MUSCLE

The **palatopharyngeus muscle** (pal-ah-to-fah-**rin**-je-us) forms the **posterior tonsillar pillar** (**ton**-sil-ar **pil**-er) in the oral cavity, a vertical fold posterior to each palatine tonsil (see Figures 4–28 and 4–29).

Origin and Insertion. The palatopharyngeus muscle originates in the soft palate and inserts in the walls of the laryngopharynx and on the thyroid cartilage.

Action. The palatopharyngeus muscle moves the palate posteroinferiorly and the posterior pharyngeal wall anterosuperiorly to help close off the nasopharynx during swallowing.

▼ LEVATOR VELI PALATINI MUSCLE

The **levator veli palatini muscle** (le-**vate**-er **vee**-lie pal-ah-**teen**-ee) is a muscle mainly situated superior to the soft palate (see Figure 4–28).

Origin and Insertion. The levator veli palatini muscle originates from the inferior surface of the temporal bone. The muscle then inserts into the **median palatine raphe** (**pal**-ah-tine **ra**-fe), a midline fibrous band of the palate (see Chapter 1).

Action. The levator veli palatini muscle raises the soft palate and helps bring it into contact with the posterior pharyngeal wall to close off the nasopharynx during speech and swallowing.

▼ TENSOR VELI PALATINI MUSCLE

The **tensor veli palatini muscle** (**ten**-ser **vee**-lie pal-ah-**teen**-ee) is a special muscle that stiffens the soft palate (see Figure 4–28). This muscle is probably active during all palatal movements. Some of its fibers are also responsible for opening the auditory tube to allow air to flow between the pharynx and middle ear cavity.

Origin and Insertion. The tensor veli palatini muscle originates from the auditory tube area and the inferior surface of the sphenoid bone. The muscle then passes inferiorly between the medial pterygoid muscle and medial pterygoid plate, forming a tendon near the pterygoid hamulus. The tendon winds around the hamulus, using it as a pulley, and then spreads out to insert in the median palatine raphe.

Action. The tensor veli palatini muscle tenses and slightly lowers the soft palate.

▼ MUSCLE OF THE UVULA

The **muscle of the uvula** (**u**-vu-lah) is the final muscle of the soft palate to be considered.

Origin and Insertion. The muscle of the uvula lies within the **uvula of the palate** (see Chapter 1 for more information). The uvula is a midline tissue structure that hangs inferiorly from the posterior margin of the soft palate (see Figures 4–28 and 4–29).

Action. The muscle of the uvula shortens and broadens the uvula, changing the contour of the posterior portion of the soft palate. This change in contour allows the soft palate to adapt closely to the posterior pharyngeal wall to help close off the nasopharynx during swallowing.

Identification Exercise

Identify the structures on the following diagrams by filling in the blanks with the correct anatomical term. You can check your answers by looking back at the figure indicated in parentheses for each identification diagram.

1. (Figure 4–4)

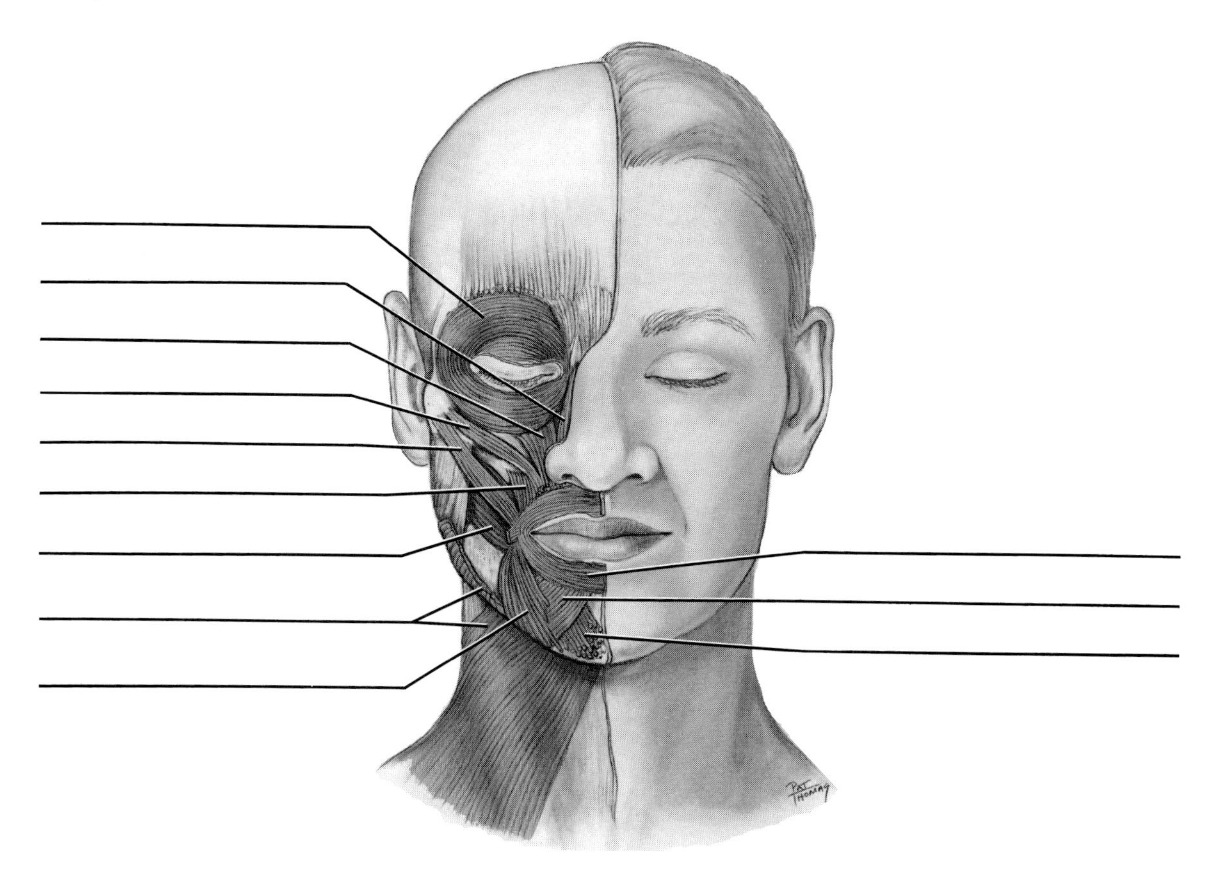

Identification Exercise *Continued*

2. (Figure 4–5)

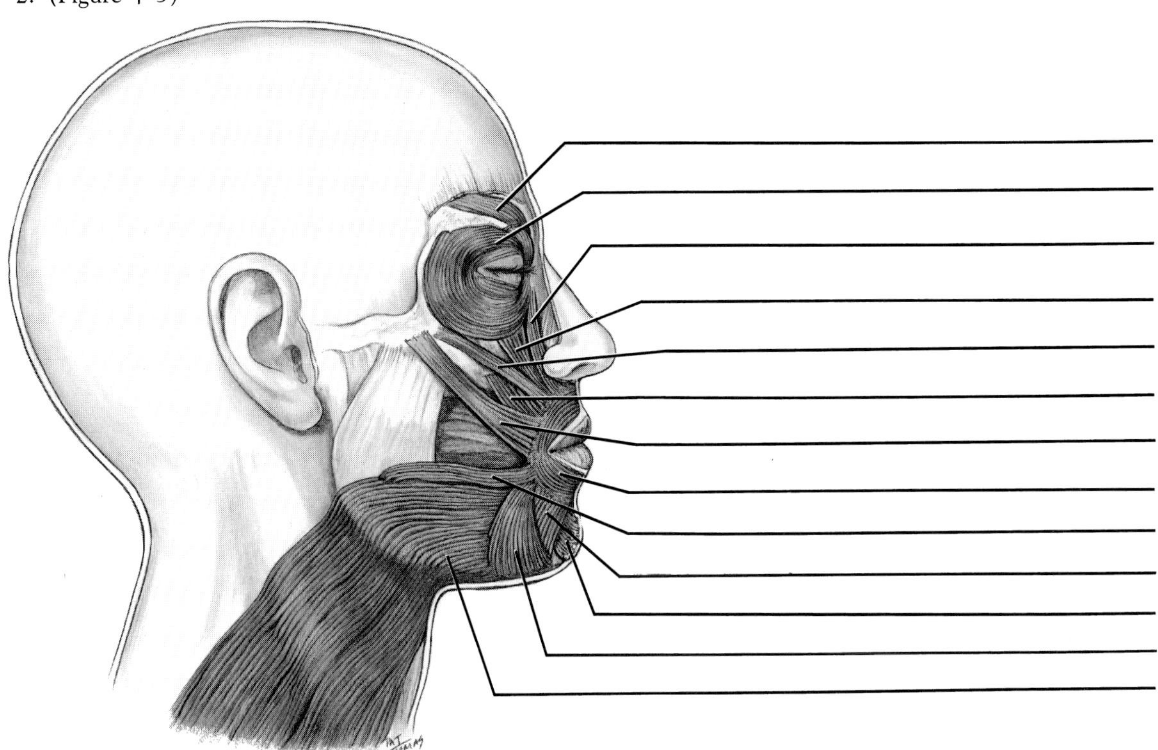

3. (Figure 4–21)

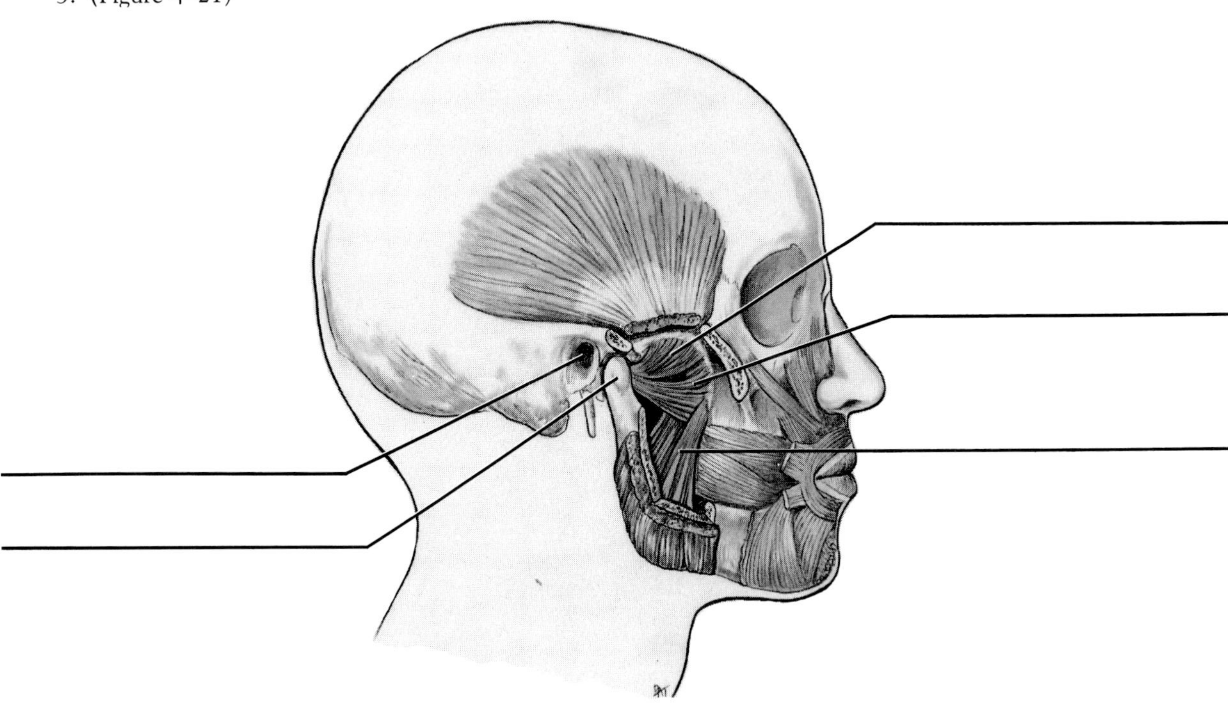

Identification Exercise *Continued*

4. (Figure 4–22)

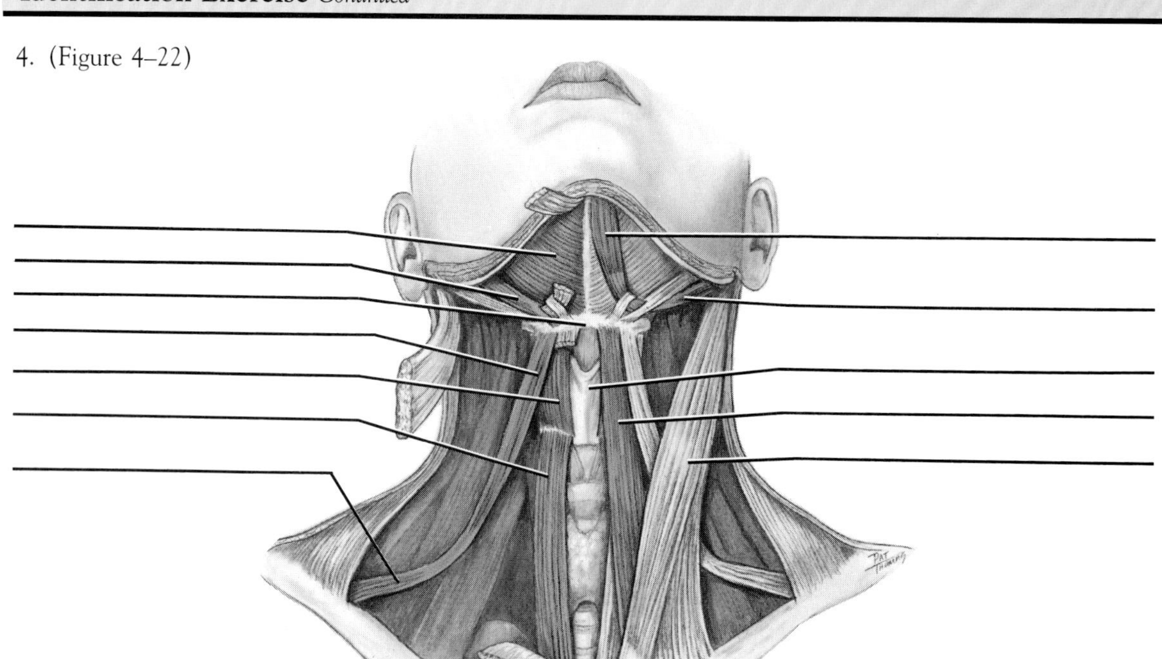

5. (Figure 4–25)

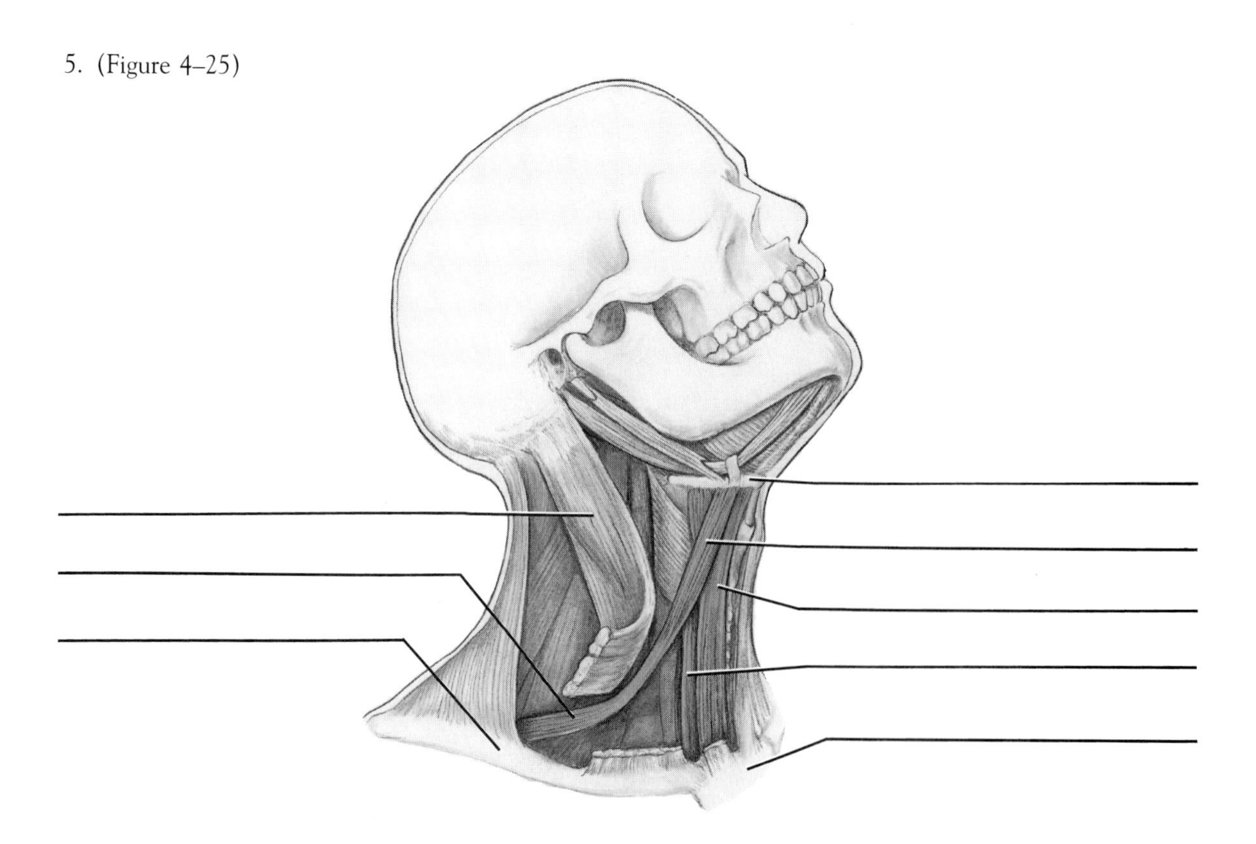

Identification Exercise *Continued*

6. (Figure 4–26)

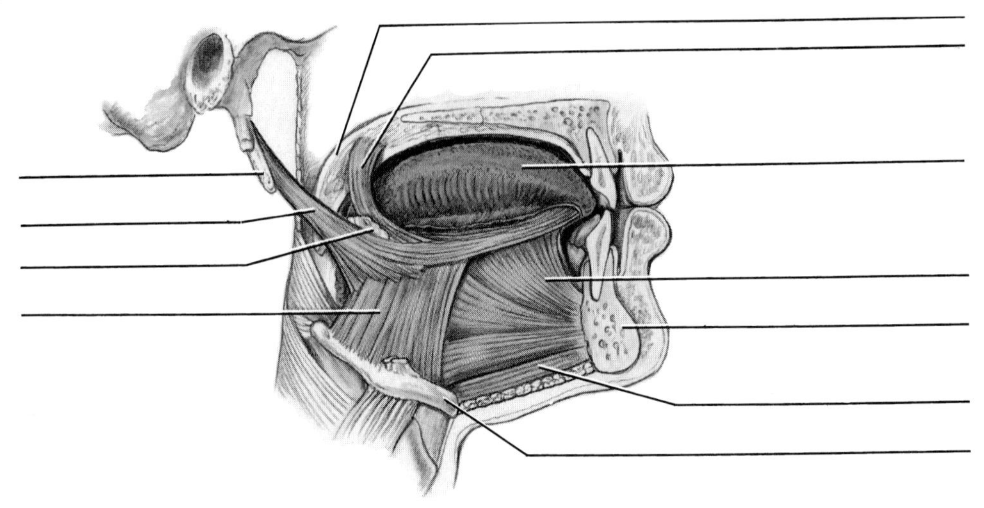

7. (Figure 4–27B)

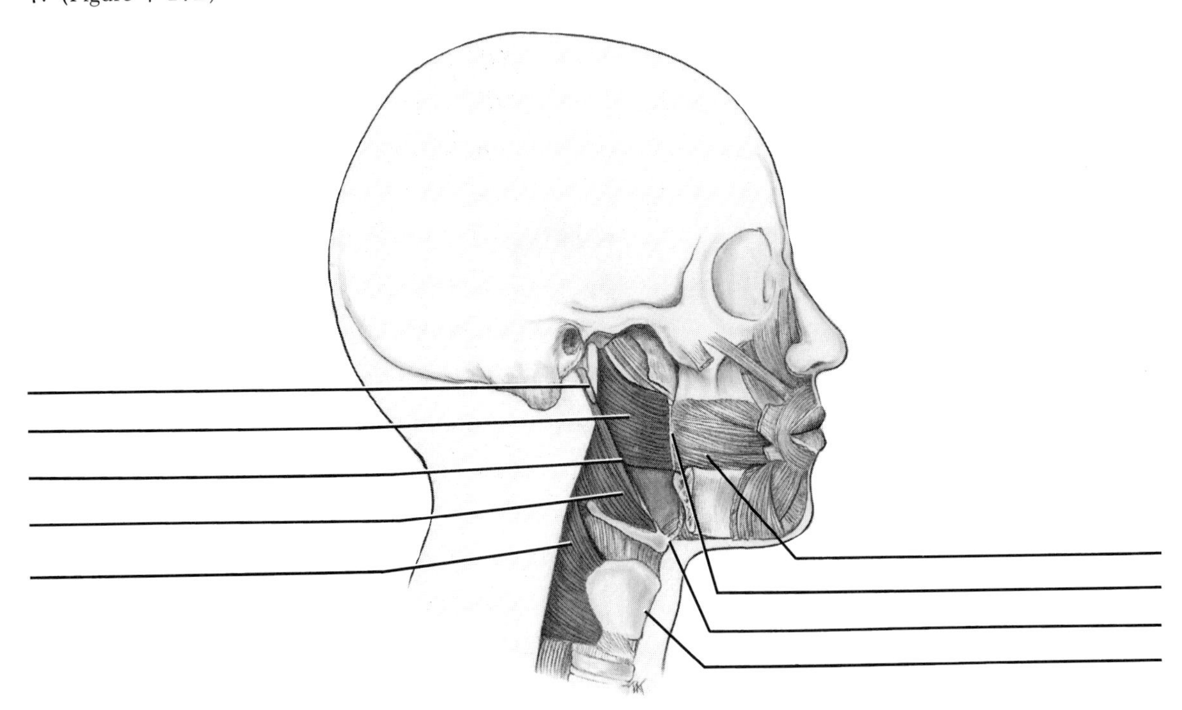

Identification Exercise *Continued*

8. (Figure 4–28)

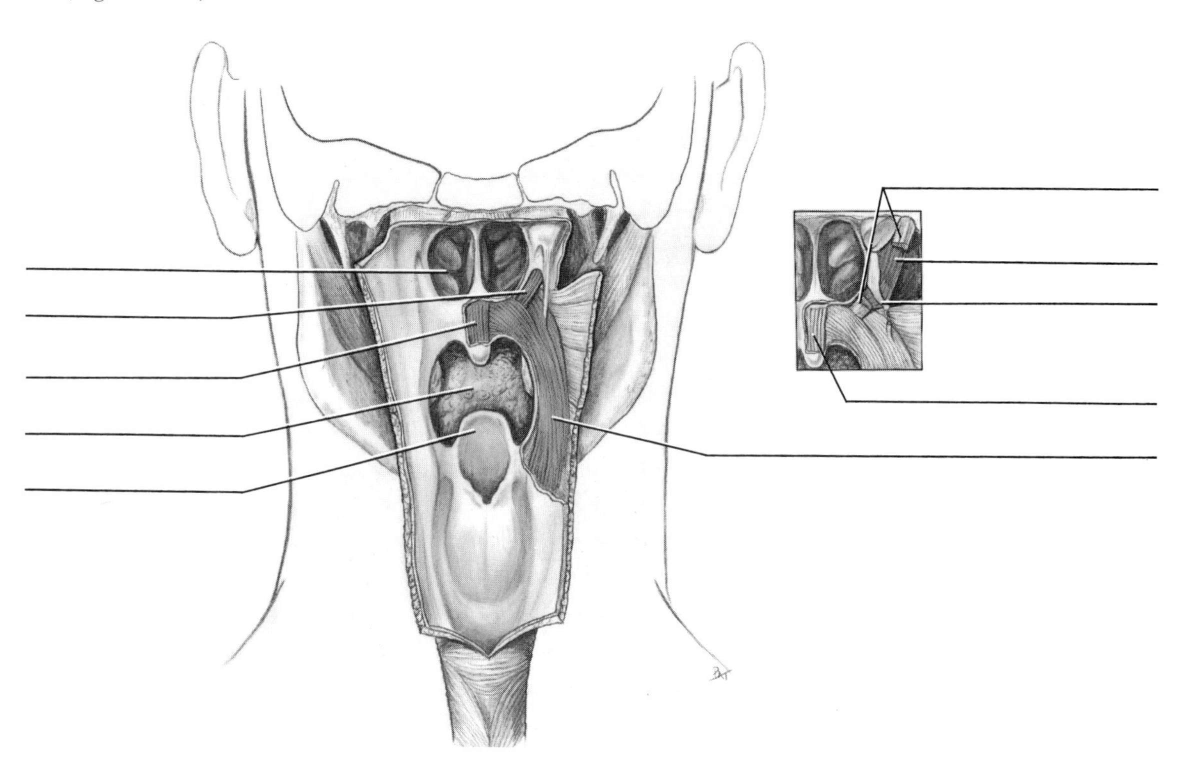

REVIEW QUESTIONS

1. Both the origin of the frontal belly of the epicranial muscle and the insertion of its occipital belly are at the
   A. Clavicle and sternum
   B. Mastoid process
   C. Epicranial aponeurosis
   D. Pterygomandibular raphe

2. Which of the following muscles is considered a muscle of mastication?
   A. Buccinator muscle
   B. Risorius muscle
   C. Mentalis muscle
   D. Masseter muscle
   E. Corrugator supercilii muscle

3. The origin of a muscle is
   A. The starting point of a muscle
   B. Where the muscle fibers join the bone tendon
   C. The muscle end attached to the least movable structure
   D. The muscle end attached to the more movable structure

4. Which of the following muscle pairs is divided by a median septum?
   A. Geniohyoid muscle
   B. Masseter muscle
   C. Digastric muscle
   D. Transverse muscle of the tongue
   E. Vertical muscle of the tongue

5. Which of the following paired muscles unites medially, forming the floor of the mouth?
   A. Geniohyoid muscle
   B. Omohyoid muscle
   C. Digastric muscle
   D. Mylohyoid muscle
   E. Transverse muscle of the tongue

6. Which of the following muscle groups depress the hyoid bone?
   A. Muscle of mastication
   B. Suprahyoid muscles
   C. Infrahyoid muscles
   D. Intrinsic tongue muscles
   E. Extrinsic tongue muscles

7. Which of the following muscles has two bellies, giving the muscle two different origins?
   A. Lateral pterygoid muscle
   B. Geniohyoid muscle
   C. Thyrohyoid muscle
   D. Stylohyoid muscle
   E. Geniohyoid muscle

8. Which of the following is the most important muscle used when the patient's lips close around the saliva ejector?
   A. Risorius muscle
   B. Mentalis muscle
   C. Mylohyoid muscle
   D. Buccinator muscle
   E. Orbicularis oris muscle

9. Which of the following muscle groups is involved in elevating the hyoid bone and depressing the mandible?
   A. Muscles of mastication
   B. Suprahyoid muscles
   C. Infrahyoid muscles
   D. Intrinsic tongue muscles
   E. Extrinsic tongue muscles

10. Which of the following muscle groups is innervated by the cervical nerves?
    A. Muscles of mastication
    B. Muscles of facial expression
    C. Suprahyoid muscles
    D. Infrahyoid muscles
    E. Intrinsic tongue muscles

11. Which muscle can make the patient's oral vestibule shallow, thereby making dental work difficult?
    A. Mentalis muscle
    B. Zygomaticus major muscle
    C. Depressor anguli oris muscle
    D. Levator anguli oris muscle

12. Which of the following muscle groups is innervated by the facial nerve?
    A. Intrinsic tongue muscles
    B. Extrinsic tongue muscles
    C. Muscles of facial expression
    D. Muscles of mastication

REVIEW QUESTIONS *Continued*

13. Which of the following muscle groups inserts directly on the hyoid bone?
    A. Geniohyoid, stylohyoid, and omohyoid muscles
    B. Masseter, stylohyoid, and digastric muscles
    C. Masseter, buccinator, and omohyoid muscles
    D. Palatopharyngeus and palatoglossus muscle and muscle of the uvula

14. Which of the following muscles is used when a patient grimaces?
    A. Epicranial muscle
    B. Corrugator supercilii
    C. Risorius muscle
    D. Mentalis muscle

15. Which of the following muscles is an extrinsic muscle of the tongue?
    A. Geniohyoid muscle
    B. Hyoglossus muscle
    C. Mylohyoid muscle
    D. Transverse muscle
    E. Vertical muscle

16. Which of the following muscles compresses the cheeks during chewing, assisting the muscles of mastication?
    A. Risorius muscle
    B. Buccinator muscle
    C. Mentalis muscle
    D. Orbicularis oris muscle

17. The superior pharyngeal constrictor muscle
    A. Originates from the larynx
    B. Inserts on the median pharyngeal raphe
    C. Overlaps the stylopharyngeus muscle
    D. Is a longitudinal muscle of the pharynx

18. Which of the following statements concerning the masseter muscle is correct?
    A. Most superficial muscle of facial expression
    B. Originates from the zygomatic arch
    C. Inserts on the medial surface of the mandible's angle
    D. Depresses the mandible during jaw movement

19. Which of the following muscles creates the anterior tonsillar pillar in the oral cavity?
    A. Palatoglossus muscle
    B. Palatopharyngeus muscle
    C. Stylopharyngeus muscle
    D. Tensor veli palatini muscle

20. Which of the following situations occurs when both sternocleidomastoid muscles are used by the patient?
    A. Neck is drawn laterally
    B. Head flexes at the neck
    C. Chin moves superiorly to the opposite side
    D. Head rotates and is drawn to the shoulders

Chapter **5**

Temporomandibular Joint

▼ *This chapter discusses the following topics:*

I. Anatomy of the Temporomandibular Joint
   A. Bones of the Joint
   B. Joint Capsule
   C. Disc of the Joint
   D. Ligaments Associated with the Joint

II. Jaw Movements with Muscle Relationships

III. Disorders of the Joint

▼ *Key words*

Depression of the mandible (de-**presh**-in) Lowering of the lower jaw.

Joint (joint) Site of a junction or union between two or more bones.

Lateral deviation of the mandible (de-vee-**ay**-shun) Shifting of the lower jaw to one side.

Ligament (**lig**-ah-mint) Band of fibrous tissue connecting bones.

Elevation of the mandible (el-eh-**vay**-shun) Raising of the lower jaw.

Protrusion of the mandible (pro-**troo**-shun) Bringing of the lower jaw forward.

Retraction of the mandible (re-**trak**-shun) Bringing of the lower jaw backward.

Subluxation (sub-luk-**ay**-shun) Acute episode of temporomandibular joint disorder where both joints become dislocated often due to excessive mandibular protrusion and depression.

Temporomandibular disorder (TMD) (tem-poh-ro-man-**dib**-you-lar) Disorder involving one or both temporomandibular joints.

Chapter 5

Temporomandibular Joint

▼ ***After studying this chapter, the reader should be able to:***

1. Define and pronounce all the key words and anatomical terms in this chapter.
2. Locate and identify the specific anatomical landmarks of the temporomandibular joint on a diagram, a skull, and a patient.
3. Describe the movements of the temporomandibular joint and their relationship to the muscles in the head and neck region.
4. Discuss the disorders of the temporomandibular joint.
5. Correctly complete the identification exercise and review questions for this chapter.
6. Integrate the knowledge about the anatomy of the temporomandibular joint into the examination of the joint for disorders.

Anatomy of the Temporomandibular Joint

A *joint* (joint) is a site of a junction or union between two or more bones. The **temporomandibular joint (TMJ)** (tem-poh-ro-man-**dib**-you-lar) is a joint on each side of the head that allows for movement of the mandible for speech and mastication. A patient may have a disease process associated with one or both of their temporomandibular joints. Thus the dental professional needs to understand the anatomy of the joint, the normal movements involved with the joint, and any possible disorders associated with the joint.

The temporomandibular joint is innervated by the mandibular division of the fifth cranial or trigeminal nerve. The blood supply to the joint is from branches of the external carotid artery.

BONES OF THE JOINT

The temporomandibular joint has two sets of articulations, one on each side of the head: the two temporal bones and the two condyles of the mandible (Figure 5–1). Both articulating bony surfaces of the joint are covered by fibrocartilage. Chapter 3 has more information on both of these bones of the joint.

Temporal Bone

The **temporal bone** (**tem**-poh-ral) is a cranial bone that articulates with the mandible at the TMJ (Figure 5–2). The articulating area on the temporal bone of the joint is located on the bone's inferior aspect. This articulating area includes the bone's articular eminence and articular fossa. The **articular eminence** (ar-**tik**-you-ler) is positioned anterior to the articular fossa and consists of a smooth rounded ridge.

The **articular fossa** (ar-**tik**-you-ler) or mandibular fossa is posterior to the articular eminence and consists of a depression on the temporal bone, posterior and medial to the zygomatic process of the temporal bone (zy-go-**mat**-ik). Posterior to the articular fossa is a sharper ridge, the postglenoid process (post-**gle**-noid).

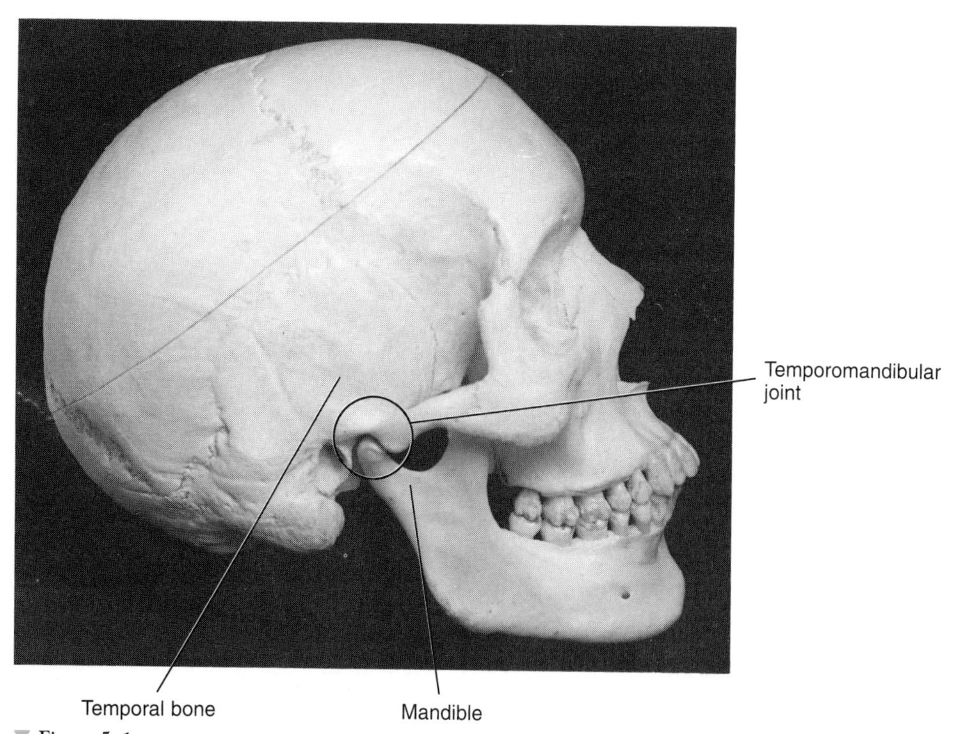

Figure 5–1
Lateral view of the skull showing the temporomandibular joint with the temporal bone and mandible.

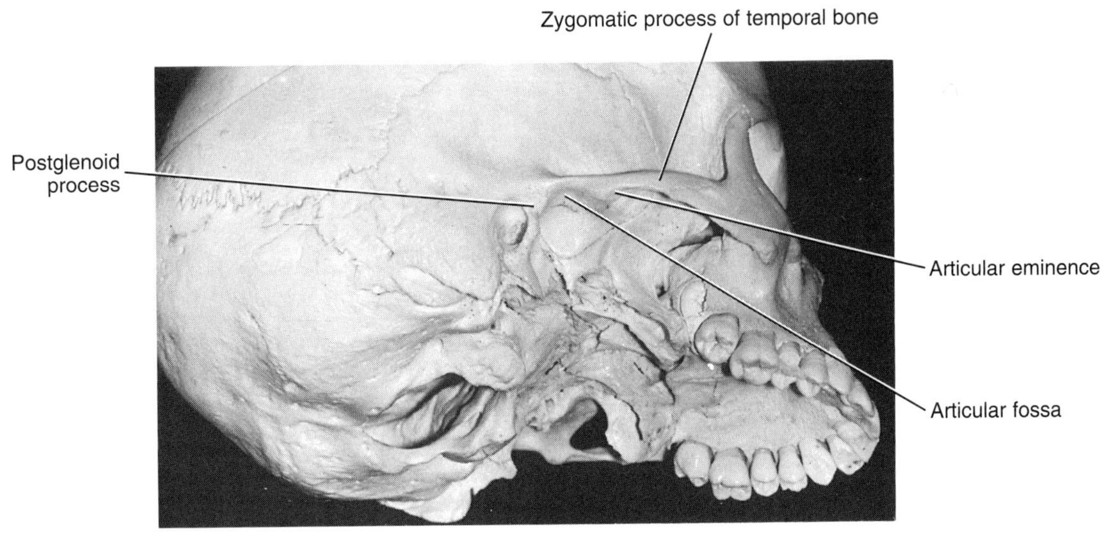

▼ **Figure 5–2**
Inferolateral view of the skull with the temporal bone.

▼ Mandible

The **mandible** (**man**-di-bl) is a facial bone that articulates with the temporal bone at the head of the **condyle of the mandible** with its **articulating surface of the condyle** (ar-**tik**-you-late-ing) (Figure 5–3). Posterior to the condyle is the **coronoid process** (**kor**-ah-noid). The depression of the **mandibular notch** (man-**dib**-you-lar) is located between the condyle and coronoid process.

▼ JOINT CAPSULE

A fibrous **joint capsule** completely encloses the TMJ (Figure 5–4). The capsule wraps around the margin of

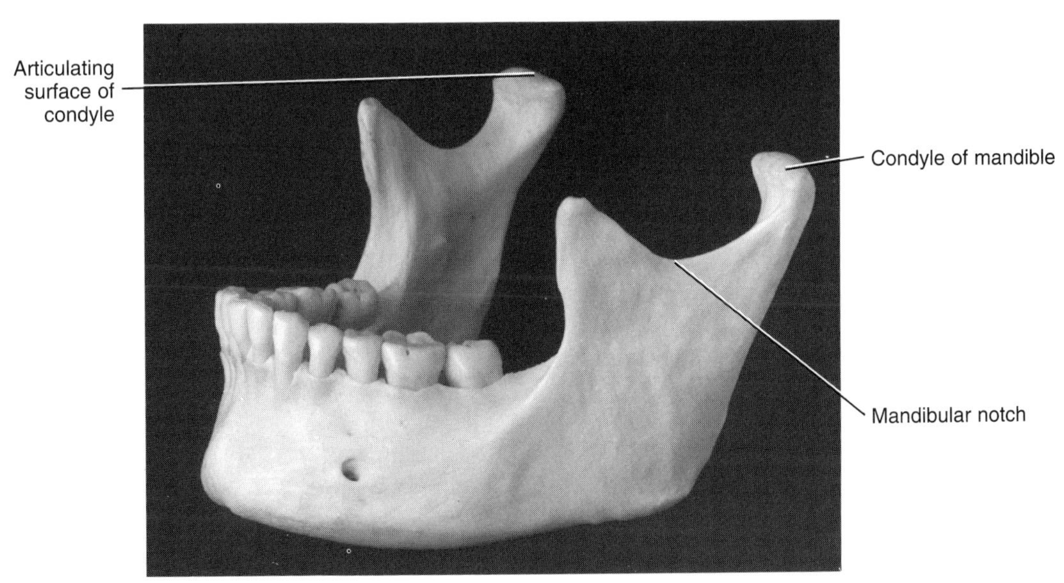

▼ **Figure 5–3**
Anterolateral view of the mandible.

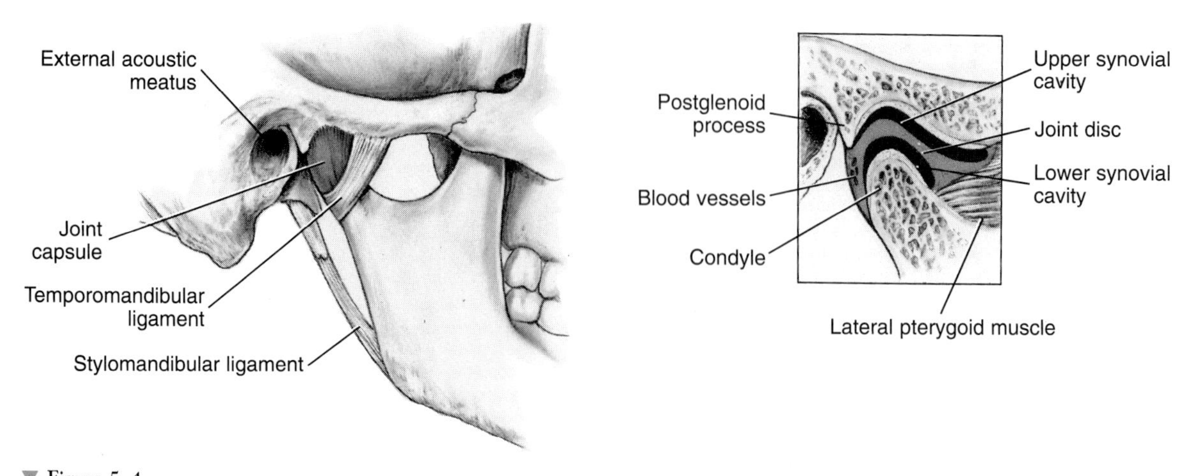

▼ **Figure 5–4**
Lateral view of the joint capsule of the temporomandibular joint and its lateral temporomandibular ligament. Note on the inset that the capsule has been removed to show the upper and lower synovial cavities and their relationship to the joint disc.

the temporal bone's articular eminence and articular fossa superiorly. Inferiorly, the capsule wraps around the circumference of the mandibular condyle, including the condyle's neck.

▼ DISC OF THE JOINT

The fibrous **disc of the joint** (disk) or meniscus of the joint is located between the temporal bone and condyle of the mandible on each side (see Figure 5–4). On parasagittal section, the disc appears caplike on the mandibular condyle, with its superior aspect concavo-convex from anterior to posterior and its inferior aspect concave. This shape of the disc conforms to the shape of the adjacent articulating bones of the TMJ and is related to normal joint movements.

The disc completely divides the TMJ into two compartments or spaces. These two compartments are **synovial cavities** (sy-**no**-vee-al **kav**-i-tees), an upper synovial cavity and a lower synovial cavity. The membranes lining the inside of the joint capsule secrete **synovial fluid** that helps lubricate the joint and fills the synovial cavities.

The disc is attached to the lateral and medial poles of the mandibular condyle. The disc is not attached to the temporal bone anteriorly, except indirectly through the capsule. Posteriorly, the disc is divided into two areas. The upper division of the posterior portion of the disc is attached to the temporal bone's postglenoid process, and the lower division attaches to the neck of the condyle. The disc blends with the capsule at these points. This posterior area of attachment of the disc to the capsule is one of the places where nerves and blood vessels enter the joint.

As a person ages, the disc can become thinner or even perforated. At any age, the disc may become dislocated forward by injury to the posterior attachment. Both perforation and displacement can lead to clinical problems (discussed later in this chapter).

▼ LIGAMENTS ASSOCIATED WITH THE JOINT

A ***ligament*** (**lig**-ah-mint) is a band of fibrous tissue connecting bones. There are three paired ligaments associated with the TMJ. These ligaments are the temporomandibular joint ligament, sphenomandibular ligament, and stylomandibular ligament (Figure 5–5).

▼ Temporomandibular Joint Ligament

The **temporomandibular joint ligament** (tem-poh-ro-man-**dib**-you-lar) is located on the lateral side of each joint and forms a reinforcement of the capsule of the TMJ. This ligament prevents the excessive retraction or moving backward of the mandible, a situation that might lead to problems with the TMJ.

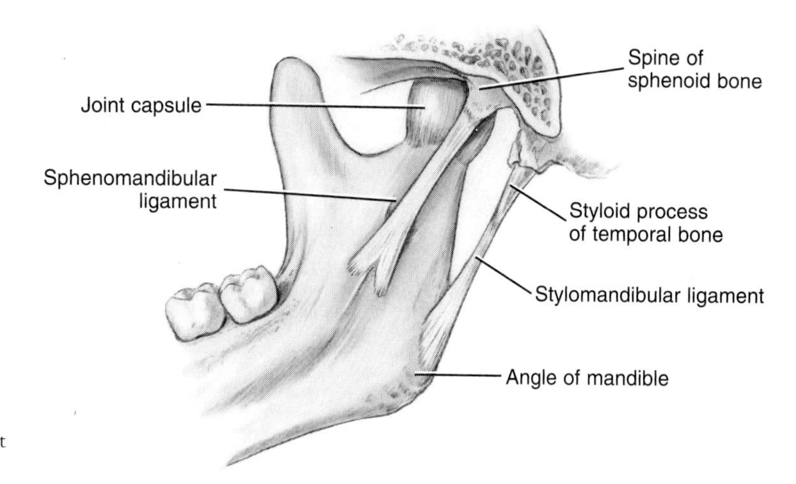

▼ Figure 5–5
Internal view of the temporomandibular joint with associated ligaments.

▼ Sphenomandibular Ligament

The **sphenomandibular ligament** (sfe-no-man-**dib**-you-lar) is not a part of the TMJ. It is located on the medial side of the mandible, some distance from the joint. This ligament runs from the angular spine of the sphenoid bone to the lingula of the mandibular foramen. It is a vestige of the embryonic lower jaw, Meckel's cartilage. The inferior alveolar nerve descends between the sphenomandibular ligament and the ramus of mandible to gain access to the mandibular foramen.

The sphenomandibular ligament becomes accentuated and taut when the mandible is protruded. The sphenomandibular ligament is a landmark for the administration of a local anesthetic for the inferior alveolar block (see Chapter 9 for more information).

▼ Stylomandibular Ligament

The **stylomandibular ligament** (sty-lo-man-**dib**-you-lar) is a variable ligament that is formed from a thickened cervical fascia in the area. This ligament runs from the styloid process of the temporal bone to the angle of the mandible. This ligament also becomes taut when the mandible is protruded.

Jaw Movements with Muscle Relationships

The TMJ allows for the movement of the mandible during speech and mastication. There are two basic types of movement performed by the joint and its associated muscles: a gliding movement and a rotational movement. To palpate the joint and its associated muscles effectively, have the patient go through all the movements of the mandible (Table 5–1). The TMJ can be palpated just anterior to the external acoustic meatus of each ear (Figure 5–6).

The gliding movement of the TMJ occurs mainly between the disc and the articular eminence of the temporal bone in the upper synovial cavity, with the disc plus the condyle moving forward or backward down and up the articular eminence. The gliding movement allows the lower jaw to move forward or backward. Bringing the lower jaw forward involves ***protrusion of the mandible*** (pro-**troo**-shun). Bringing the lower jaw backward involves ***retraction of the mandible*** (re-**trak**-shun). Protrusion involves the bilateral contraction of the lateral pterygoid muscles. The posterior portions of both temporalis muscles are involved during retraction of the mandible.

The rotational movement of the TMJ occurs mainly between the disc and the condyle of the mandible in the lower synovial cavity. The axis of rotation of the disc plus the condyle is transverse, and the movements accomplished are depression or elevation of the mandible. ***Depression of the mandible*** (de-**presh**-in) is the lowering of the lower jaw. ***Elevation of the mandible*** (el-eh-**vay**-shun) is the raising of the lower jaw.

With these two types of movement, gliding and rotation, and with the right and left TMJ working together, the finer movements of the jaw can be accomplished. These include opening and closing the jaws and shifting the lower jaw to one side.

Table 5–1

MOVEMENTS OF THE MANDIBLE AND TMJ WITH ASSOCIATED MUSCLES

Mandibular Movements	TMJ Movements	Associated Muscles
Protrusion of mandible, moving lower jaw forward	Gliding in both upper synovial cavities	Lateral pterygoid, bilateral contraction
Retraction of mandible, moving lower jaw backward	Gliding in both upper synovial cavities	Posterior portion of temporalis, bilateral contraction
Elevation and retraction of mandible, closing jaws	Gliding in both upper synovial cavities and rotation in both lower synovial cavities	Masseter, temporalis, and medial pterygoid, bilateral contraction
Depression and protrusion of mandible, opening jaws	Gliding in both upper synovial cavities and rotation in both lower synovial cavities	Suprahyoids and inferior heads of lateral pterygoid, bilateral contraction
Lateral deviation of mandible, to shift lower jaw to opposite side	Gliding in one upper synovial cavity and rotation in opposite upper synovial cavity	Lateral pterygoid, unilateral contraction

Opening the jaws during speech and mastication involves both depression and protrusion of the mandible. When the jaws close, this involves both elevation and retraction of the mandible. Thus opening and closing the jaws involve a combination of gliding and rotational movements of the TMJ in their respective joint cavities. The disc plus the condyle glides on the articular fossa in the upper synovial cavity, moving forward or backward on the articular eminence. Roughly at the same time, the condyle of the mandible rotates on the disc in the lower synovial cavity.

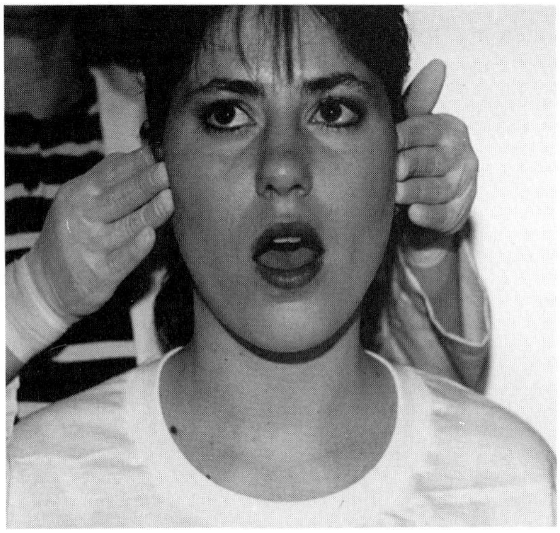

The muscles of mastication involved in elevating the mandible, during closing the jaws, are the bilateral masseter, temporalis, and medial pterygoid muscles (see Chapter 4 for more information). The anterior suprahyoid muscles are involved in depressing the mandible, during opening the jaws, when they bilaterally contract as the hyoid bone is stabilized by the other hyoid muscles (see Chapter 4 for more information). The inferior heads of the lateral pterygoid muscles may also be involved in depressing the mandible during opening the jaws.

Lateral deviation (de-vee-**ay**-shun) or lateral excursion of the mandible or shifting the lower jaw to one side occurs during mastication. Lateral deviation involves both the gliding and rotational movements of opposite TMJs in their respective joint cavities. During lateral deviation, one disc plus the condyle glides forward and medially on the articular eminence in the upper synovial cavity, while the other condyle and disc remain relatively stable in position in the articular fossa. This produces a rotation around the more stable condyle.

Contraction of one of the lateral pterygoid muscles (the one on the protruding side) is involved during

Figure 5–6
Palpation of the patient during movements of both of the temporomandibular joints.

lateral deviation. When the mandible laterally deviates to the left, the right lateral pterygoid muscle contracts, moving the right condyle forward while the left condyle stays in position, thus causing the mandible to move to the left. The reverse situation occurs when the mandible laterally deviates to the right.

During mastication, the power stroke (when the teeth crunch the food) involves a movement from a laterally deviated position back to the midline. If the food is on the right, the mandible will be deviated to the right by the left lateral pterygoid muscle. The power stroke will return the mandible to the center, so the movement is to the left and involves a retraction of the left side. This is accomplished by the left posterior portion of the temporalis muscle. At the same time, all the closing jaw muscles on the right side contract to crush the food. The reverse situation occurs if the food is on the left.

Disorders of the Joint

A patient may have a disease process associated with one or both of the temporomandibular joints or a ***temporomandibular disorder (TMD)*** (tem-poh-ro-man-**dib**-you-lar). The patient may experience chronic joint tenderness, swelling, and painful muscle spasms. Also present may be difficulties of joint movement, such as a limited or deviated mandibular opening. The dental professional provides an important role in the recognition, treatment, and maintenance of patients with this disorder.

Recognition of TMD includes palpation of the joint as the patient performs all the movements of the joint, as well as palpation of the related muscles of mastication. All signs and symptoms related to TMD, such as the amount of mandibular opening and facial pain, need to be recorded by the dental professional, as well as any parafunctional habits and related systemic diseases. The patient may have the traditional skull radiograph of the joint area, or magnetic resonance imaging (MRI) may be performed to aid in the diagnosis of TMD. MRI is a noninvasive nuclear procedure for imaging soft tissue of high fat and water content. Thus MRI can make it possible to distinguish normal tissue from diseased tissue.

There are many controversies associated with the etiology of these disorders. TMD is a heterogeneous, complex disorder involving many factors such as behavioral stressors and parafunctional habits such as clenching and bruxism. Clenching is the prolonged holding of the teeth together by the masticatory muscles. The normal condition is for the teeth to be slightly parted at rest and the muscles relaxed. Bruxism is habitual grinding of the teeth, especially at night.

Patients who clench or grind their teeth may or may not be aware of their parafunctional jaw movements. The masseter muscle may be overdeveloped in these patients, and the teeth may show abnormal wear. Clenching and bruxism are not caused by malocclusion, but both appear to be processes of the central nervous system.

Trauma to the jaw may cause TMD with the disc having adhesions to the bony surfaces, but it is not the most common etiological factor. Systemic diseases such as osteoarthritis may involve portions of the TMJ and contribute to TMD. Aging of the disc that causes wear and hardening may also be a factor in TMD, yet TMD does not become worse with age.

Most TMD cases are noted in females with the average age at 39 years, possibly because women are more likely to seek treatment. Some researchers theorize that changes in female hormones may play a role in the pathogenesis of TMD in women. More studies need to be done in this area of etiology.

Not all patients with TMD have abnormalities in the joint disc or the joint itself. Most symptoms seem to come from the muscles. Most recent studies do not support the role of TMD directly causing headaches, neck or back pain, or instability. Headaches are usually caused by muscle tension or vascular changes. Cyclic episodes of TMD and other incidents of chronic body pain are commonly encountered in the TMD population.

Joint sounds occur because of disc derangement. The posterior portion of the disc gets caught between the condyle head and the articular eminence. Joint sounds are not a reliable indicator of TMD since they can change over time in a patient. Thus the clicking, grinding, and popping of the joint during movement that is commonly present with TMD is also found in 40–60 percent of persons without TMD.

There are also many controversies in the treatment of TMD. Most recent studies have determined that malocclusion and occlusal discrepancies are not involved in most cases of TMD. Thus occlusal adjustment, repositioning the jaw, and orthodontic treatment are not the treatments of choice for all patients with TMD, nor do these treatments seem to prevent TMD. Less than half of the patients with TMD seek treatment for their disorder.

Most patients with TMD do well over time with inexpensive and reversible treatments, including patient-based or prescription pain control, relaxation therapy, stress management, habit control, moderate home-based muscular exercises, and flat plane non-repositioning oral splints. These inexpensive and reversible treatments now have the same success as more expensive and irreversible treatments. Thus only a few TMD patients (1–2 percent) require surgery or other extensive treatment. Surgery of the TMJ can now involve arthroscopy with an endoscope and lasers.

An acute episode of TMD can occur when a patient opens the mouth too wide by causing maximal depression and protrusion of the mandible, as when yawning or receiving prolonged dental care. This causes ***subluxation*** (sub-luks-**ay**-shun) or dislocation of both joints. Subluxation happens when the head of each condyle moves too far anterior on the articular eminence. When the patient tries to close and elevate the mandible, the condylar heads cannot move posteriorly because the muscles have become spastic.

Treatment of subluxation consists of relaxing these muscles and carefully moving the mandible downward and back. The condylar heads will then be able to assume the normal posterior position in relationship to the articular eminence by the muscular action of the elevating muscles of mastication. Future care of these patients involves avoidance of extreme depression of the mandible.

Identification Exercise

Identify the structures on the following diagrams by filling in the blanks with the correct anatomical term. You can check your answers by looking back at the figure indicated in parenthesis for each identification diagram.

1. (Figure 5–4)

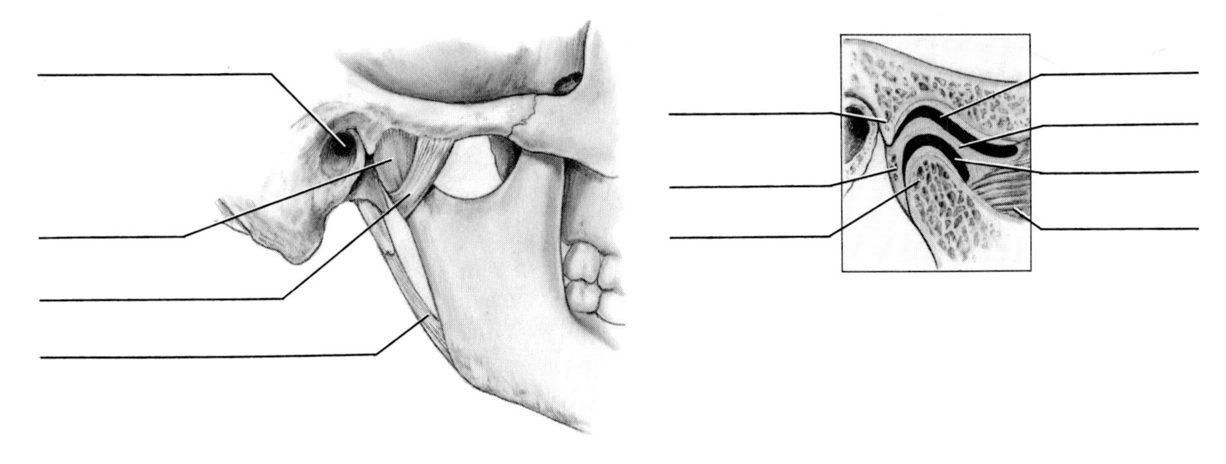

2. (Figure 5–5)

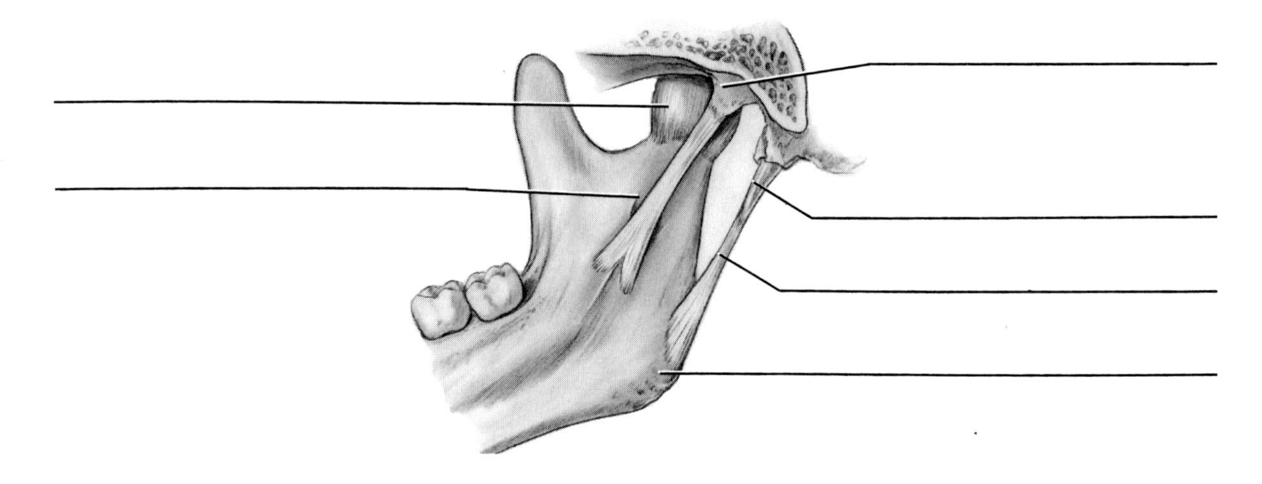

REVIEW QUESTIONS

1. Which of the following ligaments associated with the temporomandibular joint reinforces the joint capsule?
   A. Styloid ligament
   B. Stylomandibular ligament
   C. Temporomandibular ligament
   D. Sphenomandibular ligament

2. Which of the following landmarks is located on the mandible?
   A. Articular eminence
   B. Coronoid process
   C. Articular fossa
   D. Postglenoid process

3. Which of the following is a basic movement performed by the temporomandibular joint?
   A. Gliding movement only
   B. Rotational movement only
   C. Gliding and rotational movement
   D. No movement is performed

4. Which of the following muscles is involved in the lateral deviation of the mandible?
   A. Masseter muscle
   B. Medial pterygoid muscle
   C. Lateral pterygoid muscle
   D. Temporalis muscle
   E. Digastric muscle

5. Protrusion of the mandible primarily involves
   A. Opening the jaws
   B. Closing the jaws
   C. Bringing the lower jaw forward
   D. Bringing the lower jaw backward
   E. Shifting the lower jaw to one side

6. Which of the following movements is assisted by the temporalis muscle?
   A. Mandibular depression only
   B. Mandibular elevation only
   C. Mandibular retraction only
   D. Mandibular depression and elevation
   E. Mandibular elevation and retraction

7. Which of the following ligaments associated with the temporomandibular joint has the inferior alveolar nerve descend nearby to gain access to the mandibular foramen?
   A. Sphenomandibular ligament only
   B. Stylomandibular ligament only
   C. Temporomandibular ligament only
   D. Sphenomandibular and stylomandibular ligaments
   E. Stylomandibular and temporomandibular ligaments

8. Which of the following statements about the temporomandibular disc is false?
   A. The disc separates the TMJ into synovial cavities.
   B. The disc is attached anteriorly and posteriorly to the condyle.
   C. Gliding movements take place between the disc and the temporal bone.
   D. The inferior surface of the disc is concave.

9. Which area of the mandible articulates with the temporal bone at the temporomandibular joint?
   A. Lingula
   B. Mandibular notch
   C. Coronoid process
   D. Condyle

10. During both mandibular protrusion and retraction, the rotation of the articulating surface of the mandible against the disc in the lower synovial cavity is prevented by the
    A. Facial muscles
    B. Infrahyoid muscles
    C. Muscles of mastication
    D. Ligaments of the temporomandibular joint

Chapter 6

Blood Supply

Vascular System

The vascular system of the head and neck, as to the rest of the body, consists of an arterial blood supply, a capillary network, and venous drainage. The dental professional will need to be able to locate the larger blood vessels of the head and neck since these vessels may become compromised owing to a disease process or during a dental procedure, such as a local anesthetic injection. Blood vessels may also spread infection in the head and neck area (see Chapter 12 for more information). The blood vessels may also spread cancerous cells from the tumor to distant sites and at a more rapid rate than lymphatic vessels.

Blood vessels are less numerous than lymphatic vessels, yet the venous portion mainly parallels the lymphatic vessels in location (see Chapter 10 for more information). A large network of blood vessels is called a ***plexus*** (**plek**-sis). The head and neck area contains certain important venous plexuses. Blood vessels also may communicate with each other by an ***anastomosis*** (plural, ***anastomoses***) (ah-nas-tah-**moe**-sis, ah-nas-tah-**moe**-sees), a connecting channel between the vessels.

An ***artery*** (**ar**-ter-ee) is a component of the blood system that arises from the heart carrying blood away from it. Each artery starts as a large vessel and branches into smaller vessels, each one a smaller artery or an ***arteriole*** (ar-**ter**-ee-ole). Each arteriole branches into even smaller vessels until it becomes a network of capillaries. Each ***capillary*** (**kap**-i-lare-ee) is smaller than an arteriole and is able to supply blood to a large tissue area only because there are so many of them.

A ***vein*** (vane) is another component of the blood system. A vein, unlike an artery, travels to the heart carrying blood. Valves in the veins are mostly absent in the head and neck area, unlike the rest of the body. This leads to two-way flow dictated by local pressure changes, which is the reason that facial or dental infections can lead to serious complications (discussed in Chapter 12). After each smaller vein or ***venule*** (**ven**-yule) drains the capillaries of the tissue area, the venules coalesce to become larger veins. Veins are much larger and more numerous than arteries. Veins anastomose freely and have a greater variability in location in comparison to arteries.

There are also different kinds of venous networks found in the body. Superficial veins are found immediately below the skin. Deeper veins usually accompany larger arteries in a more protected location within the tissue. A ***venous sinus*** (**vee**-nus **sy**-nus) is a blood-filled space between the two layers of tissue. All these networks are connected with each other by anastomoses.

It is important to review the pathways of the arteries and veins as they exit and then enter the heart so as to understand the origins of the blood vessels of the head and neck. After the basic origins of the blood supply to the head and neck are understood, diagrams of the blood vessels overlying the skull figure are very helpful in studying this system. Relating the tissues supplied to the area blood vessels is an additional way of understanding the location of the various blood vessels.

Arterial Blood Supply to the Head and Neck

The major arteries that supply the head and neck are the common carotid and subclavian arteries. The origins from the heart to the head and neck of these two arteries are different depending on the side of the body considered. The other arteries of the head and neck are symmetrically located on each side of the body.

▼ ORIGINS TO THE HEAD AND NECK

The origins from the heart of the common carotid and subclavian arteries that supply the head and neck are different for the right and left sides of the body (Figure 6–1). For the left side of the body, the common carotid and subclavian arteries arise directly from the **aorta** (ay-**or**-tah). For the right side of the body, the common carotid and subclavian arteries are both branches from the brachiocephalic artery. The **brachiocephalic artery** (bray-kee-oo-sah-**fal**-ik) is a direct branch of the aorta.

The **common carotid artery** (**kom**-in kah-**rot**-id) is branchless and travels up the neck, lateral to the trachea and larynx, to the upper border of the thyroid cartilage (see Figure 6–1). The common carotid artery travels in a sheath beneath the sternocleidomastoid muscle. This sheath also contains the internal jugular vein and the tenth cranial or vagus nerve. The common carotid artery ends by dividing into the internal and external carotid arteries at about the level of the larynx (Figure 6–2).

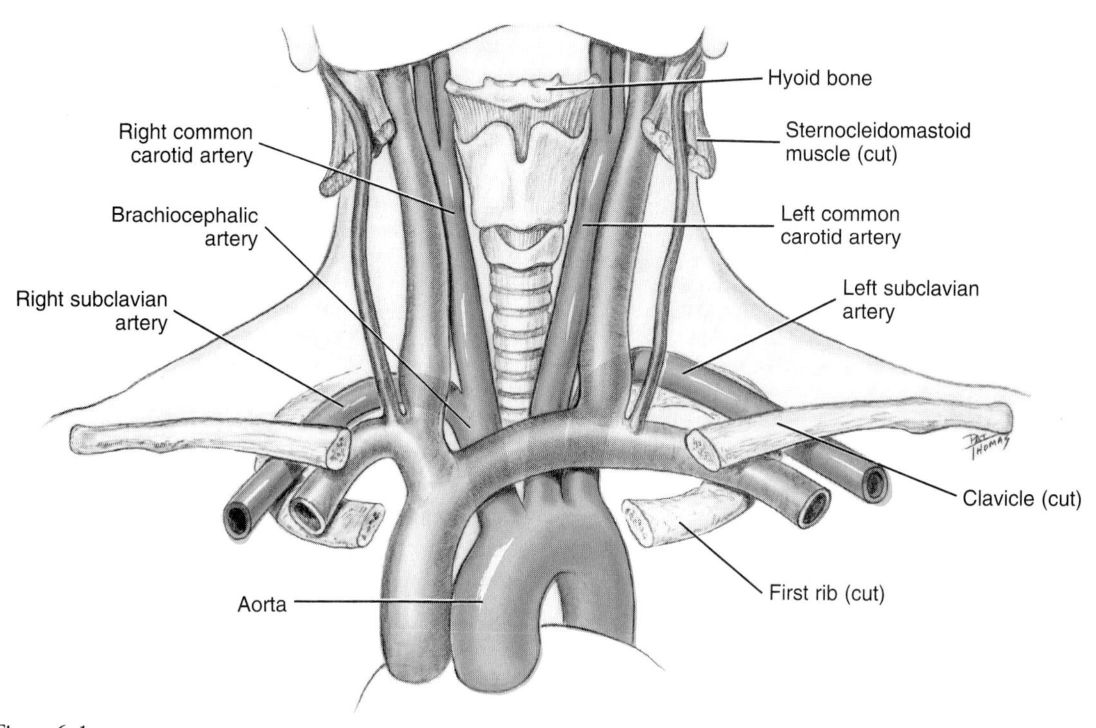

Figure 6–1
Origins from the heart of the arterial blood supply for the head and neck outlining the pathway of the common carotid and subclavian arteries. Note that the pathway is different for the right and left sides of the body.

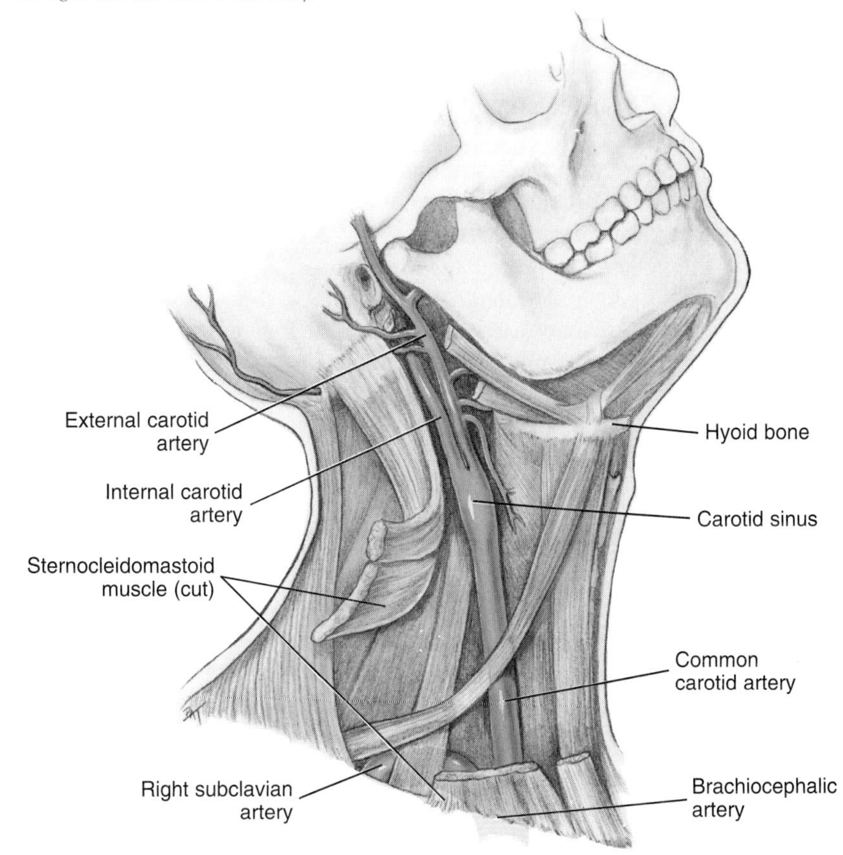

Figure 6–2
Pathway of the internal carotid artery after branching off the common carotid artery.

Just before the common carotid artery bifurcates into the internal and external carotid arteries, it exhibits a swelling called the **carotid sinus** (kah-**rot**-id **sy**-nus) (see Figure 6–2). When the common carotid artery is palpated against the larynx, the most reliable arterial pulse of the body can be monitored. If the anterior border of the sternocleidomastoid muscle is rolled posteriorly at the level of the thyroid cartilage of the larynx (Adam's apple), the ***carotid pulse*** can be felt in the groove of tissue produced. This pulse is most reliable since the common carotid is a major artery supplying the brain and therefore in an emergency situation (cardiopulmonary resuscitation) remains palpable when peripheral arteries are not, such as the radial artery. The carotid pulse also is easily accessible during dental treatment.

The **subclavian artery** (sub-**klay**-vee-an) arises lateral to the common carotid artery (see Figure 6–1). The subclavian artery gives off branches to supply both intracranial and extracranial structures, but its major destination is the upper extremity (arm).

▼ INTERNAL CAROTID ARTERY

The **internal carotid artery** (kah-**rot**-id) is a division that travels upward in a slightly lateral position (in relationship to the external carotid artery) after leaving the common carotid artery (see Figure 6–2). This artery is hidden by the sternocleidomastoid muscle of the neck. The internal carotid artery has no branches in the neck but continues adjacent to the internal jugular vein within the carotid sheath to the skull base, where it enters the cranium. The internal carotid artery supplies intracranial structures and is the source of the **ophthalmic artery** (of-**thal**-mic), which supplies the eye, orbit, and lacrimal gland.

▼ EXTERNAL CAROTID ARTERY

The **external carotid artery** (kah-**rot**-id) travels upward in a more medial position (to the internal carotid artery) after arising from the common carotid artery (Figure 6–3). The external carotid artery supplies the extracranial tissues of the head and neck, including the oral cavity. The external carotid artery has four sets of branches grouped according to their location to the main artery: the anterior, medial, posterior, and terminal branches (Table 6–1).

▼ Anterior Branches of the External Carotid Artery

There are three anterior branches from the external carotid artery: the superior thyroid, lingual, and facial arteries (Figure 6–4). The lingual and facial arteries continue to divide to serve areas of the head and neck of interest to dental professionals.

▼ SUPERIOR THYROID ARTERY

The **superior thyroid artery** (**thy**-roid) is an anterior branch from the external carotid artery (see Figure 6–4). The superior thyroid artery has branches: the infrahyoid branch (in-frah-**hi**-oid), sternocleidomastoid branch (stir-no-klii-do-**mass**-toid), superior laryngeal artery (lah-**rin**-je-al), and superior and inferior thyroid. These branches supply the tissues inferior to the hyoid bone, including the infrahyoid muscles, sternocleidomastoid muscle, muscles of the larynx, and thyroid gland.

▼ LINGUAL ARTERY

The **lingual artery** is an anterior branch from the external carotid artery and arises above the superior thyroid artery at the level of the hyoid bone (see Figure 6–4). The lingual artery travels anteriorly to the apex of the tongue by way of its inferior surface. The lingual artery supplies the tissues superior to the hyoid bone, including the suprahyoid muscles and floor of the mouth by the dorsal lingual, deep lingual, sublingual, and suprahyoid branches.

The tongue is also supplied by branches of the lingual artery, including several small dorsal lingual branches to the base and body, and the deep lingual artery, the terminal part of the lingual artery, to the apex.

The **sublingual artery** (sub-**ling**-gwal) supplies the mylohyoid muscle, sublingual salivary gland, and mucous membranes of the floor of the mouth. The small suprahyoid branch (soo-prah-**hi**-oid) supplies the suprahyoid muscles.

▼ FACIAL ARTERY

The **facial artery** is the final anterior branch from the external carotid artery (Figures 6–4 and 6–5). The facial artery arises slightly superior to the lingual artery as it branches off anteriorly. Sometimes the

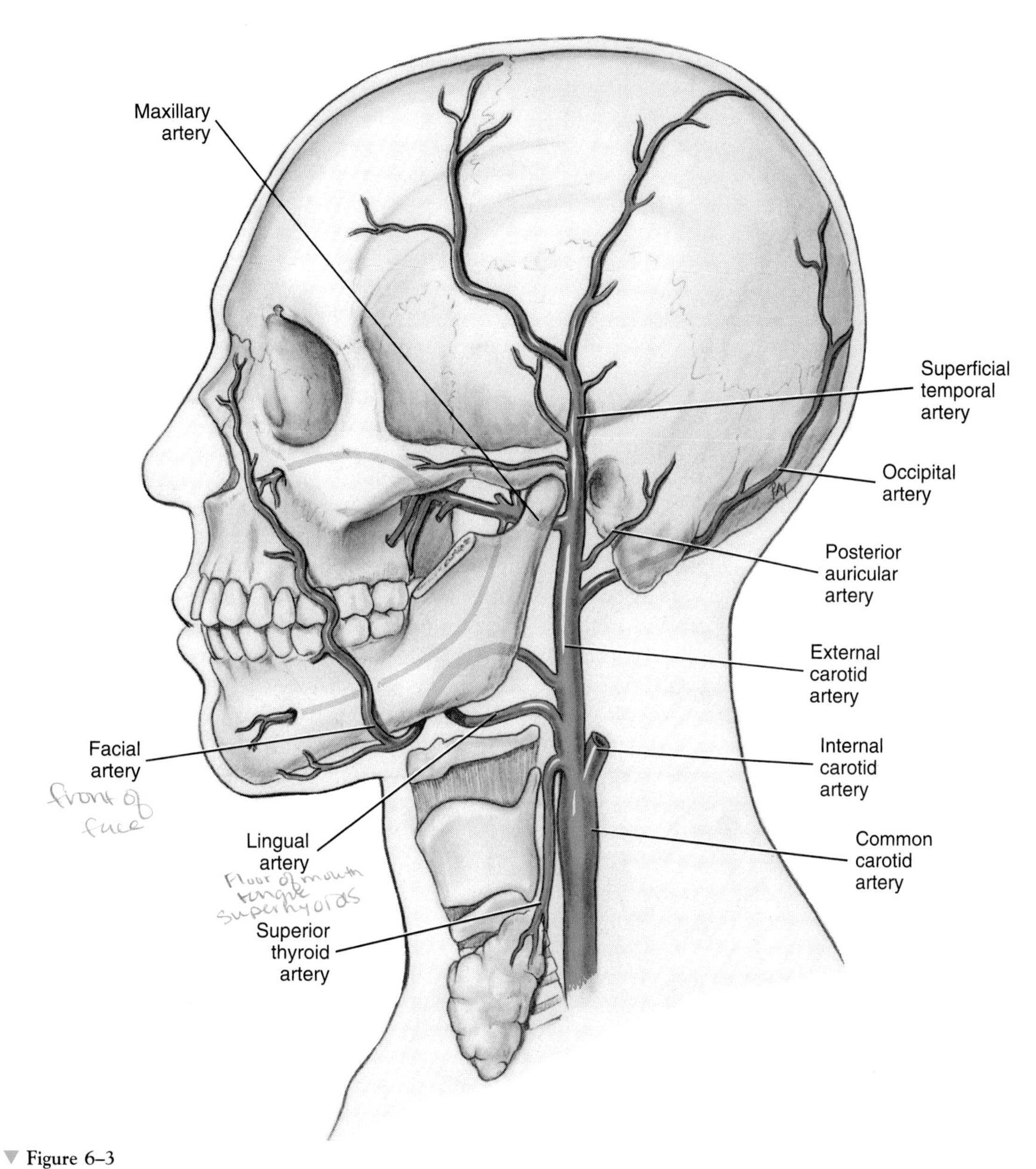

▼ **Figure 6–3**
Pathway of the external carotid artery after branching off the common carotid artery. Note that the medial branch of the external carotid artery, the ascending pharyngeal artery, cannot be seen in this view.

Table 6–1

BRANCHES OF THE EXTERNAL CAROTID ARTERY

Branches of External Carotid Artery	Position of Branches	Further Branches
Superior thyroid	Anterior	Superior and inferior thyroid, infrahyoid, superior laryngeal and sternocleidomastoid
Lingual	Anterior	Dorsal lingual, deep lingual, sublingual, and suprahyoid
Facial	Anterior	Ascending palatine, glandular, submental, inferior labial, superior labial, and angular
Ascending pharyngeal	Medial	Pharyngeal and meningeal
Occipital	Posterior	Muscular, sternocleidomastoid, auricular, and meningeal
Posterior auricular	Posterior	Auricular and stylomastoid
Superficial temporal	Terminal	Transverse facial, middle temporal, frontal, and parietal
Maxillary	Terminal	See Table 6–2

facial and lingual arteries share a common trunk. The facial artery has a complicated path as it runs medial to the mandible, over the submandibular salivary gland, and then around the mandible's lower border to its lateral side.

From the lower border of the mandible, the facial artery runs anteriorly and superiorly near the angle of the mouth and along the side of the nose. The facial artery terminates at the medial canthus of the eye. Thus the facial artery supplies the face in the oral, buccal, zygomatic, nasal, infraorbital, and orbital regions.

The facial artery is paralleled by the facial vein in the head area, although they do not run together. In the neck, the artery is separated from the vein by the posterior belly of the digastric muscle, stylohyoid muscle, and submandibular salivary gland. The facial artery's major branches include the ascending palatine, submandibular, submental, inferior labial, superior labial, and angular arteries.

The **ascending palatine artery** (ah-**send**-ing **pal**-ah-tine) is the first branch from the facial artery (see Figure 6–5). The ascending palatine artery supplies the soft palate, palatine muscles, and palatine tonsils and can be the source of the serious blood loss or hemorrhage during its removal by a tonsillectomy (blood vessel lesions are discussed later in this chapter).

The glandular branches and **submental artery** (sub-**men**-tal) are branches from the facial artery that supply the submandibular lymph nodes, submandibular salivary gland, and mylohyoid and digastric muscles.

The **inferior labial artery** is another branch from the facial artery that supplies the lower lip tissues,

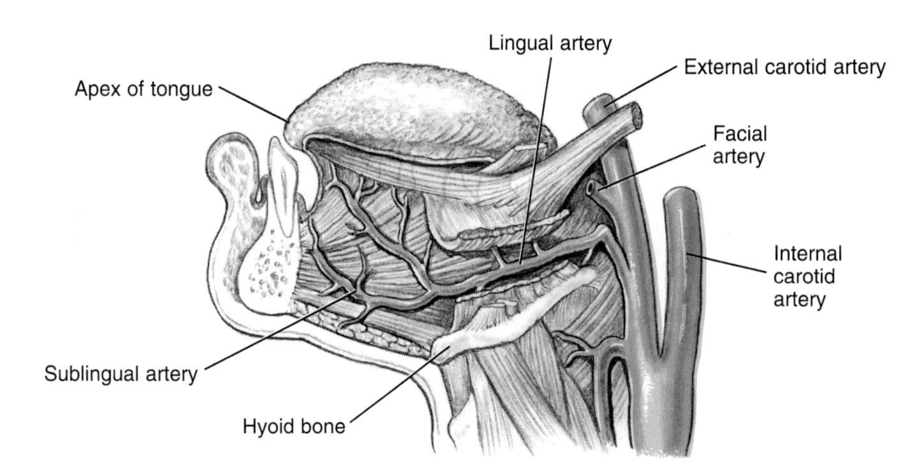

Figure 6–4
Pathways of the lingual artery and superior thyroid artery.

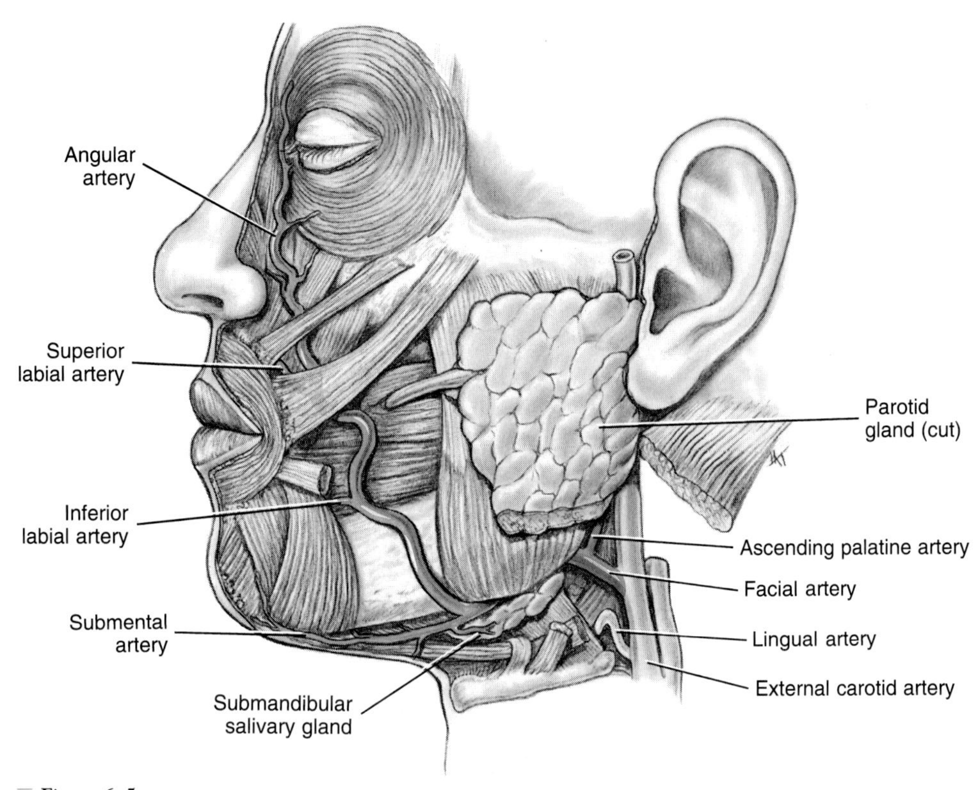

Figure 6–5
Pathway of the facial artery.

including the muscles of facial expression, such as the depressor anguli oris muscle. The **superior labial artery** is also a branch from the facial artery that supplies the upper lip tissues.

The **angular artery** (**ang**-u-lar) is the termination of the facial artery and supplies the tissues along the side of the nose (see Figure 6–5).

Medial Branch of the External Carotid Artery

There is only one medial branch from the external carotid artery, the small **ascending pharyngeal artery** (ah-**send**-ing fah-**rin**-je-al), that arises close to the origin of the external carotid artery (see Figure 6–5). The ascending pharyngeal artery has many small branches, such as the pharyngeal branches (fah-**rin**-je-al) and meningeal branches (me-**nin**-je-al) which supply the pharyngeal walls (where they anastomose with the ascending palatine artery), soft palate, and meninges of the brain.

Posterior Branches of the External Carotid Artery

There are two posterior branches from the external carotid artery: the occipital artery and posterior auricular artery (Figure 6–6).

▼ OCCIPITAL ARTERY

The **occipital artery** (ok-**sip**-it-al), a posterior branch from the external carotid artery, arises from the external carotid artery as it passes upward behind the ascending ramus of the mandible, and travels to the posterior portion of the scalp (see Figure 6–6). The occipital artery supplies the suprahyoid and sternocleidomastoid muscles, as well as the scalp and meningeal tissues in the occipital region. The artery supplies these regions by the muscular branches, sternocleidomastoid branches (stir-no-klii-do-**mass**-toid), auricular (aw-**rik**-yule-lar), and meningeal branches (me-**nin**-je-al). At its origin, the occipital artery is closely related to the twelfth cranial or hypoglossal nerve.

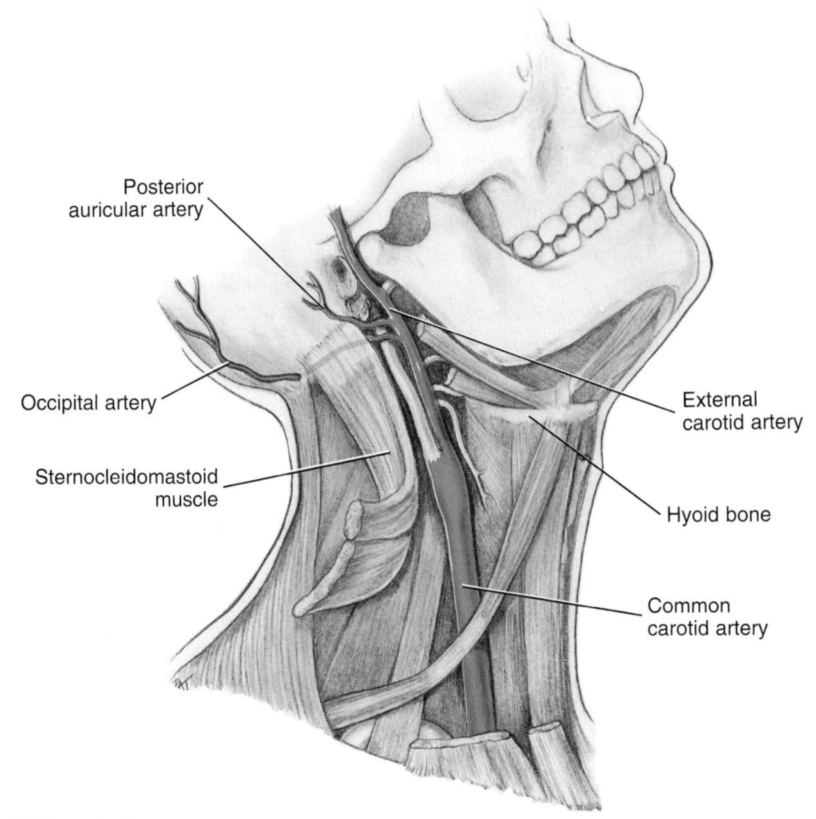

▼ Figure 6–6
Pathways of the occipital artery and posterior auricular artery.

▼ POSTERIOR AURICULAR ARTERY

The small **posterior auricular artery** (aw-**rik**-yule-lar) is also a posterior branch from the external carotid artery (see Figure 6–6). The posterior auricular artery arises superior to the occipital artery and stylohyoid muscle about the level of the tip of the styloid process. The posterior auricular artery supplies the internal ear by its auricular branch and the mastoid air cells by the **stylomastoid artery** (sty-lo-**mass**-toid).

▼ Terminal Branches of the External Carotid Artery

There are two terminal branches of the external carotid artery, the superficial temporal artery and maxillary artery (Figures 6–7, 6–8, and 6–9). The external carotid artery splits into these terminal branches within the parotid salivary gland. In addition, both terminal branches give rise to many important arteries in the head and neck area.

▼ SUPERFICIAL TEMPORAL ARTERY

The **superficial temporal artery** (**tem**-poh-ral) is the smaller terminal branch of the external carotid artery (see Figure 6–7). The artery arises within the parotid salivary gland. This artery can sometimes be visible in patients under the skin of the temporal region. The superficial temporal artery has several branches, including the transverse facial artery, middle temporal artery, frontal branch, and parietal branch.

The small **transverse facial artery** supplies the parotid salivary gland duct and nearby facial tissues. The equally small **middle temporal artery** (**tem**-poh-ral) supplies the temporalis muscle. The frontal branch (**frunt**-il) and parietal branch (pah-**ry**-it-il) both supply portions of the scalp in the frontal and parietal regions.

▼ MAXILLARY ARTERY

The **maxillary artery** (**mak**-sil-lare-ee) is the larger terminal branch of the external carotid artery (Figures

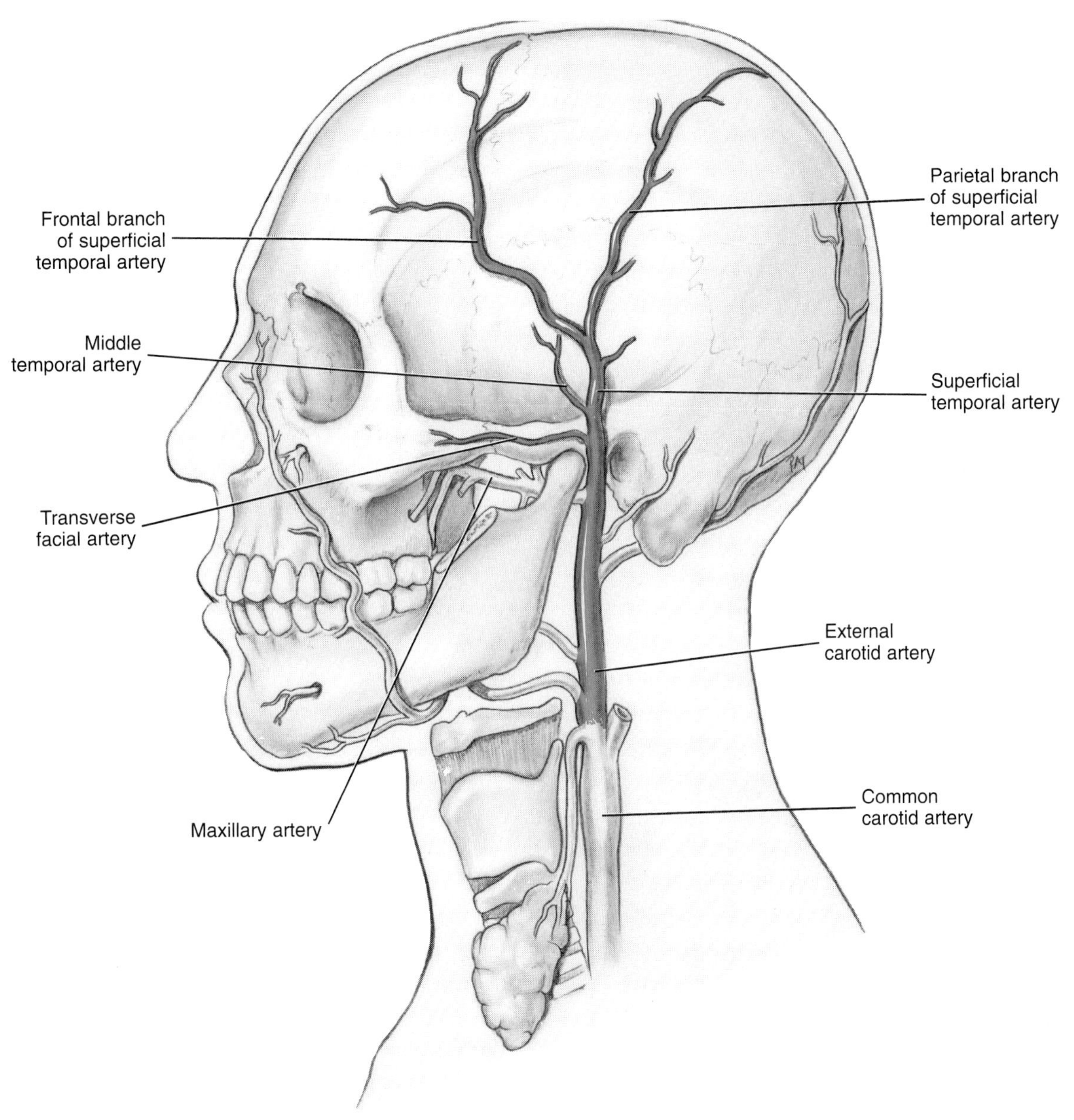

▼ **Figure 6–7**
Pathway of the superficial temporal artery.

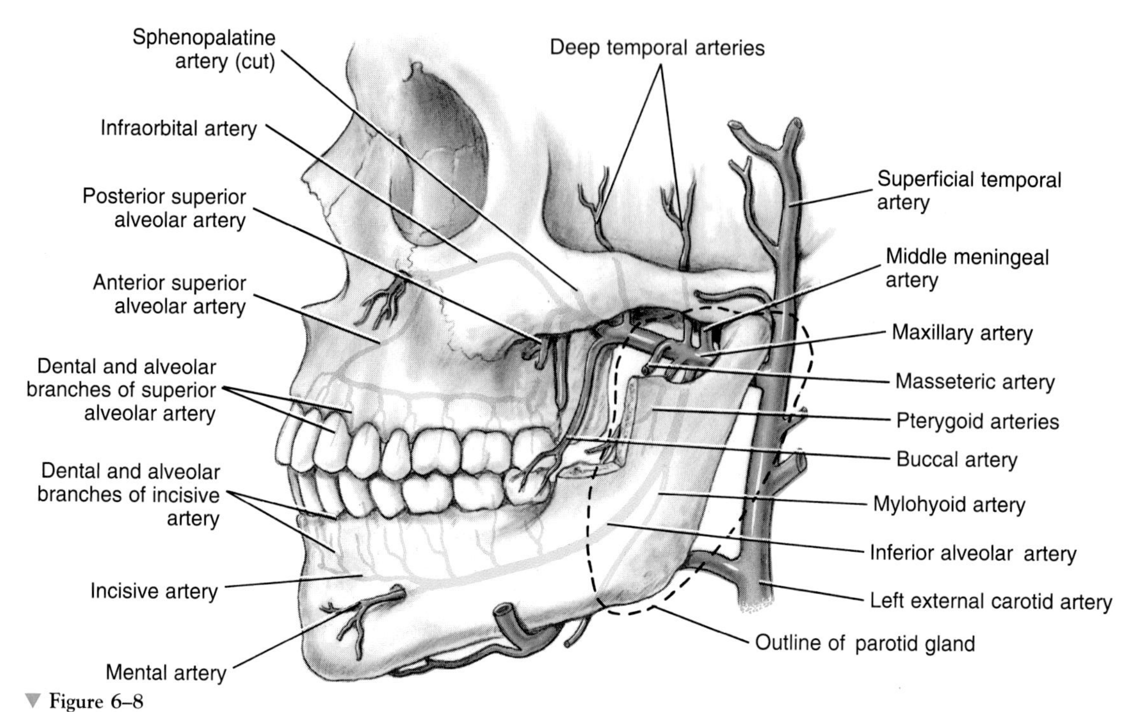

▼ **Figure 6–8**
Pathway of the maxillary artery (except those branches to the nasal cavity and palate).

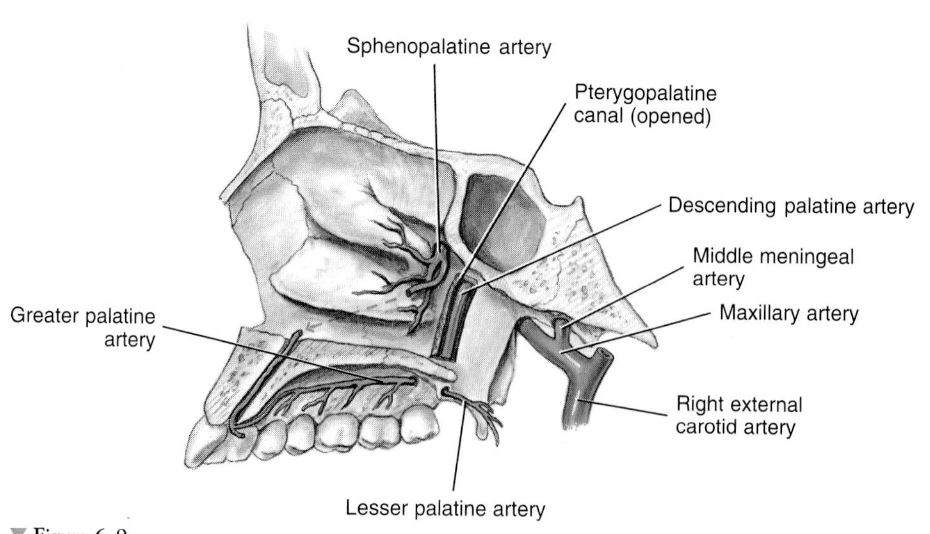

▼ **Figure 6–9**
Pathways of the greater palatine artery, lesser palatine artery, and sphenopalatine artery.

6–8 and 6–9). The maxillary artery begins at the neck of the mandibular condyle within the parotid salivary gland. The maxillary artery runs between the mandible and the sphenomandibular ligament anteriorly and superiorly through the infratemporal fossa. The artery may run either superficial or deep to the lateral pterygoid muscle.

After traversing the infratemporal fossa, the maxillary artery enters the pterygopalatine fossa. The pterygopalatine fossa is behind and below the eye (see Chapter 3 for more information). Within the infratemporal and pterygopalatine fossae, the maxillary artery gives off many branches. The branches within the infratemporal fossa include the middle meningeal and inferior alveolar arteries and several arteries to muscles (Table 6–2).

The **middle meningeal artery** (meh-**nin**-je-al) supplies the meninges of the brain by way of the foramen spinosum, located on the inferior surface of the skull, as well as the skull bones (see Figure 6–8).

The **inferior alveolar artery** (al-**ve**-o-lar) also arises from the maxillary artery in the infratemporal fossa (see Figure 6–8). The artery turns inferiorly to enter the mandibular foramen and then the mandibular canal, along with the inferior alveolar nerve. The mylohyoid artery branches from the inferior alveolar artery before it enters the canal.

The **mylohyoid artery** (my-lo-**hi**-oid) arises from the inferior alveolar artery before the main artery enters the mandibular canal by way of the mandibular foramen (see Figure 6–8). The mylohyoid artery travels in the mylohyoid groove on the inner surface of the mandible and supplies the floor of the mouth and the mylohyoid muscle.

Within the mandibular canal, the inferior alveolar artery gives off the dental and alveolar branches (see Figure 6–8). The dental branches of the inferior alveolar artery supply the pulp tissue of the mandibular posterior teeth by way of each tooth's apical foramen. The alveolar branches of the inferior alveolar artery supply the periodontium of the mandibular posterior teeth, including the gingiva. The inferior alveolar artery then branches into two arteries within the mandibular canal: the mental and incisive arteries.

The **mental artery** (**ment**-il) arises from the inferior alveolar artery and exits the mandibular canal by way of the mental foramen (see Figure 6–8). The mental foramen is located on the outer surface of the mandible, usually beneath the apices of the first and second mandibular premolar teeth. After the mental artery exits the canal, the artery supplies the tissues of the chin and anastomoses with the inferior labial artery.

The **incisive artery** (in-**sy**-ziv) branches off the inferior alveolar artery and remains in the mandibular canal to divide into dental and alveolar branches (see Figure 6–8). The dental branches of the incisive artery supply the pulp tissue of the mandibular anterior teeth by way of each tooth's apical foramen. The alveolar branches of the incisive artery supply the

Table 6–2

BRANCHES OF THE MAXILLARY ARTERY

Major Branches of Maxillary Artery	Further Branches	Tissues Supplied
Middle meningeal		Meninges of brain and bones of skull
Inferior alveolar	Mylohyoid, mental, and incisive	Mandibular teeth, mouth floor, and mental region
Deep temporal(s)		Temporalis muscle
Pterygoid(s)		Lateral and medial pterygoid muscles
Masseteric		Masseter muscle
Buccal		Buccinator muscle and buccal region
Posterior superior alveolar		Posterior maxillary teeth and maxillary sinus
Infraorbital	Orbital and anterior superior alveolar	Orbital region, face, and anterior maxillary teeth
Greater palatine	Lesser palatine(s)	Hard and soft palates
Sphenopalatine	Lateral nasal, septal, and nasopalatine	Nasal cavity and anterior hard palate

periodontium of the mandibular anterior teeth, including the gingiva, and anastomose with the alveolar branches of the incisive artery on the other side.

The maxillary artery also has branches that are located near the muscle they supply (see Figure 6–8). These arteries all accompany branches of the mandibular division of the fifth cranial or trigeminal nerve. The **deep temporal arteries** (**tem**-poh-ral) supply the anterior and posterior portions of the temporalis muscle. The **pterygoid arteries** (**teh**-re-goid) supply the lateral and medial pterygoid muscles. The **masseteric artery** (mass-et-**tehr**-ik) supplies the masseter muscle. The **buccal artery** supplies the buccinator muscle and other soft tissues of the cheek.

Just as the maxillary artery leaves the infratemporal fossa and enters the pterygopalatine fossa, it gives off the **posterior superior alveolar artery** (al-**ve**-o-lar) (see Figure 6–8). This artery enters the posterior superior alveolar foramina on the maxillary tuberosity and then gives off dental branches and alveolar branches. The posterior alveolar superior alveolar artery also anastomoses with the anterior superior alveolar artery.

The dental branches of the posterior superior alveolar artery supply the pulp tissue of the posterior maxillary teeth by way of each tooth's apical foramen. The alveolar branches of the posterior superior alveolar artery supply the periodontium of the posterior maxillary teeth, including the gingiva. Some branches also supply the maxillary sinus.

The **infraorbital artery** (in-frah-**or**-bit-al) branches from the maxillary artery in the pterygopalatine fossa and may share a common trunk with the posterior superior alveolar artery (see Figure 6–8). The infraorbital artery then enters the orbit through the inferior orbital fissure. While in the orbit, the artery travels in the infraorbital canal. Within the canal, the infraorbital artery provides orbital branches (**or**-bit-al) to the orbit and gives off the anterior superior alveolar artery.

The **anterior superior alveolar artery** (al-**ve**-o-lar) arises from the infraorbital artery and gives off dental and alveolar branches (see Figure 6–8). The anterior superior alveolar artery also anastomoses with the posterior superior alveolar artery.

The dental branches of the anterior superior alveolar artery supply the pulp tissue of the anterior maxillary teeth by way of each tooth's apical foramen. The alveolar branches of the anterior superior alveolar artery supply the periodontium of the anterior maxillary teeth, including the gingiva.

After giving off these branches in the infraorbital canal, the infraorbital artery emerges onto the face from the infraorbital foramen (see Figure 6–8). The artery's terminal branches supply parts of the infraorbital region of the face and anastomose with the facial artery.

Also in the pterygopalatine fossa, the maxillary artery gives rise to the **greater palatine artery** and **lesser palatine arteries** (**pal**-ah-tine), which travel to the palate through the pterygopalatine canal and the greater and lesser palatine foramina to supply the hard and soft palates, respectively (see Figure 6–9). The maxillary artery ends by becoming the **sphenopalatine artery** (sfe-no-**pal**-ah-tine), which supplies the nasal cavity. The sphenopalatine artery gives rise to the posterior lateral nasal branches and septal branches, including a nasopalatine branch (nay-zo-**pal**-ah-tine) that accompanies the nasopalatine nerve through the incisive foramen on the maxilla (see Figure 6–9).

Venous Drainage of the Head and Neck

The veins of the head and neck start out as small venules and become larger as they near the base of the neck on their way to the heart. The veins of the head and upper neck are usually symmetrically located but have a greater variability in location than do the arteries, anastomosing freely. Veins are also generally larger and more numerous than arteries in the same tissue area.

The internal jugular vein drains the brain, as well as most of the other tissues of the head and neck (Table 6–3), whereas the external jugular vein drains only a small portion of the extracranial tissues. However, the two veins have many anastomoses. The beginnings of both veins are discussed initially, and then later their route to the heart.

FACIAL VEIN

The **facial vein** drains into the internal jugular vein, which is discussed later in this chapter (Figure 6–10). The facial vein begins at the medial corner of the eye, where it begins by the junction of two veins from the frontal region, the **supratrochlear vein** (soo-prah-**trok**-lere) and **supraorbital vein** (soo-prah-**or**-bit-al). The supraorbital vein also anastomoses with the ophthalmic veins. The **ophthalmic**

Table 6–3

VEINS OF THE HEAD

Region or Tributaries Drained	Drainage Veins	Major Veins
Meninges of brain	Middle meningeal	Pterygoid plexus
Lateral scalp area	Superficial temporal and posterior auricular	Retromandibular and external jugular
Frontal region	Supratrochlear and supraorbital	Facial and ophthalmic
Orbital region	Ophthalmic(s)	Cavernous sinus and pterygoid plexus
Superficial temporal and maxillary veins	Retromandibular	External jugular
Upper lip area	Superior labial	Facial
Maxillary teeth	Posterior superior alveolar	Pterygoid plexus
Lower lip area	Inferior labial	Facial
Mandibular teeth and submental region	Inferior alveolar	Pterygoid plexus
Submental region	Submental	Facial
Lingual and sublingual regions	Lingual	Facial or internal jugular
Deep facial areas and posterior superior alveolar and inferior alveolar veins	Pterygoid plexus	Maxillary
Pterygoid plexus of veins	Maxillary	Retromandibular

veins (of-**thal**-mic) drain the tissues of the orbit. This anastomosis provides a communication with the cavernous venous sinus, which may become fatally infected through the spread of dental infection (discussed further in this chapter and Chapter 12). This is especially significant since the facial vein, like other veins of the head, has no valves to control the direction of blood flow.

The facial vein receives branches from the same areas of the face that are supplied by the facial artery. This vein anastomoses with the deep veins, such as the pterygoid plexus in the infratemporal fossa, and with the large retromandibular vein before joining the internal jugular vein at the level of the hyoid bone (discussed later in this chapter).

The facial vein has some important tributaries in the oral region (see Figure 6–10). The **superior labial vein** drains the upper lip. The **inferior labial vein** drains the lower lip. The **submental vein** (sub-**men**-tal) drains the tissues of the chin, as well as the submandibular region.

One excellent example of the venous variability concerns the **lingual veins.** These include the dorsal lingual veins that drain the dorsal surface of the tongue, the highly visible deep lingual veins that drain the ventral surface of the tongue, and the sublingual veins that drain the floor of the mouth. These lingual veins may join to form a single vessel or may empty into larger vessels separately. They also may drain indirectly into the facial vein or directly into the internal jugular vein.

RETROMANDIBULAR VEIN

The **retromandibular vein** (reh-tro-man-**dib**-you-lar) will form the external jugular vein from a portion of its route. The retromandibular vein is formed by the merger of the superficial temporal vein and maxillary vein (Figure 6–11). The retromandibular vein emerges from the parotid salivary gland and courses inferiorly. This vein and its beginning venules drain areas similar to those supplied by the superficial temporal and maxillary arteries.

Below the parotid gland, the retromandibular vein typically divides (see Figure 6–11). The anterior division joins the facial vein, and the posterior division continues its downward course on the surface of the sternocleidomastoid muscle. After being joined by the posterior auricular vein (aw-**rik**-you-lar), which drains the lateral scalp behind the ear, this posterior division of the retromandibular veins becomes the external jugular vein. The external jugular vein is discussed later in this chapter.

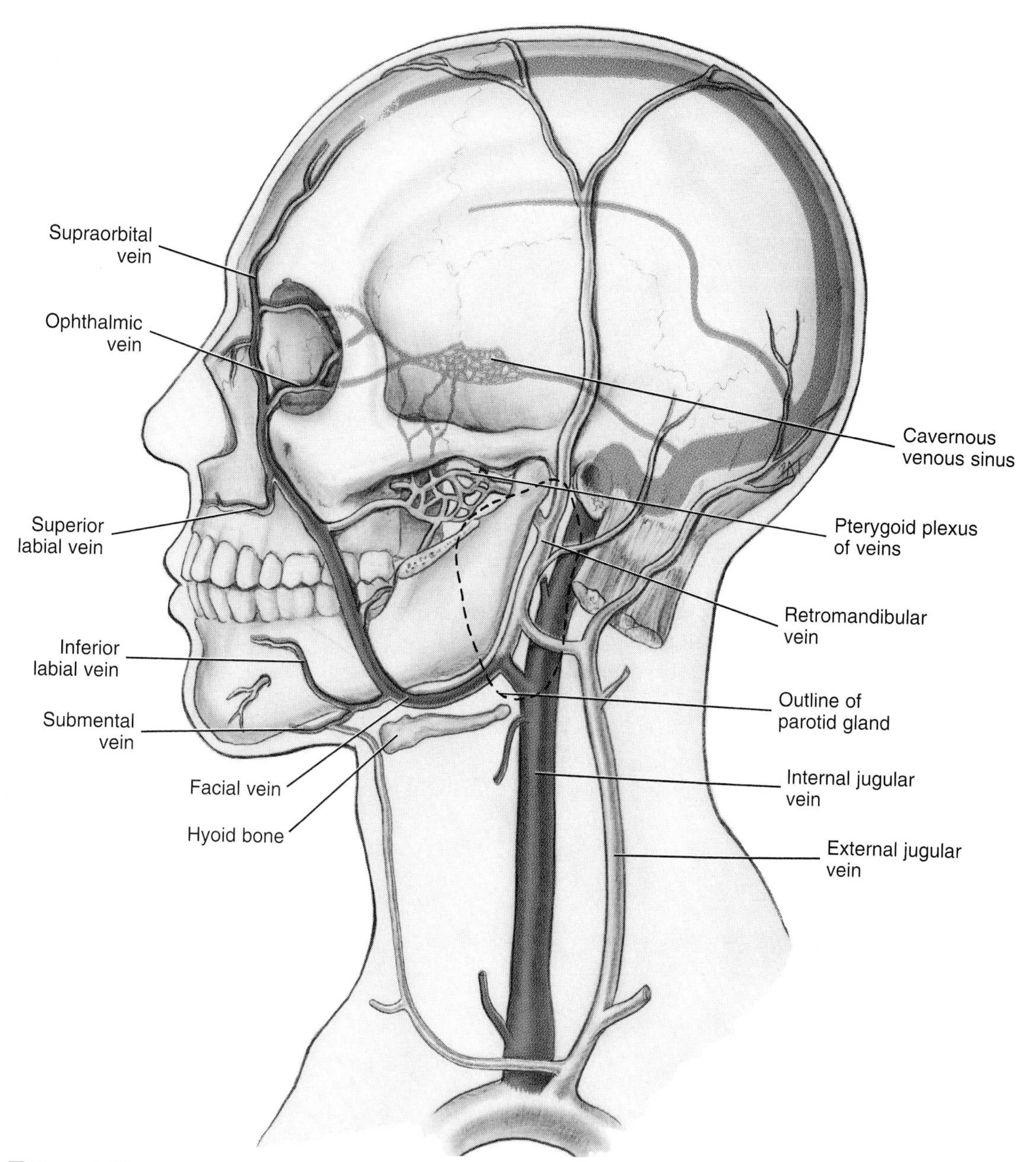

Figure 6–10
Pathways of the internal jugular vein and facial vein, as well as the location of the cavernous venous sinus.

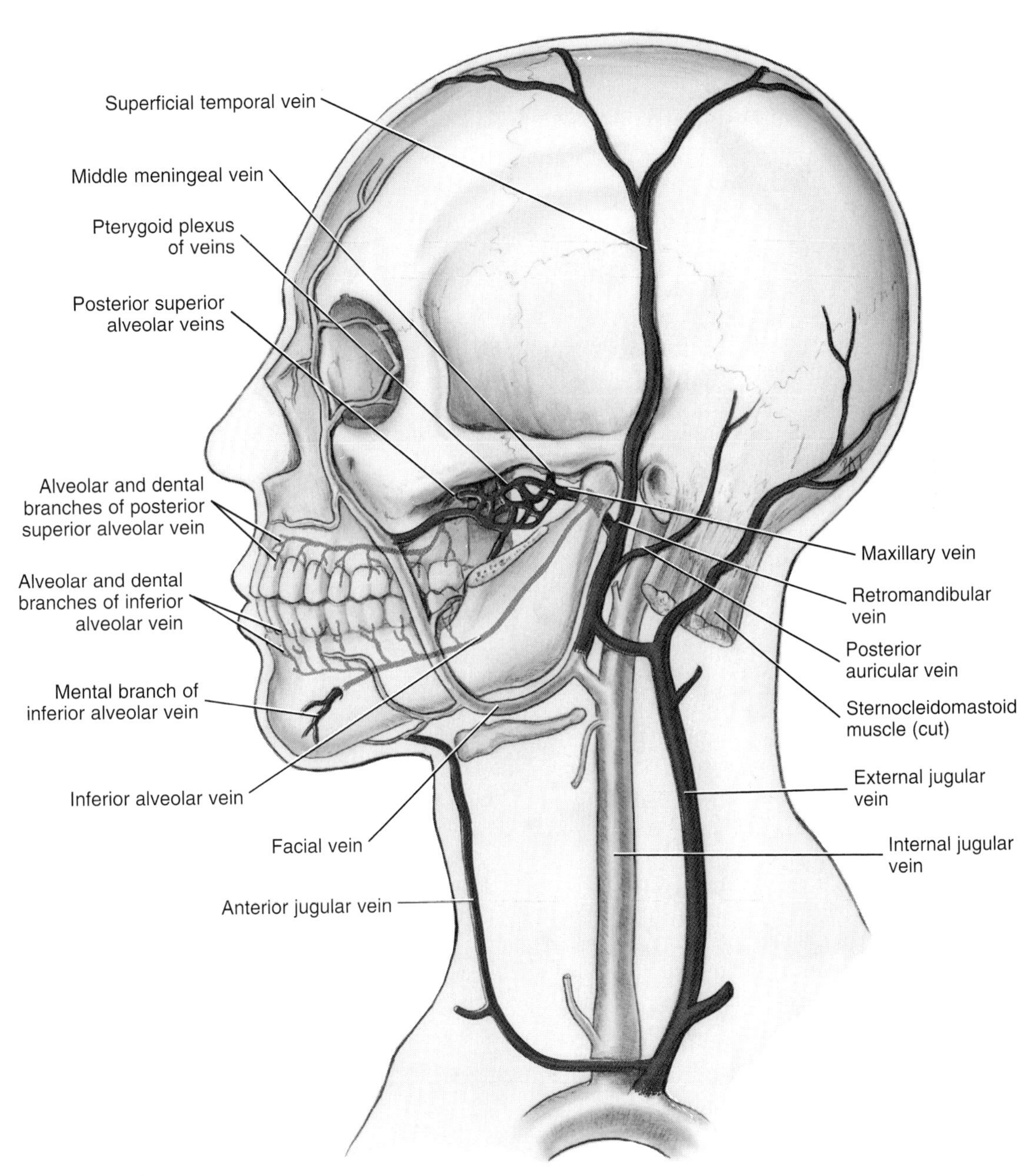

▼ **Figure 6–11**
Pathways of the retromandibular vein and external jugular vein, including the anterior jugular vein.

Superficial Temporal Vein

The **superficial temporal vein** (**tem**-poh-ral) drains the lateral scalp and is superficially located (see Figure 6–11). The superficial temporal vein goes on to drain into and form the retromandibular vein, along with the deeper maxillary vein.

Maxillary Vein

The **maxillary vein** (**mak**-sil-lare-ee) is deeper than the superficial temporal vein and begins in the infratemporal fossa by collecting blood from the pterygoid plexus, near the maxillary artery (see Figure 6–11). Through the pterygoid plexus, the maxillary vein receives the middle meningeal, posterior superior alveolar, inferior alveolar, and other veins, such as those from the nose and palate (those areas served by the maxillary artery). After receiving these veins, the maxillary vein merges with the superficial temporal vein to drain into and form the retromandibular vein.

PTERYGOID PLEXUS OF VEINS

The **pterygoid plexus of veins** (**teh**-ri-goid) is a collection of small anastomosing vessels located around the pterygoid muscles and surrounding the maxillary artery on each side of the face in the infratemporal fossa (see Figure 6–11). This plexus anastomoses with both the facial and retromandibular veins. The pterygoid plexus protects the maxillary artery from being compressed during mastication. By either filling or emptying, the pterygoid plexus can accommodate changes in volume of the infratemporal fossa that occur when the mandible moves.

The pterygoid plexus drains the veins from the deep portions of the face and then drains into the maxillary vein. The **middle meningeal vein** (meh-**nin**-je-al) also drains the blood from the meninges of the brain into the pterygoid plexus of veins.

Some portions of the pterygoid plexus of veins are very close to the maxillary tuberosity, reflecting the drainage of dental tissues into the plexus. Thus there is a possibility of piercing the pterygoid plexus when a posterior superior alveolar anesthetic block is performed if the angulation of the needle is not correct (see Chapter 9 for more information). When the pterygoid plexus of veins is pierced, a small amount of the blood escapes and enters the tissues, causing tissue tenderness, swelling, and the discoloration of a hematoma (discussed later in this chapter).

A spread of infection along the needle tract deep into the tissues can also occur when the posterior superior alveolar anesthetic block is incorrectly administered (see Chapter 12 for more information). The pterygoid plexus of veins may also be involved in the spread of infection to the cavernous venous sinus (discussed later in this chapter and in Chapter 12).

POSTERIOR SUPERIOR ALVEOLAR VEIN

The pterygoid plexus of veins also drains the **posterior superior alveolar vein** (al-**ve**-o-lar), which is formed by the merging of its dental and alveolar branches (see Figure 6–11). The dental branches of the posterior superior alveolar vein drain the pulp tissue of the maxillary teeth by way of each tooth's apical foramen. The alveolar branches of the posterior alveolar vein drain the periodontium of the maxillary teeth, including the gingiva.

INFERIOR ALVEOLAR VEIN

The **inferior alveolar vein** (al-**ve**-o-lar) forms from the merging of its dental branches, alveolar branches, and mental branches in the lower arch, where they also drain into the pterygoid plexus (see Figure 6–11). The dental branches of the inferior alveolar vein drain the pulp tissue of the mandibular teeth by way of each tooth's apical foramen. The alveolar branches of the inferior alveolar vein drain the periodontium of the mandibular teeth, including the gingiva.

The mental branches of the inferior alveolar vein (**ment**-il) enter the mental foramen after draining the chin area, on the outer surface of the mandible, where they anastomose with branches of the facial vein. The mental foramen is on the surface of the mandible, usually between the apices of the mandibular first and second premolars.

VENOUS SINUSES

There are **venous sinuses** located in the meninges of the brain. Specifically, these sinuses are within the dura mater of the brain, a dense connective tissue that lines the inside of the cranium. These dural sinuses are channels by which blood is conveyed from the cerebral veins into the veins of the neck, particularly the internal jugular vein.

The most important venous sinus to dental professionals is the **cavernous venous sinus** (**kav**-er-nus) that is located on each side of the body of the sphenoid bone (see Figure 6–10). Each cavernous venous sinus communicates with the one on the opposite side and also with the pterygoid plexus of veins and superior ophthalmic vein, which anastomoses with the facial vein. The cavernous venous sinus may be involved with the spread of infection from the teeth or periodontium, which can lead to fatal results (discussed further in Chapter 12).

INTERNAL JUGULAR VEIN

The **internal jugular vein** (**jug**-you-lar) drains most of the tissues of the head and neck (Figures 6–10 and 6–12). As mentioned before, the internal jugular vein, unlike many veins in other parts of the body, does not have any one-way valves, nor does any head and neck vein (with the minor exception of the external jugular vein, discussed later). Not having any valves to prevent the flow of blood backward, this vein may become involved with the spread of infection (discussed in Chapter 12).

The internal jugular vein originates in the cranial cavity and leaves the skull through the jugular foramen. It receives many tributaries, including the veins from the lingual, sublingual, and pharyngeal areas, as well as the facial vein. The internal jugular vein runs with the common carotid artery and vagus nerve in the carotid sheath. The internal jugular vein descends in the neck to merge with the subclavian vein.

EXTERNAL JUGULAR VEIN

As mentioned before, the posterior division of the retromandibular vein becomes the **external jugular vein** (**jug**-you-lar). The external jugular vein continues the descent down the neck, terminating in the subclavian vein (see Figures 6–11 and 6–12). Alone among the veins of the head and neck, the external jugular vein has valves near its entry into the subclavian vein. Usually the external jugular vein is visible as it crosses

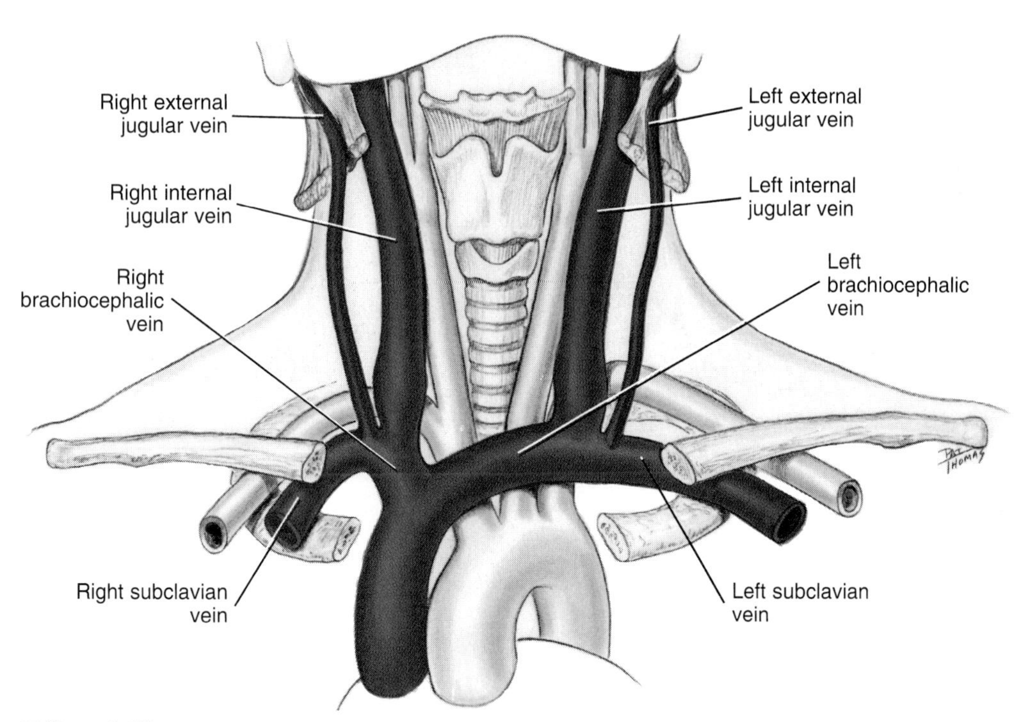

Figure 6–12
Pathways to the heart from the head and neck, including the external and internal jugular veins, subclavian vein, brachiocephalic veins, and superior vena cava.

the sternocleidomastoid muscle; to increase its visibility, it can be distended by gentle supraclavicular digital pressure to block outflow.

The **anterior jugular vein** drains into the external jugular vein (or directly into the subclavian vein) before it joins the subclavian vein (see Figure 6–11). The anterior jugular vein begins below the chin, communicating with veins in the area, and descends near the midline, within the superficial fascia, receiving branches from the superficial cervical structures. Only one anterior jugular vein may be present, but usually two veins are present, anastomosing with each other through a jugular venous arch.

▼ PATHWAYS TO THE HEART FROM THE HEAD AND NECK

On each side of the body, the external jugular vein joins the subclavian vein from the arm, and then the internal jugular vein merges with the **subclavian vein** (sub-**klay**-vee-an) to form the **brachiocephalic vein** (bray-kee-oo-sah-**fal**-ik) (see Figure 6–12). The brachiocephalic veins unite to form the **superior vena cava** (**vee**-na **kay**-va) and then travel to the heart. Because the superior vena cava is on the right side of the heart, the brachiocephalic veins are asymmetrical. The right brachiocephalic vein is short and vertical, and the left brachiocephalic vein is long and horizontal.

Blood Vessel Lesions

Blood vessels may become compromised in certain disease processes, such as high blood pressure, infection, trauma, or endocrine pathology. These disease processes may lead to blood vessel lesions. One of these lesions is a clot or ***thrombus*** (plural, ***thrombi***) (**throm**-bus, **throm**-by) that forms on the inner vessel wall (Figure 6–13).

A thrombus may dislodge from the inner vessel wall and travel as an ***embolus*** (plural, ***emboli***) (**em**-bol-us, **em**-bol-eye), foreign material in the blood (Figure 6–14). Both of these blood vessel lesions can cause occlusion of the vessel, where the blood flow is blocked either partially or fully. Bacteria traveling the blood can also cause a ***bacteremia*** (bak-ter-**ee**-me-ah). A transient bacteremia can occur with dental treatment and is very serious in certain medically compromised patients (discussed further in Chapter 12).

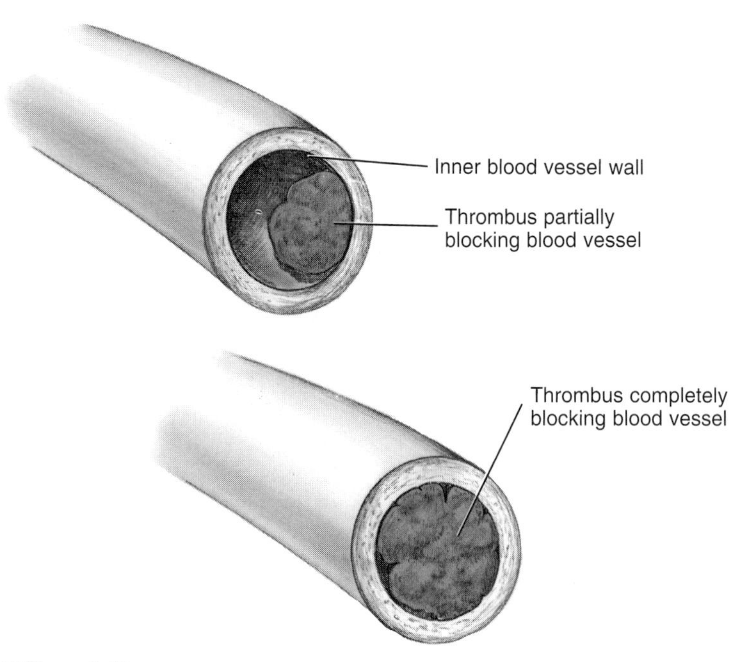

▼ Figure 6–13
Diagrams of a thrombus formed on the inner blood vessel wall and partially and completely blocking the blood vessel.

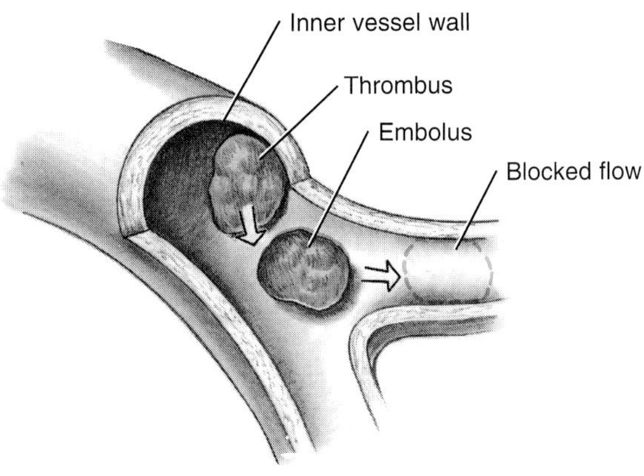

Figure 6–14
Diagram of a dislodged thrombus forming an embolus and then traveling in a blood vessel.

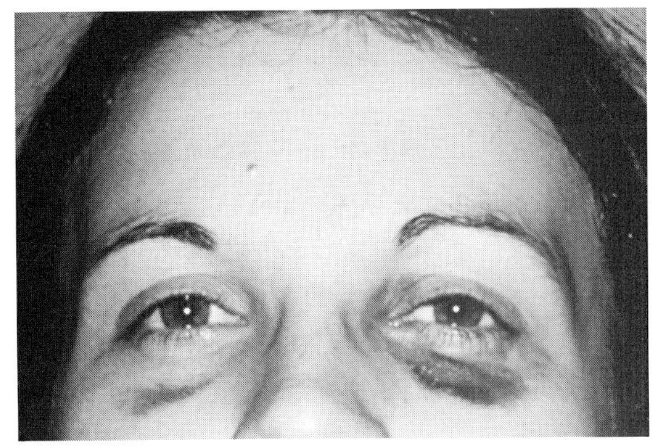

Figure 6–15
Two extraoral hematomas formed in a patient after administration of a local anesthetic injection in the area of the pterygoid plexus on the left side and an injection near the infraorbital foramen on the right side. (Courtesy of Dr. Mark Egbert.)

This occlusion of the blood vessel can hamper blood circulation and cause further complications, such as a stroke (cerebrovascular accident), a heart attack (myocardial infarction), or tissue destruction (gangrene) depending on the lesion's location. These thrombi may also be infected and spread infection by way of embolus formation to such areas as the cavernous venous sinus (see Chapter 12 for more information). A dental professional needs to keep the possibility of blood vessel lesions in mind when treating a patient with blood vessel disease or dentally related infections.

When a blood vessel is seriously traumatized, large amounts of the blood can escape into the surrounding tissue without clotting, causing a ***hemorrhage*** (**hem**-ah-rij). This is a serious, life-threatening blood vessel lesion. Other blood vessel lesions can involve tumorous or abnormal developmental growth of blood vessel tissues. A dental professional needs to be aware of the patient's health history with regard to these serious blood vessel diseases.

Blood vessels may also undergo localized trauma that results in a bruise. A bruise or ***hematoma*** (hee-mah-**toe**-mah) results when a blood vessel is injured and a small amount of the blood escapes into the surrounding tissue and then clots (Figures 6–15 and 6–16). This escaped blood causes tissue tenderness, swelling, and discoloration that will last until the blood is broken down by the body.

A hematoma may result during a local anesthetic injection, especially when giving a posterior superior alveolar block near the pterygoid plexus of veins (see Chapter 9 for more information) or other blocks, such as an infraorbital block or inferior alveolar block. Thus a dental professional needs to be aware of the location of the blood vessels to prevent their injury during dental treatment.

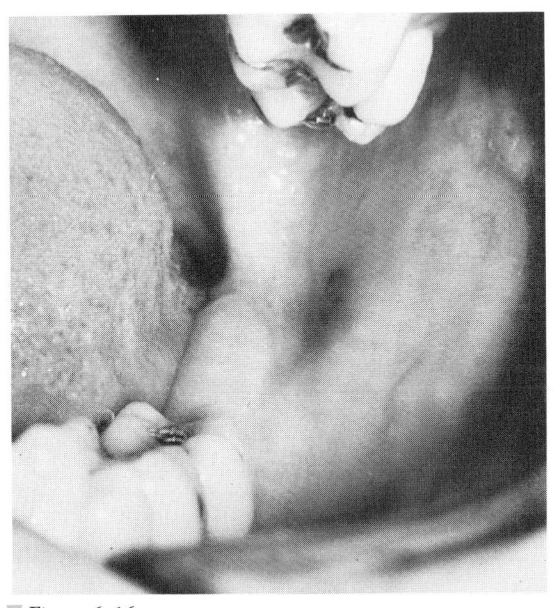

Figure 6–16
An intraoral hematoma in the tissues on the medial surface of the mandibular ramus after an inferior alveolar nerve block.

Identification Exercises

Identify the structures on the following diagrams by filling in the blanks with the correct anatomical term. You can check your answers by looking back at the figure indicated in parenthesis for each identification diagram.

1. (Figure 6–1)

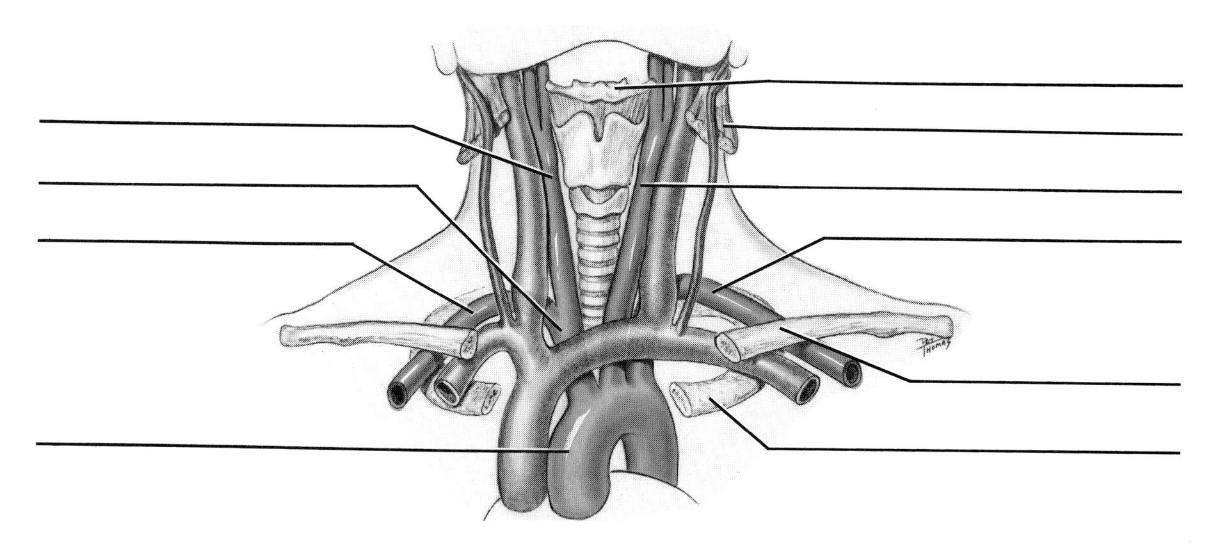

2. (Figure 6–3)

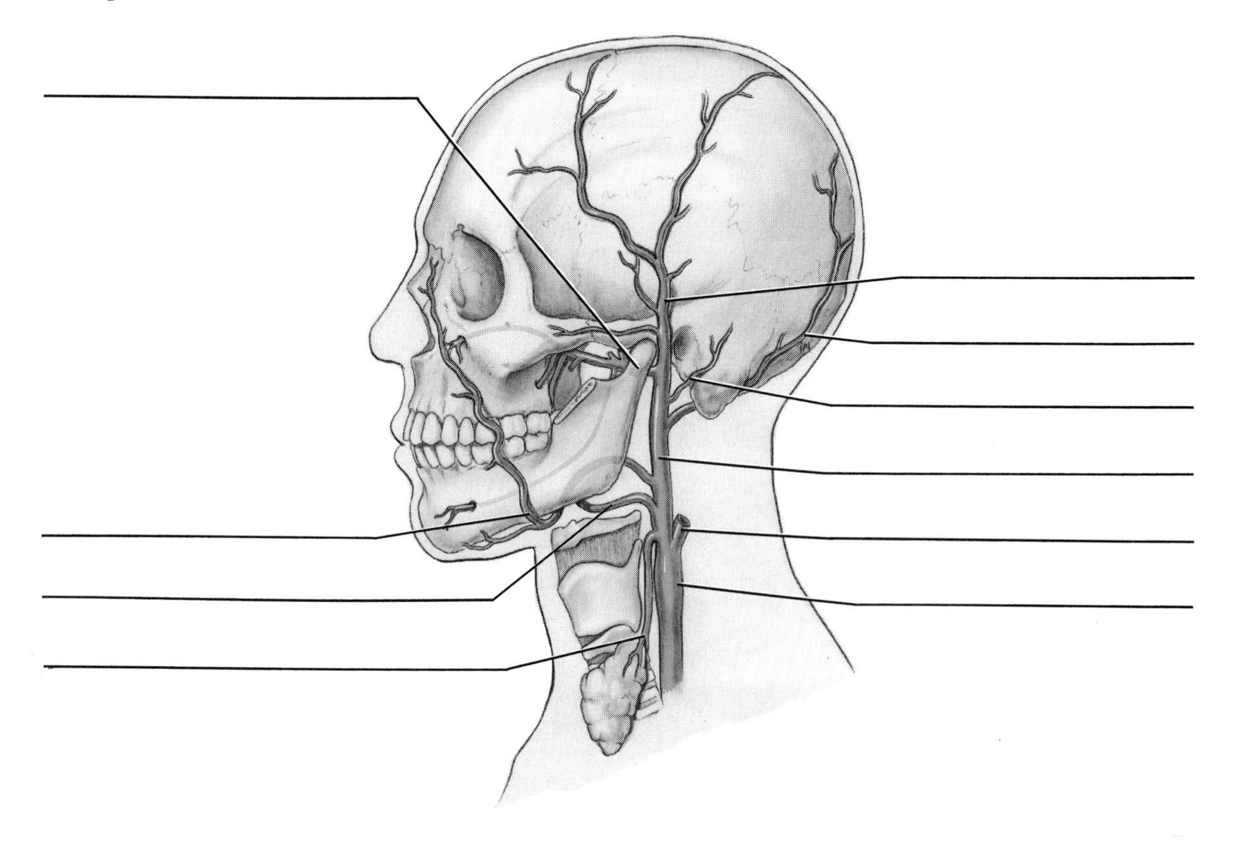

Identification Exercises *Continued*

3. (Figure 6–4)

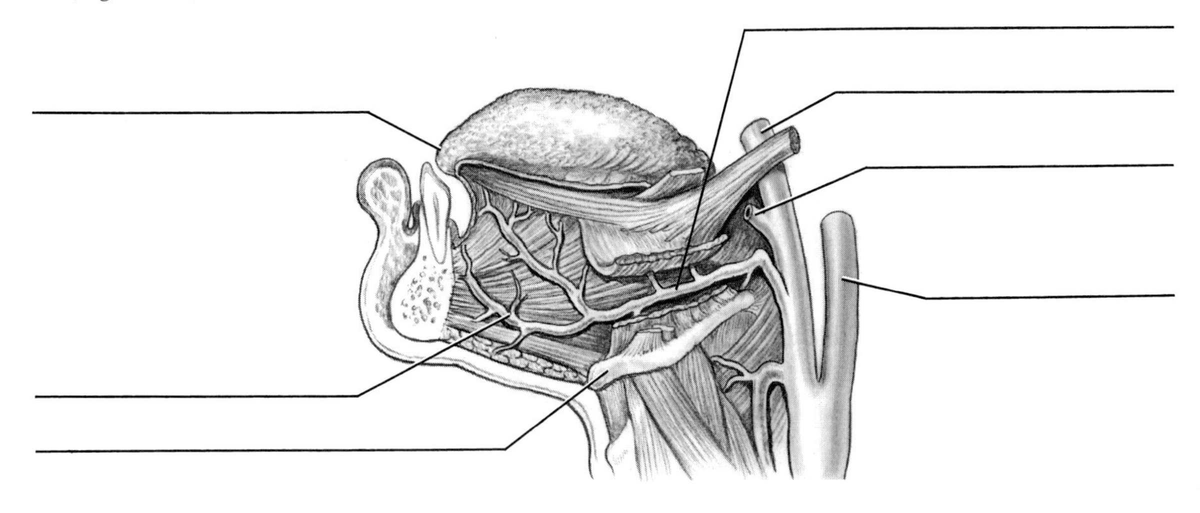

4. (Figure 6–5)

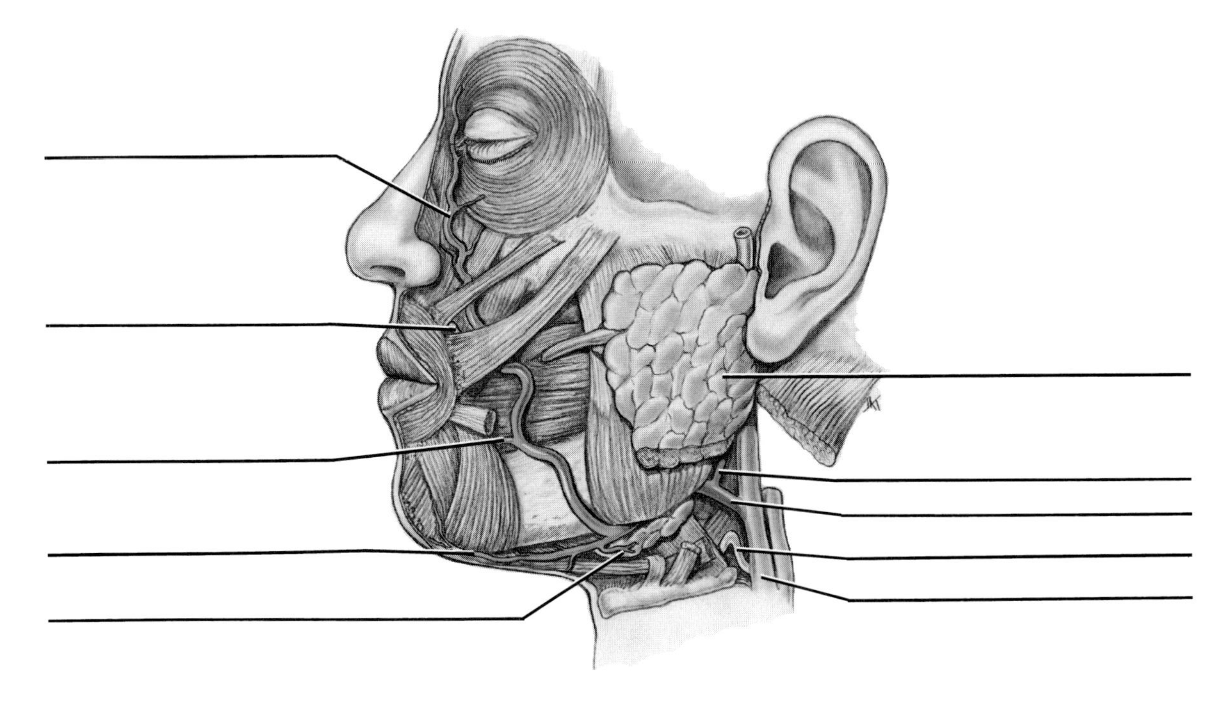

Identification Exercises *Continued*

5. (Figure 6–8)

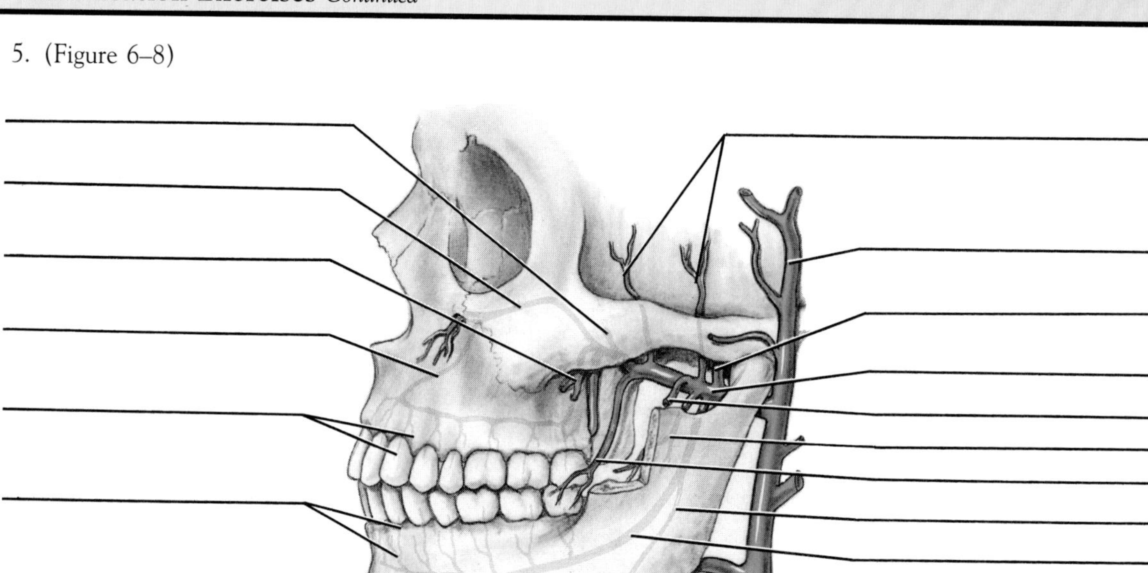

6. (Figure 6–9)

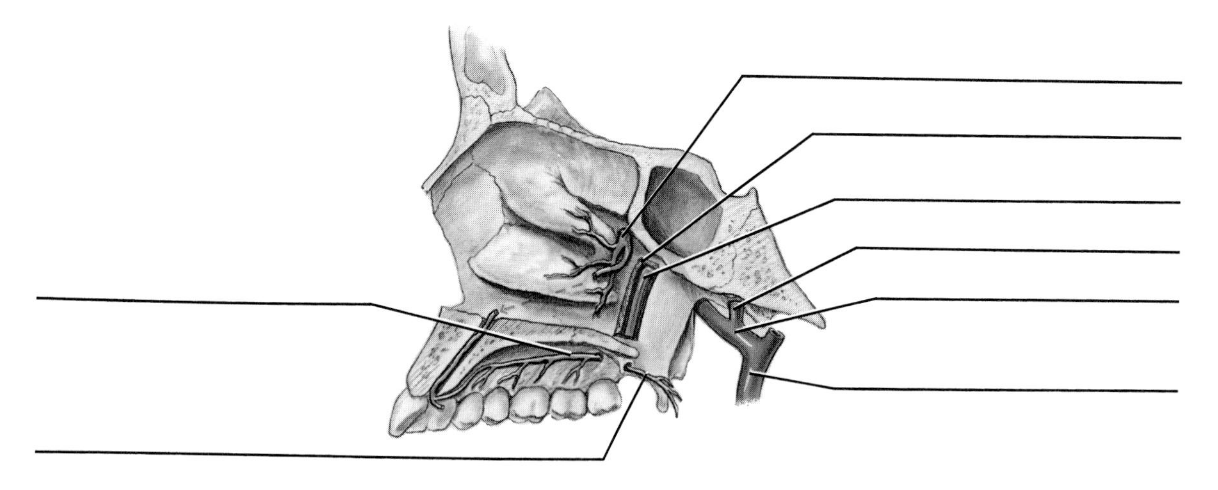

Identification Exercises *Continued*

7. (Figure 6–10)

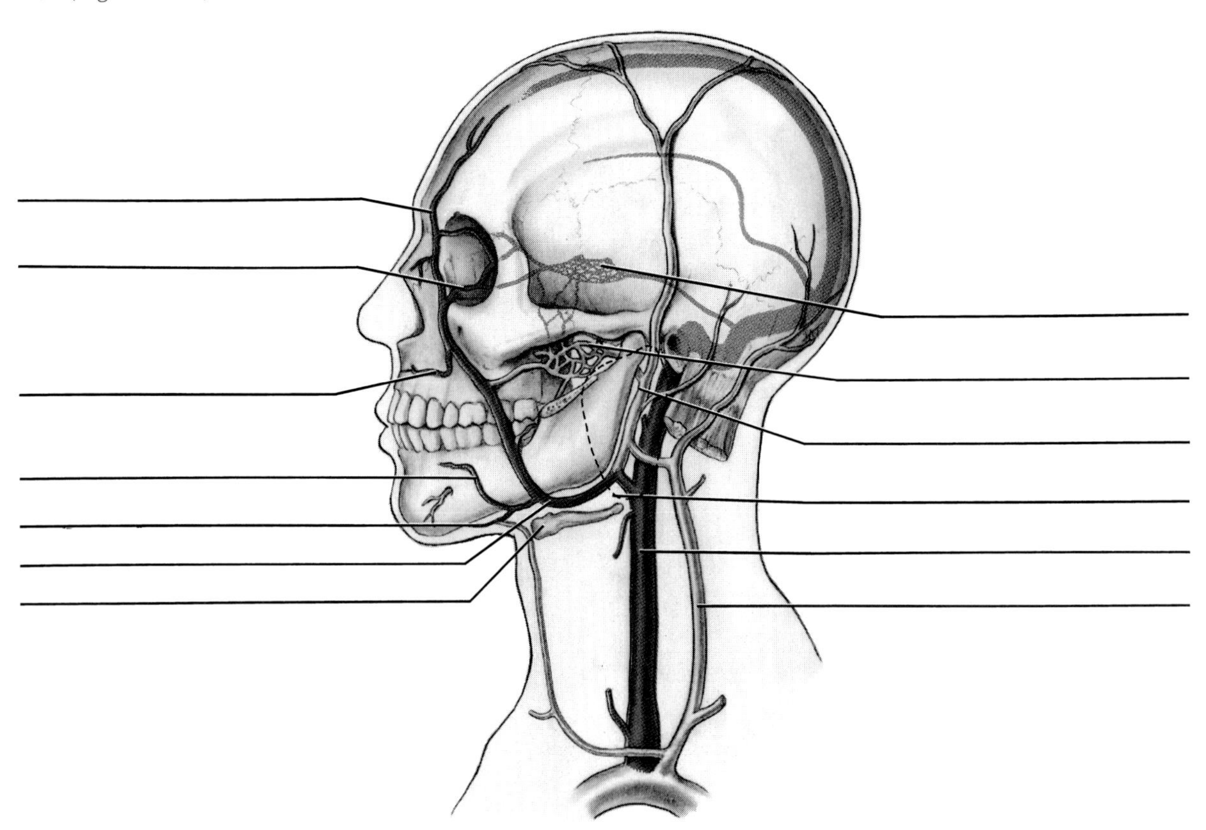

Identification Exercises *Continued*

8. (Figure 6–11)

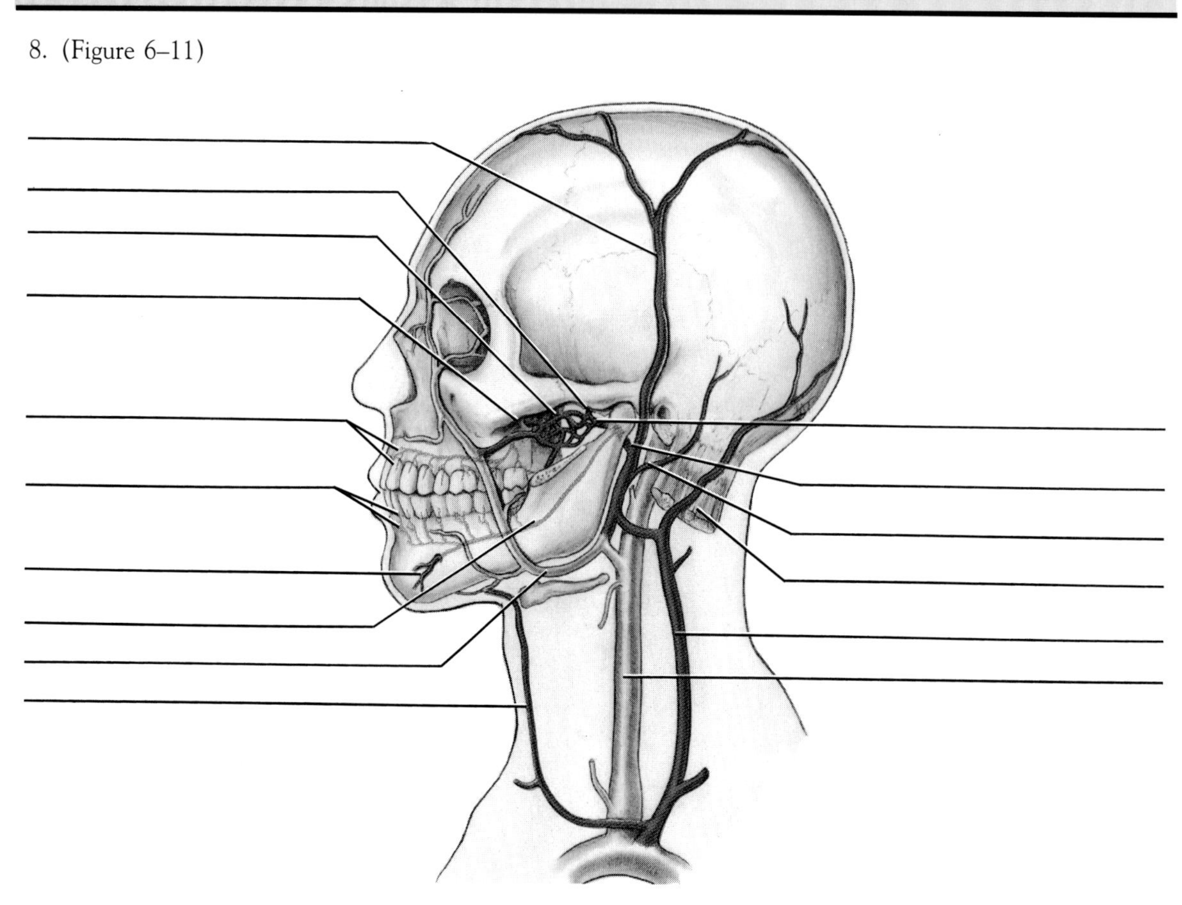

9. (Figure 6–12)

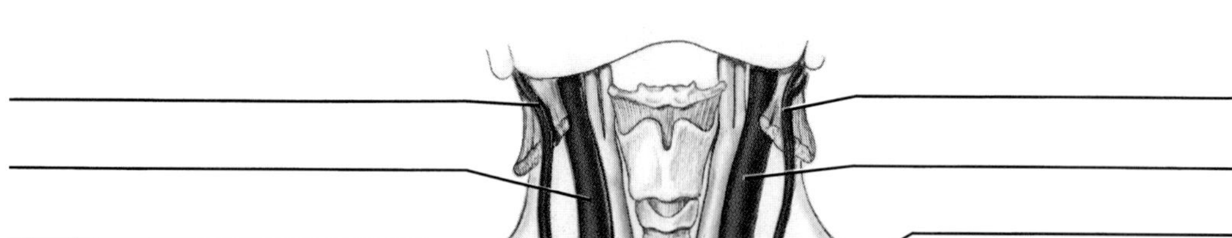

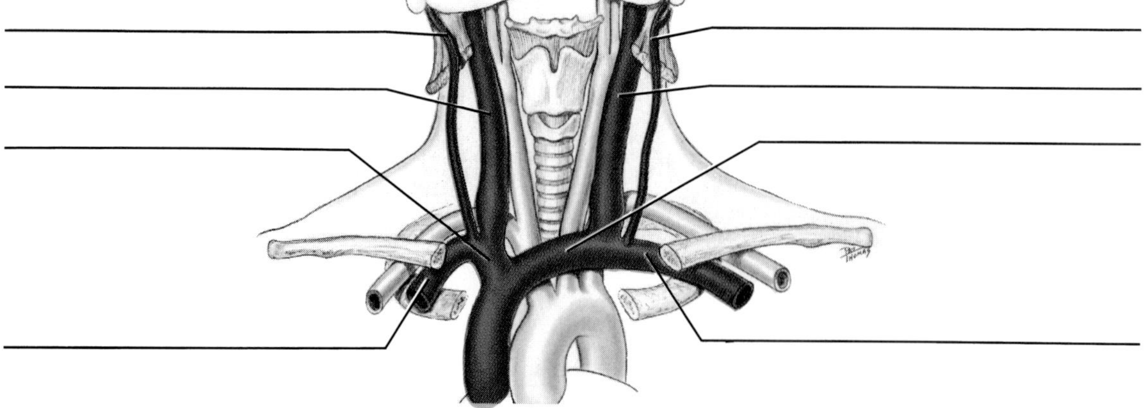

REVIEW QUESTIONS

1. The posterior superior alveolar artery and its branches supply the
   A. Maxillary posterior teeth and periodontium
   B. Mandibular posterior teeth and periodontium
   C. Sternocleidomastoid muscle and thyroid gland
   D. Temporalis muscle and parotid salivary gland

2. Which of the following statements concerning the pterygoid plexus is correct?
   A. Located around the infrahyoid muscles
   B. Protects the superficial temporal artery
   C. Drains the maxillary and mandibular dental tissues
   D. Contains valves to prevent backflow of blood

3. Which of the following veins results from the merger of the superficial temporal vein and maxillary vein?
   A. Facial vein
   B. Retromandibular vein
   C. Internal jugular vein
   D. External jugular vein
   E. Brachiocephalic vein

4. Which of the following arteries arises from the inferior alveolar artery before the artery enters the mandibular canal?
   A. Mylohyoid artery
   B. Incisive artery
   C. Mental artery
   D. Posterior superior alveolar artery
   E. Submental artery

5. Which of the following artery and foramen pairs below is matched correctly?
   A. Buccal artery-infraorbital foramen
   B. Middle meningeal artery-foramen spinosum
   C. Incisive artery-mental foramen
   D. Inferior labial artery-mandibular foramen
   E. Submental artery-mental foramen

6. Which of the following arteries listed below supply the mucous membranes and glands of the hard and soft palates?
   A. Greater and lesser palatine arteries
   B. Posterior superior alveolar artery
   C. Anterior superior alveolar artery
   D. Infraorbital artery

7. Which of the following blood vessel lesions may result when a clot on the inner blood vessel wall becomes dislodged and travels in the vessel?
   A. Hematoma
   B. Venous sinus
   C. Embolus
   D. Hemorrhage

8. Which of the following statements concerning the maxillary artery is correct?
   A. Arises from the internal carotid artery
   B. Enters the pterygopalatine fossa and forms terminal branches
   C. Arises from the zygomaticofacial foramen to emerge on the face
   D. Has mandibular, maxillary, nasal, palatine, and occipital branches

9. A venous sinus of the vascular system is a:
   A. Network of blood vessels
   B. Clot on the inner vessel wall
   C. Blood-filled space between two tissue layers
   D. Smaller vein or venule

10. Which of the following is a branch from the facial artery?
    A. Superior labial artery
    B. Ascending pharyngeal artery
    C. Posterior auricular artery
    D. Transverse facial artery

▼ *This chapter discusses the following topics:*

I. Glandular Tissue
II. Lacrimal Glands
III. Salivary Glands
  A. Major Salivary Glands
  B. Minor Salivary Glands
IV. Thyroid Gland
V. Parathyroid Glands
VI. Thymus Gland

▼ *Key Words*

Duct (dukt) Passageway to carry the secretion from the exocrine gland to the location where it will be used.

Endocrine gland (**en**-dah-krin) Type of gland without a duct, with the secretion being poured directly into the blood, which then carries the secretion to the region being used.

Exocrine gland (**ek**-sah-krin) Type of gland with an associated duct that serves as a passageway for the secretion to be emptied directly into the location where the secretion is to be used.

Gland (gland) Structure that produces a chemical secretion necessary for normal body functioning.

Goiter (**goit**-er) Enlarged thyroid gland due to a disease process.

Chapter 7

Glandular Tissue

▼ ***After reading and studying this chapter, the reader should be able to:***

1. Define and pronounce all the key words and anatomical terms in this chapter.
2. Locate and identify all the glandular tissue and associated structures in the head and neck region on a diagram, a skull, and a patient.
3. Correctly complete the identification exercise and review questions for this chapter.
4. Integrate the knowledge about the head and neck glands during clinical practice when these glands may be involved in a disease process.

Glandular Tissue

The glandular tissue in the head and neck area includes the lacrimal, salivary, thyroid, parathyroid, and thymus glands. A dental professional needs to be able to locate and identify these glands and their innervation, lymphatic drainage, and blood supply (Table 7–1). This information will help the dental professional determine if the glands are involved in a disease process and the extent of that involvement.

A ***gland*** (gland) is a structure that produces a chemical secretion necessary for normal body functioning. An ***exocrine gland*** (**ek**-sah-krin) is a gland having a duct associated with it. A ***duct*** (dukt) is a passageway that allows the secretion to be emptied directly into the location where the secretion is to be used. An ***endocrine gland*** (**en**-dah-krin) is a ductless gland, with the secretion being poured directly into the blood, which then carries the secretion to the region to be used. Motor nerves associated with both types of glands help regulate the flow of the secretion, and sensory nerves are also present.

Lacrimal Glands

The **lacrimal glands** (**lak**-ri-mal) are paired exocrine glands that secrete **lacrimal fluid** or tears. Lacrimal fluid is a watery fluid that lubricates the conjunctiva lining the inside of the eyelids and the front of the eyeball. The fluid leaves the gland through 8–12 fine tubules. After passing over the eyeball, the lacrimal fluid is drained through a small hole in each eyelid, ending in the nasolacrimal sac (nay-so-**lak**-rim-al), a thin-walled structure behind the medial canthus. From the nasolacrimal sac, the lacrimal fluid continues into the **nasolacrimal duct,** ultimately draining into the inferior nasal meatus. This connection explains why crying leads to a runny nose.

Location. Each lacrimal gland is located in the lacrimal fossa of the frontal bone (see Chapter 3 and

Table 7–1

GLANDULAR TISSUE: LOCATION, INNERVATION, LYMPHATIC DRAINAGE, AND BLOOD SUPPLY

Glandular Tissue	Location	Innervation	Lymphatics	Blood Supply
Lacrimal gland with nasolacrimal duct	Lacrimal fossa of frontal bone	Greater petrosal and lacrimal nerves	Superficial parotid	Lacrimal and ophthalmic arteries
Parotid gland with parotid duct	Parotid space behind mandibular ramus, anterior and inferior to ear	Ninth and fifth cranial nerves	Deep parotid nodes	Branches of external carotid artery
Submandibular gland with submandibular duct	Submandibular space: below and behind body of mandible	Chorda tympani and seventh cranial nerve	Submandibular nodes	Facial and lingual arteries
Sublingual gland with sublingual duct(s)	Sublingual space: floor of mouth, medial to body of mandible	Chorda tympani and seventh cranial nerve	Submandibular nodes	Sublingual and submental arteries
Minor salivary glands with ducts	Buccal, labial, and lingual mucosa, soft and hard palate, floor of mouth, and base of circumvallate lingual papillae	Seventh cranial nerve	Various nodes depending on location	Various arteries depending on location
Thyroid gland	Inferior to hyoid bone, junction of larynx and trachea	Cervical ganglion	Superior deep cervical nodes	Superior and inferior thyroid arteries
Parathyroid gland	Close to or within thyroid	Cervical ganglion	Superior deep cervical nodes	Inferior thyroid artery
Thymus gland	In thorax, inferior to hyoid bone, deep to sternum and superficial and lateral to trachea	Tenth and cervical nerves	Within substance of gland	Inferior thyroid and internal thoracic arteries

Figure 2–4). The lacrimal fossa is located just inside the lateral portion of the supraorbital ridge inside the orbit. The nasolacrimal duct is formed at the junction of the lacrimal and maxillary bones.

Innervation. The lacrimal glands are innervated by parasympathetic fibers from the greater petrosal nerve, a branch of the facial or seventh cranial nerve. These preganglionic fibers synapse at the pterygopalatine ganglion, and postganglionic fibers reach the gland through branches of the trigeminal nerve. The lacrimal nerve serves as an afferent nerve for the lacrimal gland.

Lymphatics. The lacrimal glands drain into the superior parotid lymph nodes.

Blood Supply. The lacrimal glands are supplied by the lacrimal artery, a branch of the ophthalmic artery of the internal carotid artery.

Salivary Glands

The **salivary glands** (**sal**-i-ver-ee) produce **saliva** (sah-**li**-vah) that lubricates and cleanses the oral cavity and helps in digestion. These glands are controlled by the autonomic nervous system. There are major salivary glands and minor salivary glands as defined by their size. Both the major and minor salivary glands are exocrine glands and thus have ducts associated with them. These ducts help drain the saliva directly into the oral cavity where the saliva is used.

The salivary glands may become enlarged, tender, and possibly firmer due to various disease processes. The salivary glands may also become involved in salivary stone (sialolith) formation, blocking the drainage of saliva from the duct and causing gland enlargement and tenderness. Certain medications or disease processes may result in decreased or increased production of saliva by these glands.

MAJOR SALIVARY GLANDS

The **major salivary glands** are large paired glands and have named ducts associated with them. There are three major salivary glands: the parotid, submandibular, and sublingual glands.

Parotid Salivary Gland

The **parotid salivary gland** (pah-**rot**-id) is the largest encapsulated major salivary gland but provides only 25 percent of the total salivary volume. The saliva from the parotid is a purely serous type of secretion, a watery protein fluid. The parotid gland is divided into two lobes, a superficial lobe and a deep lobe. The duct associated with the parotid gland is the **parotid duct** or Stensen's duct. This long duct emerges from the anterior border of the gland, superficial to the masseter muscle. The duct pierces the buccinator muscle. The duct then opens up into the oral cavity upon the inner surface of the cheek, usually opposite the second maxillary molar. The **parotid papilla** (pah-**pil**-ah) is a small elevation of tissue that marks the opening of the parotid duct on the inner surface of the cheek.

The parotid salivary gland becomes enlarged and tender when the patient has mumps. This enlargement and tenderness usually involves the gland bilaterally, first one side and then the other side. Mumps is a viral infection of the parotid gland that is being prevented by a childhood vaccination. The parotid gland can also be involved in tumorous growth that can also change the consistency of the gland and cause facial paralysis on the involved side (the facial nerve travels through the gland).

Location. The parotid gland occupies the parotid fascial space, an area behind the mandibular ramus, anterior and inferior to the ear (Figures 7–1 and 7–5). The gland extends irregularly from the zygomatic arch to the angle of the mandible. This gland is effectively palpated on the sides of the face. Start in front of each ear and move to the cheek area and down to the angle of the mandible (Figure 7–2).

Innervation. The parotid gland is innervated by the motor or efferent (parasympathetic) nerves of the otic ganglion of the ninth cranial or glossopharyngeal nerve, as well as the afferent nerves from the auriculotemporal branch of the fifth cranial or trigeminal nerve. The seventh cranial or facial nerve and its branches travel through the parotid gland between its lobes but are not involved in its innervation.

Lymphatics. The parotid gland drains into the deep parotid lymph nodes.

Blood Supply. The parotid gland is supplied by branches of the external carotid artery.

Submandibular Salivary Gland

The **submandibular salivary gland** (sub-man-**dib**-you-lar) is the second largest encapsulated major salivary gland, yet provides 60–65 percent of total salivary volume. The saliva from the submandibular

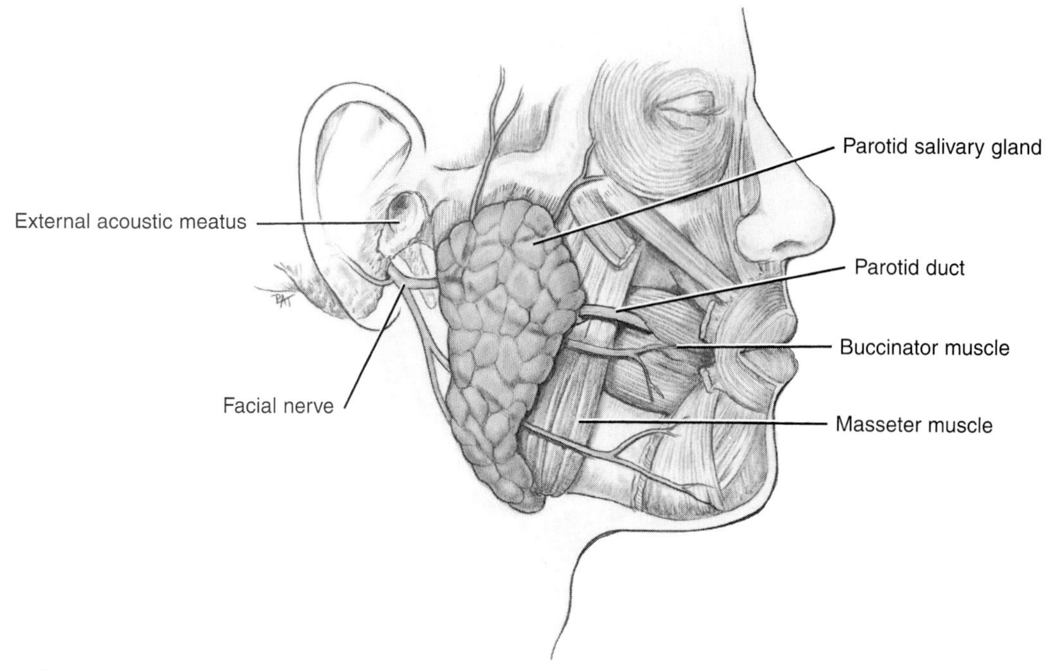

▼ Figure 7–1
The parotid salivary gland and associated structures.

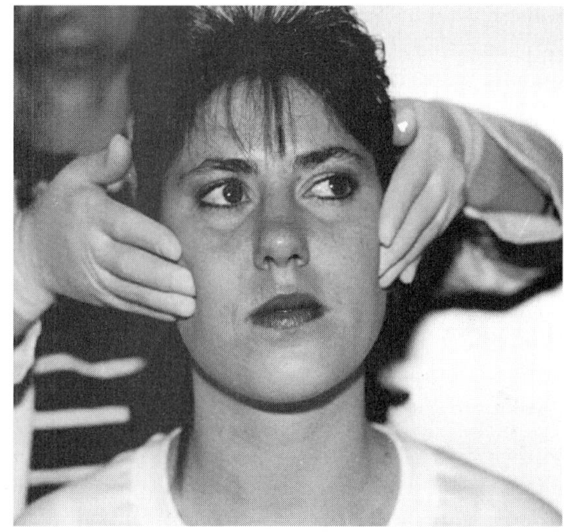

▼ Figure 7–2
Palpating the parotid salivary gland by starting in front of each ear and then moving to the cheek area and down to the angle of the mandible.

gland is a mixed secretion of both serous and mucous types, having both thick carbohydrate and watery protein fluids. The duct associated with the submandibular gland is the submandibular duct or Wharton's duct. This long duct travels along the anterior floor of the mouth. The duct then opens into the oral cavity at the **sublingual caruncle** (sub-**ling**-gwal **kar**-unk-el), a small papilla near the midline of the mouth floor on each side of the lingual frenum. The duct's tortuous travel for a considerable upward distance in its course may be the reason the submandibular gland is the most common salivary gland involved in salivary stone formation.

Location. The submandibular gland occupies the submandibular fossa in the submandibular fascial space, mainly the posterior portion (Figures 7–3 and 7–5). Most of the gland is a lobe superficial to the mylohyoid muscle, but a small deep lobe wraps around the posterior border of the muscle. The duct arises from this deep lobe and stays on the inside of the mylohyoid muscle.

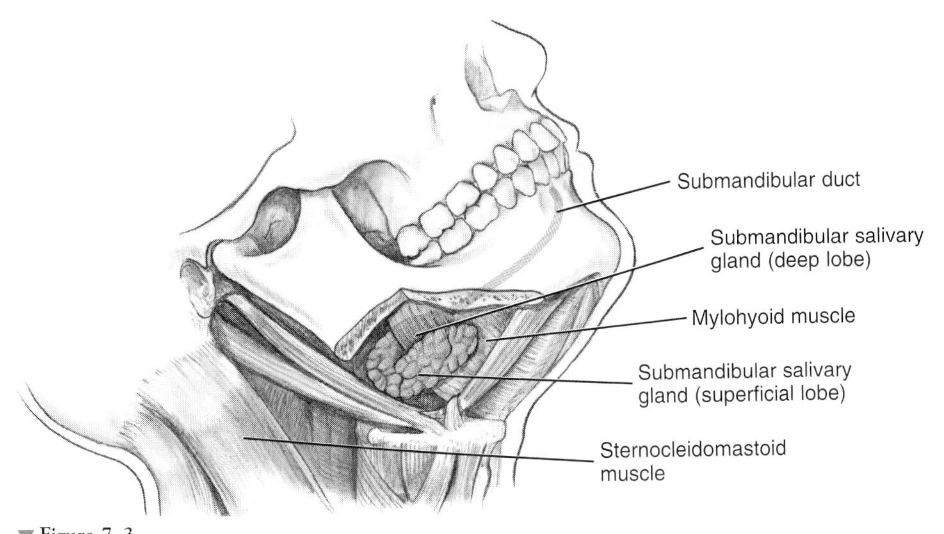

Figure 7–3
The submandibular salivary gland and associated structures.

The duct lies very close to the large lingual nerve, a branch of the fifth cranial or trigeminal nerve, which is sometimes injured in surgery to remove salivary stones from the duct. The submandibular gland is posterior to the sublingual salivary gland. The submandibular gland is effectively palpated below and behind the body of the mandible bilaterally. Start by moving inward from the lower border of the mandible near its angle as the patient lowers the head (Figure 7–4).

Innervation. The submandibular gland is innervated by the efferent (parasympathetic) fibers of the chorda tympani and the submandibular ganglion of the seventh cranial or facial nerve.

Lymphatics. The submandibular gland drains into the submandibular lymph nodes.

Blood Supply. The submandibular gland is supplied by branches of the facial and lingual arteries.

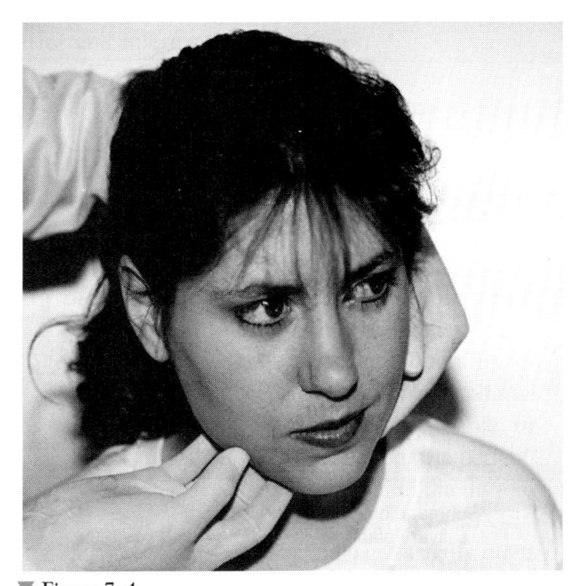

Figure 7–4
Palpating the submandibular salivary gland by palpating inward from the lower border of the mandible near its angle as the patient lowers the head.

Sublingual Salivary Gland

The **sublingual salivary gland** (sub-**ling**-gwal) is the smallest, most diffuse, and only unencapsulated major

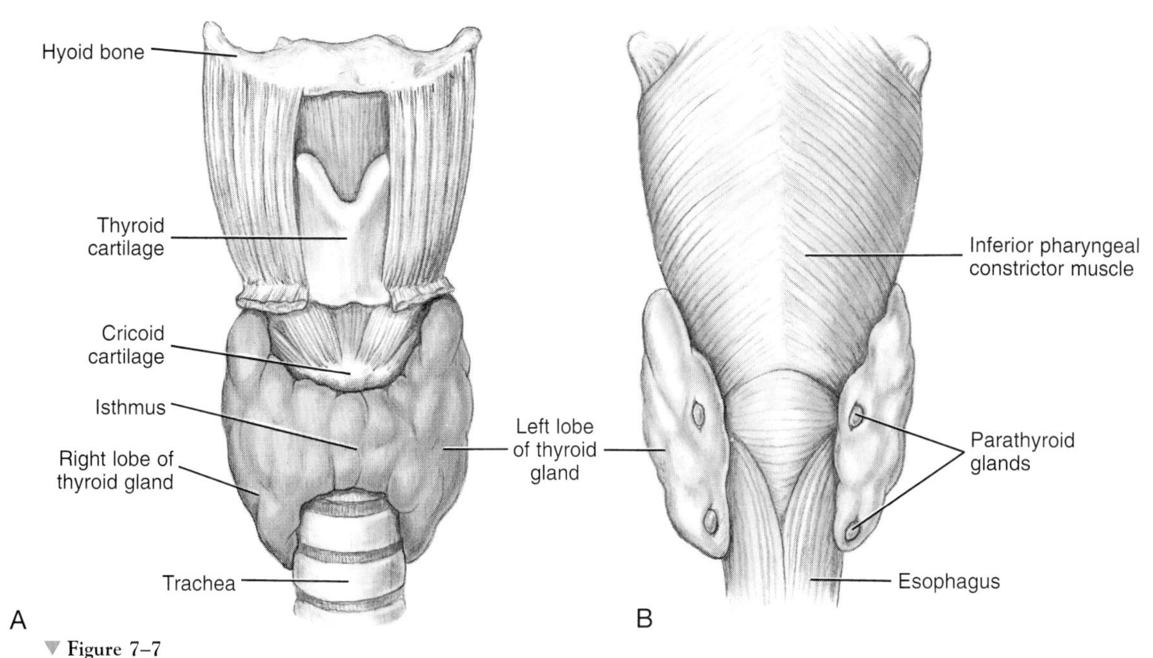

Figure 7–7
A: Anterior view of the thyroid gland and associated structures. **B:** Posterior view of the parathyroid glands and associated structures.

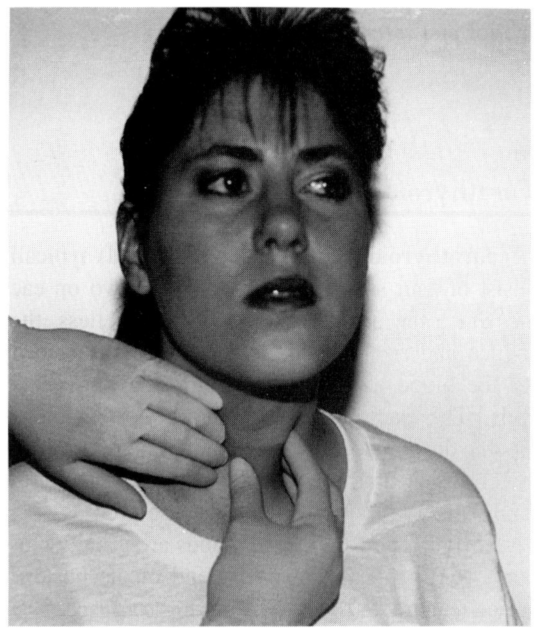

Figure 7–8
Palpating the thyroid gland with one hand by placing the other hand on one side of the trachea while gently displacing the thyroid tissue to that side of the neck.

Blood Supply. The parathyroid glands are primarily supplied by the inferior thyroid arteries.

Thymus Gland

The **thymus gland** (**thy**-mus) is an endocrine gland and therefore is ductless. The thymus gland is a part of the immune system that fights disease processes. The **T-cell lymphocytes,** white blood cells of the immune system, mature in the gland in response to stimulation by thymus hormones. The gland grows in size from birth to puberty while performing this task. After puberty, the gland stops growing and starts to shrink.

By adulthood, the gland has almost disappeared, making it mainly a temporary structure. The adult gland consists of two lateral lobes in close contact at the midline. The thymus gland is not easily palpated, but its involvement in various disease processes may alter the treatment of the patient. Older patients may have had radiation therapy to the thymus in childhood to shrink the gland so as to prevent sudden infant death. It is now known that the large thymus of a child is not related to suffocation and sudden infant death.

The radiation levels used on these patients in the past may result in thyroid cancer.

Location. The thymus gland is located in the thorax and the anterior region of the base of the neck, inferior to the thyroid gland (Figure 7–9). The gland is deep to the sternum and the sternohyoid and sternothyroid muscles. The gland is also superficial and lateral to the trachea.

Innervation. The thymus gland's innervation consists of branches of the tenth cranial or vagus nerve and cervical nerves.

Lymphatics. The thymus gland's lymphatics arise within the substance of the gland and terminate in the internal jugular vein.

Blood Supply. The thymus gland is supplied by the inferior thyroid and internal thoracic arteries.

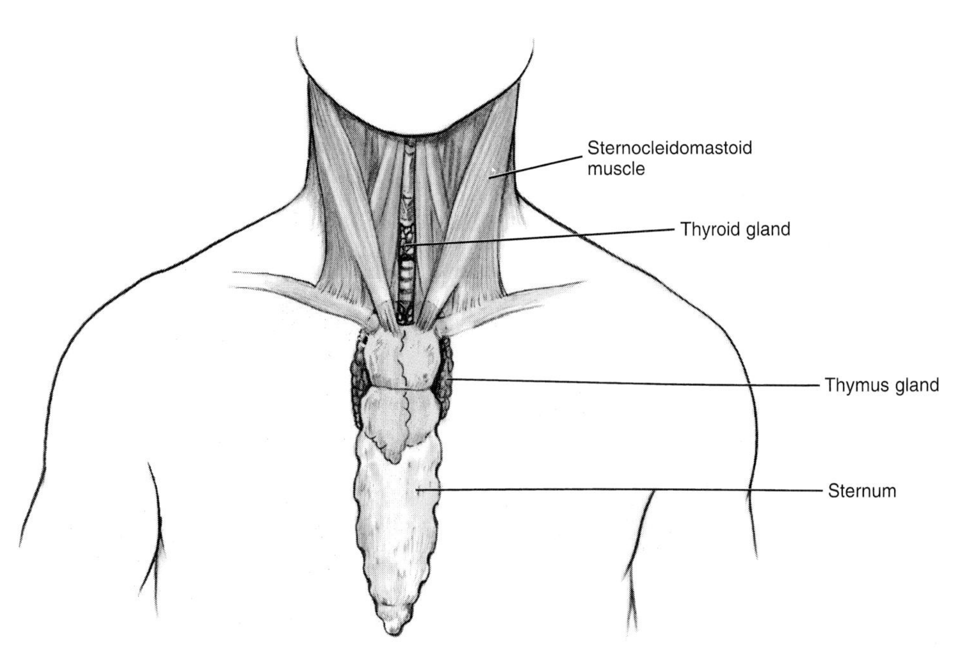

Figure 7–9
The thymus gland and associated structures.

Identification Exercise

Identify the structures on the following diagrams by filling in the blanks with the correct anatomical term. You can check your answers by looking back at the figure indicated in parenthesis for each identification diagram.

1. (Figure 7–1)

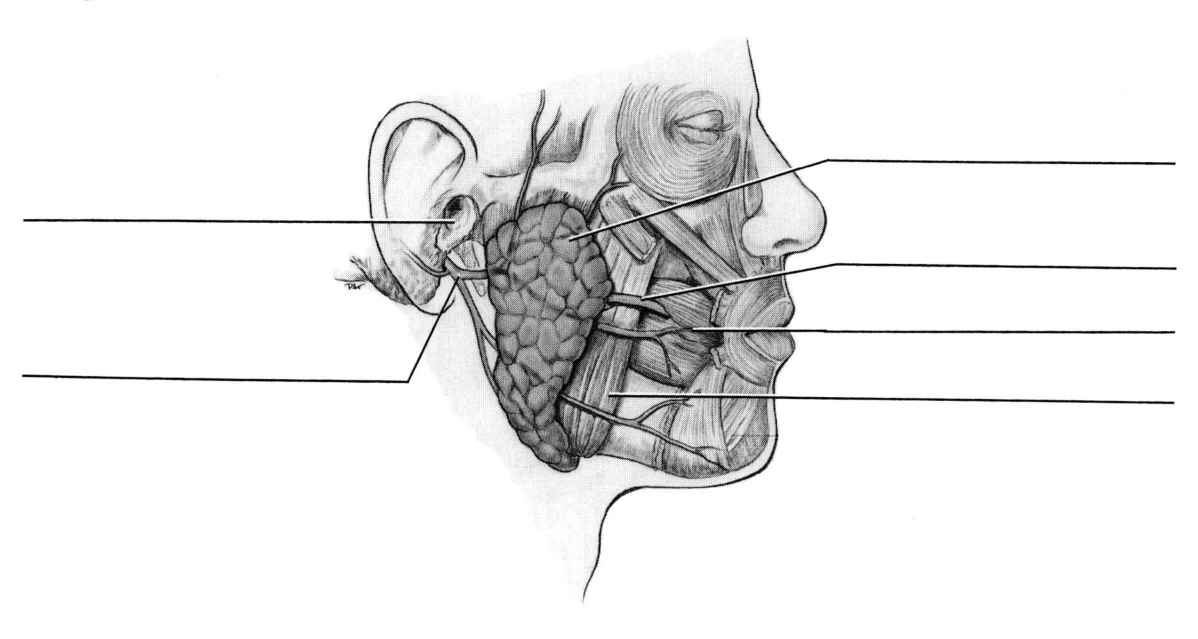

2. (Figure 7–3)

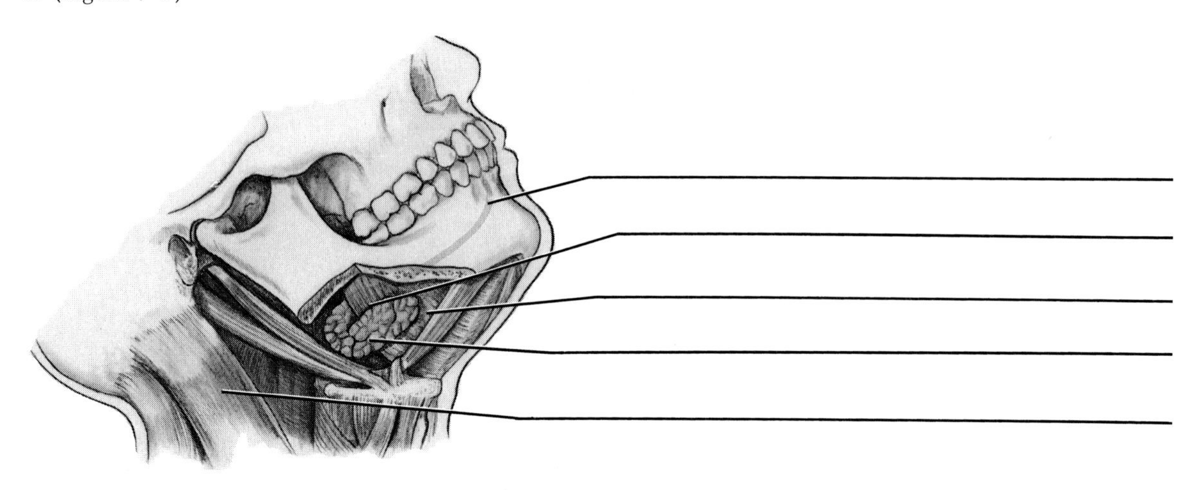

Identification Exercise *Continued*

3. (Figure 7–5)

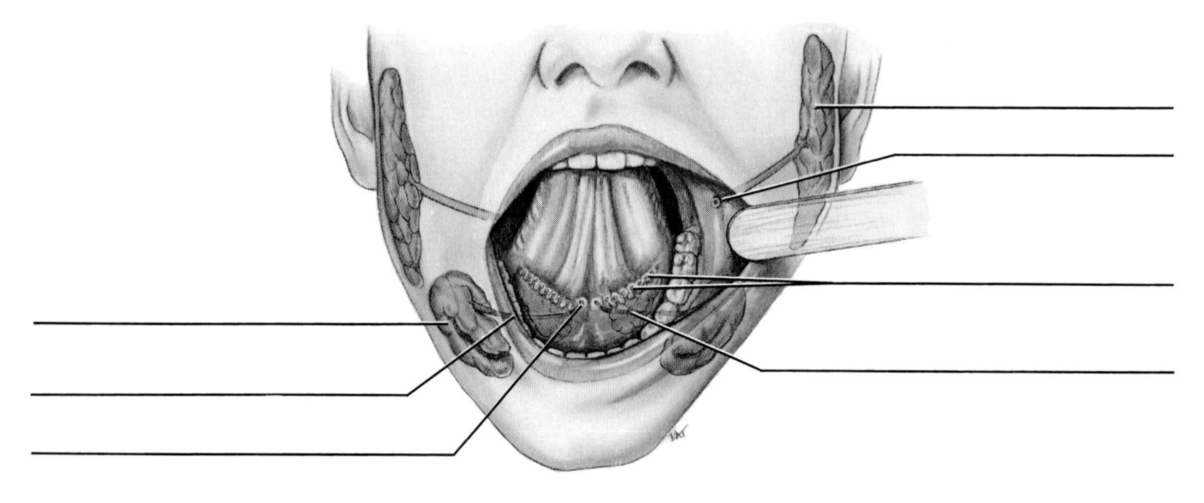

4. (Figure 7–7A)

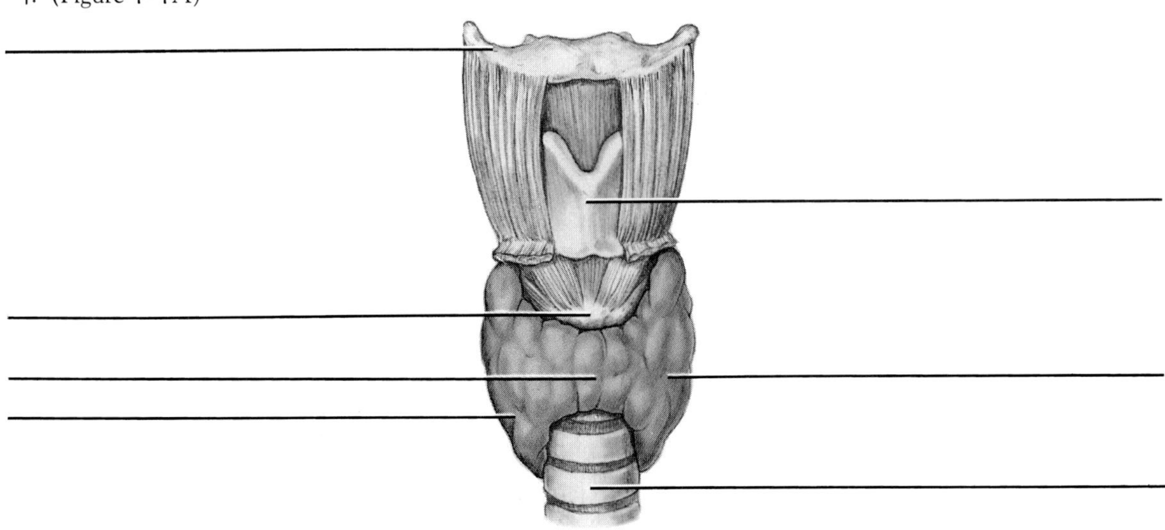

REVIEW QUESTIONS

1. The sublingual salivary gland is located
   A. Anterior to the submandibular gland
   B. Inferior to the mylohyoid muscle
   C. Lateral to the body of the mandible
   D. In the mandibular vestibule area

2. Which of the following glands has both a superficial and deep lobe?
   A. Thymus gland
   B. Parotid gland
   C. Thyroid gland
   D. Sublingual gland
   E. Lacrimal gland

3. Which of the following nerves innervates both the submandibular and sublingual salivary glands?
   A. Trigeminal nerve
   B. Chorda tympani
   C. Hypoglossal nerve
   D. Vagus nerve

4. Which of the following nerves travels through the parotid salivary gland but is not involved in its innervation?
   A. Trigeminal nerve
   B. Facial nerve
   C. Vagus nerve
   D. Glossopharyngeal nerve

5. Which of the following glands shrinks as a person matures?
   A. Thymus gland
   B. Parotid gland
   C. Thyroid gland
   D. Sublingual gland
   E. Submandibular gland

6. Which gland has a duct that usually opens on the inner surface of the cheek, opposite the second maxillary molar?
   A. Thymus gland
   B. Parotid gland
   C. Thyroid gland
   D. Sublingual gland
   E. Submandibular gland

7. Which oral landmark marks the opening of the submandibular duct?
   A. Parotid raphe
   B. Lingual frenum
   C. Parotid papilla
   D. Sublingual caruncle
   E. Nasolacrimal duct

8. The thyroid gland is located
   A. Anterior to the larynx
   B. Superior to the hyoid bone
   C. Posterior to the surrounding pharynx
   D. In the posterior and medial neck region

9. Which of the following blood vessels supplies the parotid salivary gland?
   A. Facial artery
   B. Lingual artery
   C. Internal carotid artery
   D. External carotid artery

10. As endocrine glands, the parathyroid glands are known to
    A. Have one primary duct
    B. Have multiple secondary ducts
    C. Drain directly into blood vessels
    D. Drain directly into the thyroid gland

Chapter 8

Nervous System

▼ *This chapter discusses the following topics:*

I. Nervous System
   A. Autonomic Nervous System
   B. Cranial Nerves

II. Nerves to the Oral Cavity and Associated Structures
   A. Trigeminal Nerve
   B. Facial Nerve

III. Nerve Lesions of the Head and Neck

▼ *Key words*

Afferent nerve (**af**-er-int) Sensory nerve that carries information from the periphery of the body to the brain or spinal cord.

Bell's palsy (belz **pawl**-ze) Type of unilateral facial paralysis involving the facial nerve.

Efferent nerve (**ef**-er-ent) Motor nerve that carries information away from the brain or spinal cord to the periphery of the body.

Facial paralysis (pah-**ral**-i-sis) Loss of action of the facial muscles.

Ganglion/ganglia (**gang**-gle-in, **gang**-gle-ah) Accumulation of neuron cell bodies outside the central nervous system.

Innervation (in-er-**vay**-shin) Supply of nerves to tissues or organs.

Nerve (nurv) Bundle of neural processes outside the central nervous system, part of the peripheral nervous system.

Neuron (**noor**-on) Cellular component of the nervous system that is individually composed of a cell body and neural processes.

Synapse (**sin**-aps) Junction between two neurons or between a neuron and an effector organ, where neural impulses are transmitted by electrical or chemical means.

Trigeminal neuralgia (try-**jem**-i-nal noor-**al**-je-ah) Type of lesion of the trigeminal nerve involving facial pain.

Chapter 8

Nervous System

▼ *After studying this chapter, the reader should be able to:*

1. Define and pronounce all the key words and anatomical terms in this chapter.
2. Identify and trace the route of the cranial nerves from the skull on a series of diagrams.
3. Briefly discuss the general function of each of the cranial nerves.
4. Identify and trace the route of the nerves to the oral cavity and associated structures of the head and neck on a diagram, a skull, and a patient.
5. Describe the tissues innervated by each of the nerves of the head and neck.
6. Discuss certain nerve lesions associated with the head and neck region.
7. Correctly complete the identification exercise and review questions for this chapter.
8. Integrate the knowledge about the head and neck nerves into the functioning of the region's muscles, joints, and glands, the concepts of pain management with local anesthesia, and also the related nervous system disorders.

Nervous System

The nervous system causes muscles to contract, resulting in facial expressions and joint movements, such as in mastication and speech. The system stimulates glands to secrete and regulates many other systems of the body, such as the blood system. The nervous system also allows for sensation to be perceived, such as pain and touch.

The nervous system and its components are important for the dental professional to know since they allow for the function of the muscles, temporomandibular joint, and glands of the head and neck (see Chapters 4, 5, and 7 for more information). A thorough understanding of certain nerves also is important in pain management using local anesthesia during dental treatment (see Chapter 9 for more information). Finally, there are certain related nervous system disorders of the head and neck that need to be known by the dental professional (discussed later in this chapter).

The nervous system has two main divisions, the central nervous system and peripheral nervous system. These two parts of the nervous system are constantly interacting. The **central nervous system** consists of the spinal cord and the brain and will not be discussed in this textbook.

The **peripheral nervous system** (per-**if**-er-al) consists of the spinal and cranial nerves and includes the autonomic nervous system. The spinal nerves extend from the spinal cord to the periphery of the body but are not discussed in this textbook. The cranial nerves and autonomic nervous system are discussed in this chapter.

The ***neuron*** (**noor**-on) is the cellular component of the nervous system and is composed of a cell body and neural processes (sometimes called nerve fibers). A ***nerve*** (nurv) is a bundle of neural processes outside the central nervous system and in the peripheral nervous system. A ***synapse*** (**sin**-aps) is the junction between two neurons or between a neuron and an effector organ, where neural impulses are transmitted by electrical or chemical means.

In order to function, most tissues or organs have ***innervation*** (in-er-**vay**-shin), a supply of nerves to the part. A nerve allows information to be carried to and from the brain, which is the central informational center. An accumulation of neuron cell bodies outside the central nervous system is a ***ganglion*** (plural, ***ganglia***) (**gang**-gle-in, **gang**-gle-ah).

There are two types of nerves, afferent and efferent. Many other discussions of these nerves use the terms sensory or motor because afferent and efferent sound similar to many students. In this text, the internationally recognized terms afferent and efferent are used.

An ***afferent nerve*** (**af**-er-int) or sensory nerve carries information from the periphery of the body to the brain (or spinal cord). Thus an afferent nerve carries sensory information such as taste, pain, or proprioception to the brain. Proprioception is information concerning the movement and position of the body. This information is sent to the brain to be analyzed, acted upon, associated with other information, and stored as memory.

An ***efferent nerve*** (**ef**-er-ent) or motor nerve carries information away from the brain to the periphery of the body. Thus an efferent nerve carries information to the muscles in order to activate them, often in response to information received by way of the afferent nerves. One motor neuron with its branching fibers may control hundreds of muscle fibers. Autonomic nerves are (by definition) always efferent. Recent research in nonhuman primates shows that most efferent nerves may also carry some fibers for proprioception, which is likely to occur in humans also.

AUTONOMIC NERVOUS SYSTEM

The **autonomic nervous system** (awt-o-**nom**-ik) is a part of the peripheral nervous system. This system operates without conscious control as the caretaker of the body. Autonomic fibers are efferent nerves and are always in two-nerve chains. The first nerve carries autonomic fibers to a ganglion where they end near the cell bodies of the second nerve. The autonomic nervous system has two parts, the sympathetic and parasympathetic systems. Most tissues or organ systems are supplied by both parts of the autonomic nervous system.

The **sympathetic nervous system** (sim-pah-**thet**-ik) is involved in "fight or flight" responses, such as the shutdown of salivary gland secretion. Thus such a response by the sympathetic system leads to a dry mouth (xerostomia). Sympathetic nerves arise in the spinal cord and relay in ganglia arranged like a chain running up the neck close to the vertebral column on both sides. Therefore, all the sympathetic neurons in the head have already relayed in a ganglion. Sympathetic fibers reach the cranial tissues they supply by traveling with the arteries.

The **parasympathetic nervous system** (pare-ah-sim-pah-**thet**-ik) is involved in "rest or digest" responses, such as the stimulation of salivary gland secretion. Thus such a response by the parasympathetic system leads to salivary flow to aid in digestion.

Parasympathetic fibers associated with glands of the head and neck region are carried in various cranial nerves and are briefly described here, as well as in more detail later in this chapter. Their ganglia are located in the head, and therefore parasympathetic neurons in this region may be either preganglionic neurons (before relaying in the ganglion) or postganglionic neurons (after relaying in the ganglion).

The principal parasympathetic outflow for glands in the head and neck is carried in the seventh and ninth cranial nerves. The seventh cranial or facial nerve has two branches involved in glandular secretion. The greater petrosal nerve is associated with the pterygopalatine ganglion, with the lacrimal gland being the major target organ. The chorda tympani nerve is associated with the submandibular ganglion, and the target organs are the submandibular and sublingual salivary glands. The lesser petrosal nerve, a branch of the ninth cranial or glossopharyngeal nerve, is associated with the otic ganglion, and the target organ is the parotid salivary gland.

CRANIAL NERVES

The **cranial nerves** (**kray**-nee-al) are an important part of the peripheral nervous system. There are 12 paired cranial nerves, and all are connected to the brain at its base and pass through the skull by way of fissures or foramina (Figures 8–1 and 8–2, Table 8–1) (see Chapter 3).

Some cranial nerves are either afferent or efferent, and others have both types of neural processes. A general background of all 12 cranial nerves is discussed in this chapter (Figure 8–1, Table 8–1). Roman numerals and anatomical terms are used to designate the cranial nerves.

Cranial Nerve I

The first cranial nerve, cranial nerve I or the **olfactory nerve** (ol-**fak**-ter-ee), transmits smell from the nasal mucosa to the brain and thus functions as an afferent nerve. The olfactory nerve enters the skull through the perforations in the cribriform plate of the ethmoid bone to join the olfactory bulb in the brain.

Cranial Nerve II

The second cranial nerve, cranial nerve II or **optic nerve** (**op**-tik), transmits sight from the retina of the eye to the brain and thus functions as an afferent nerve. The optic nerve enters the skull through the optic canal of the sphenoid bone on its way from the retina. In the skull, the right and left optic nerves join at the optic chiasma, where many of the fibers cross to the opposite side before continuing into the brain as the optic tracts.

Cranial Nerve III

The third cranial nerve, cranial nerve III or **oculomotor nerve** (ok-yule-oh-**mote**-er), serves as an efferent nerve to some of the eye muscles that move the eyeball. The oculomotor nerve also carries preganglionic parasympathetic fibers to the ciliary ganglion near the eyeball. The postganglionic fibers innervate small muscles inside the eyeball. The oculomotor nerve lies in the lateral wall of the cavernous sinus and exits the skull through the superior orbital fissure of the sphenoid bone on its way to the orbit.

Cranial Nerve IV

The very small fourth cranial nerve, cranial nerve IV or **trochlear nerve** (**trok**-lere), also serves as an efferent nerve for one eye muscle, as well as proprioception, similar to the oculomotor nerve but without any parasympathetic fibers. Similar to the oculomotor nerve, the trochlear nerve runs in the lateral wall of the cavernous sinus and exits the skull through the superior orbital fissure of the sphenoid bone on its way to the orbit.

Cranial Nerve V

The fifth cranial nerve, cranial nerve V or **trigeminal nerve** (try-**jem**-i-nal), has both an efferent component for the muscles of mastication, as well as some other cranial muscles, and an afferent component for the teeth, tongue, and oral cavity, as well as most of the skin of the face and head. Although the trigeminal has

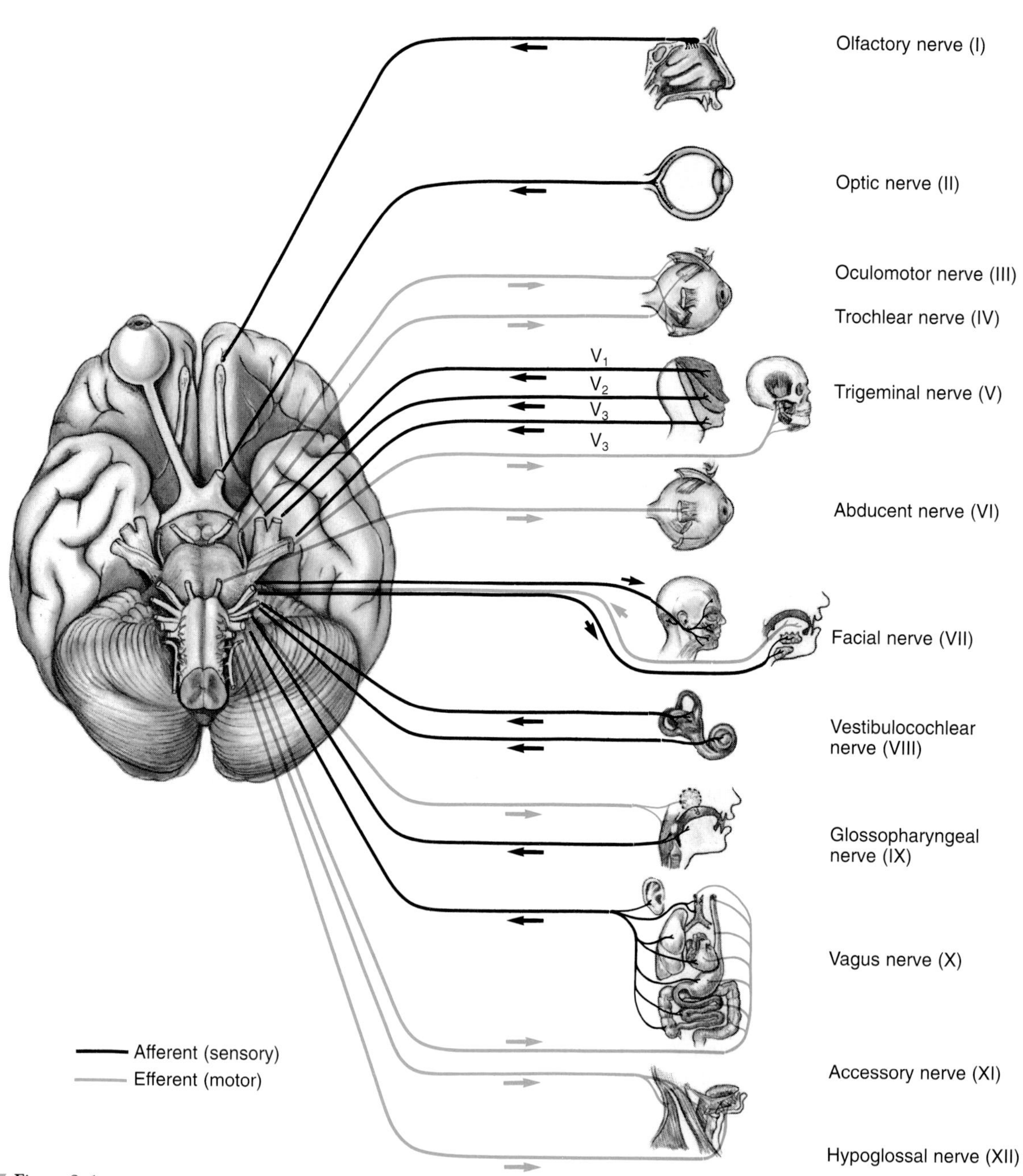

Figure 8–1
Inferior view of the brain showing cranial nerves and the organs and tissues they innervate.

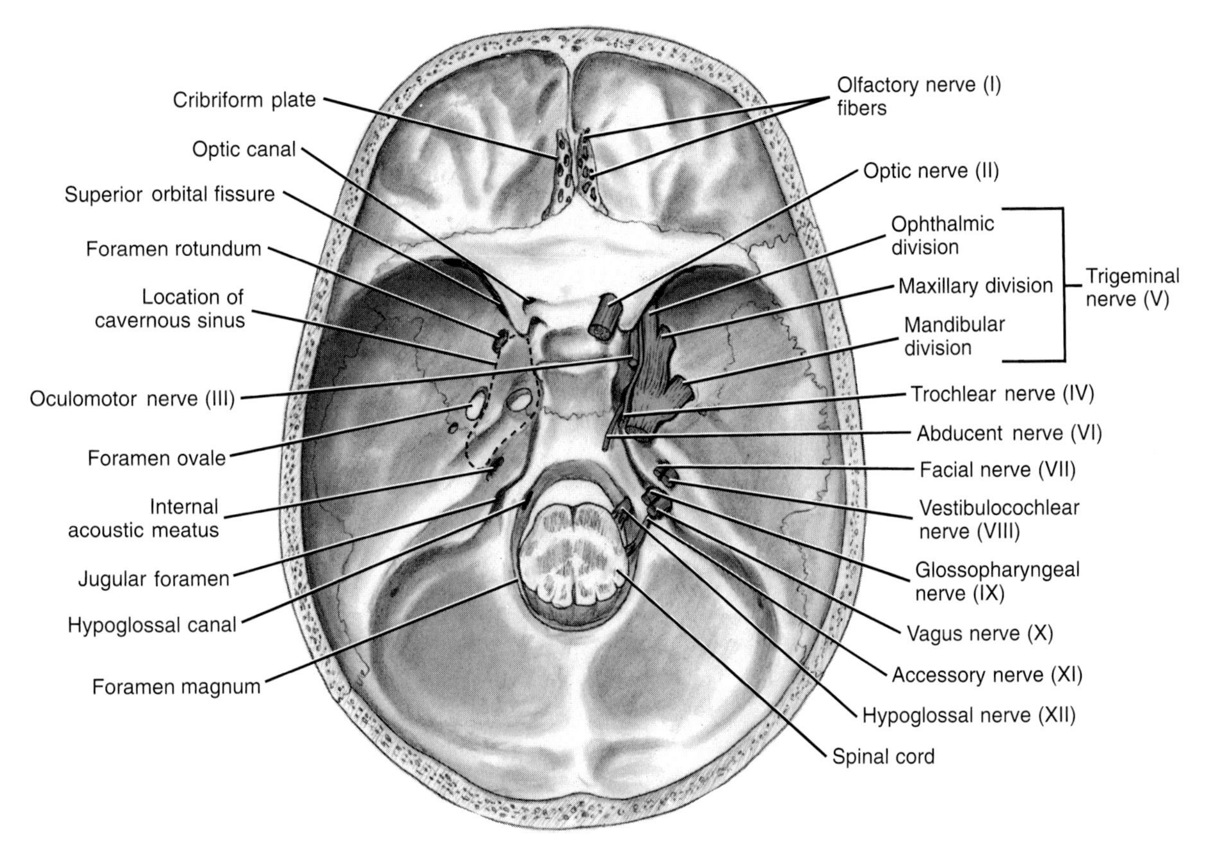

Figure 8–2
Internal view of the base of the skull showing cranial nerves exiting or entering the skull.

Table 8–1

TWELVE CRANIAL NERVES WITH THE NERVE TYPE AND TISSUES INNERVATED

Nerve	Nerve Type and Tissues Innervated
I: Olfactory	Afferent: nasal mucosa
II: Optic	Afferent: retina of the eye
III: Oculomotor	Efferent: eye muscles
IV: Trochlear	Efferent: eye muscles
V: Trigeminal	Efferent: muscles of mastication and other cranial muscles Afferent: face and head skin, teeth, oral cavity, and tongue (general sensation)
VI: Abducens	Efferent: eye muscles
VII: Facial	Efferent: muscles of facial expression, other cranial muscles, and lacrimal, submandibular, sublingual, and minor glands (parasympathetic) Afferent: skin around ear and tongue (taste sensation)
VIII: Vestibulocochlear	Afferent: inner ear
IX: Glossopharyngeal	Efferent: stylopharyngeus muscle and parotid gland (parasympathetic) Afferent: skin around ear and tongue (taste and general sensation)
X: Vagus	Efferent: muscles of soft palate, pharynx, and larynx and thorax and abdominal organs (parasympathetic) Afferent: skin around ear and epiglottis (taste sensation)
XI: Accessory	Efferent: muscles of neck, soft palate, and pharynx
XII: Hypoglossal	Efferent: tongue muscles

no preganglionic parasympathetic fibers, many postganglionic parasympathetic fibers travel along with its branches.

The trigeminal nerve is the largest cranial nerve and has two roots, sensory and motor roots (Figure 8–3). The **sensory root of the trigeminal nerve** has three divisions: the ophthalmic, maxillary, and mandibular divisions. The ophthalmic division provides sensation to the upper face and scalp. The maxillary and mandibular divisions provide the middle and lower face, respectively.

Each division of the sensory root of the nerve enters the skull in a different location in the sphenoid bone. The ophthalmic division enters through the superior orbital fissure. The maxillary division enters by way of the foramen rotundum. The mandibular division passes through the skull by way of the foramen ovale.

The **motor root of the trigeminal nerve** accompanies the mandibular division of the sensory root and also exits the skull through the foramen ovale of the sphenoid bone.

The trigeminal nerve is the most important cranial nerve to the dental professional since it innervates many relevant tissues of the head and neck (Table 8–2). The trigeminal nerve is discussed in greater detail later in this chapter.

Cranial Nerve VI

The sixth cranial nerve, cranial nerve VI or abducent nerve, **abducens** (ab-**doo**-senz), serves as an efferent nerve to one of the muscles that moves the eyeball, similar to the oculomotor and trochlear nerves. Similar to both of those cranial nerves, the abducens exits the skull through the superior orbital fissure of the sphenoid bone on its way to the orbit.

However, the abducens has a somewhat different intracranial course. Rather than lying in the wall of the cavernous sinus, the nerve runs through the sinus, close to the internal carotid artery, and is often the first

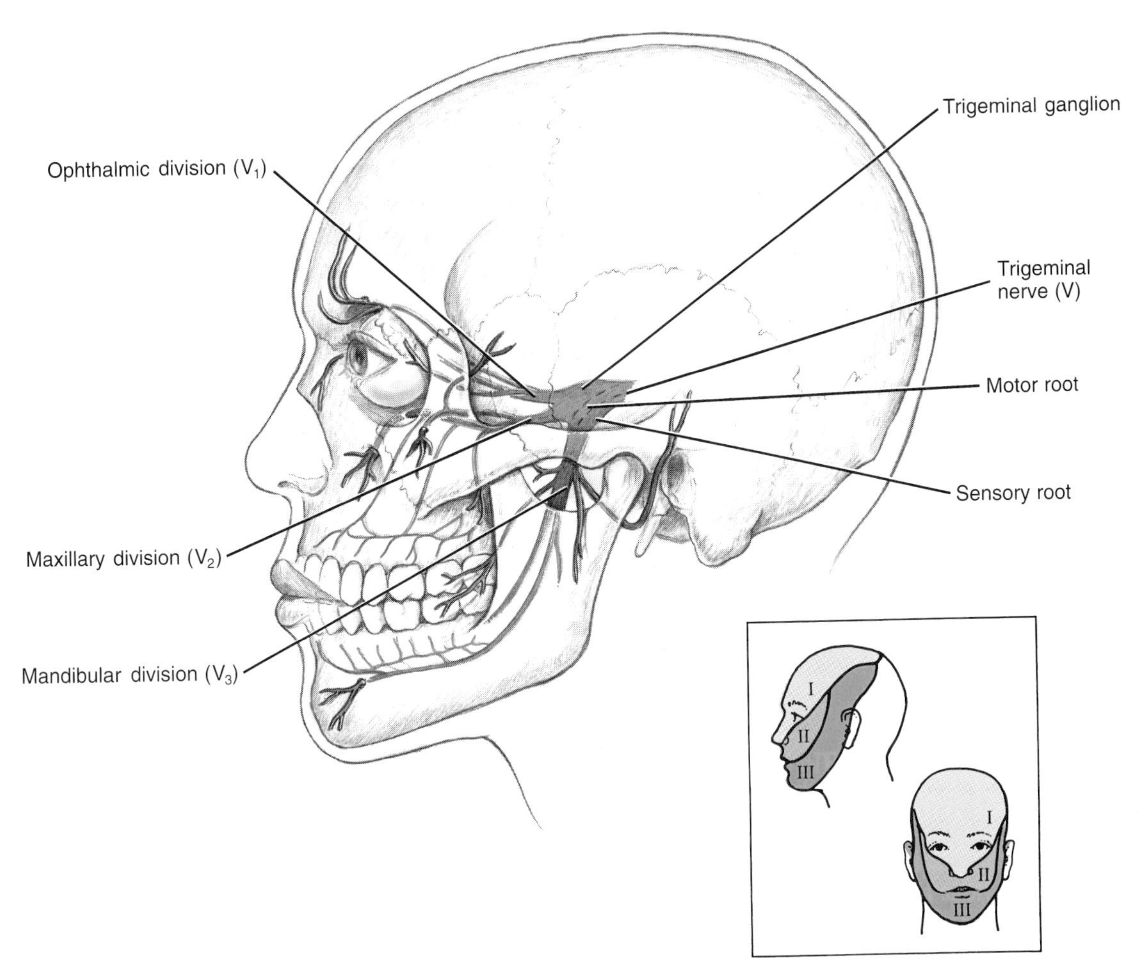

▼ **Figure 8–3**
The general pathway of the trigeminal or fifth cranial nerve and its motor and sensory roots and three divisions (inset shows the pattern of innervation for each division).

nerve affected by infection of the sinus (see Chapter 12 for more information).

▼ Cranial Nerve VII

The seventh cranial nerve, cranial nerve VII or **facial nerve,** carries both efferent and afferent components. The facial nerve carries an efferent component for the muscles of facial expression and for the preganglionic parasympathetic innervation of the lacrimal gland (relaying in the pterygopalatine ganglion) and submandibular and sublingual salivary glands (relaying in the submandibular ganglion). The afferent component serves a tiny patch of skin behind the ear, taste sensation, and the body of the tongue.

The facial nerve leaves the cranial cavity by passing through the internal acoustic meatus, which leads to the facial canal inside the temporal bone. Finally, the facial nerve exits the skull by way of the stylomastoid foramen of the temporal bone. The facial nerve is also important to dental professionals since it innervates many relevant tissues of the head and neck (Table 8–2). The facial nerve is discussed in more detail later in this chapter.

▼ Cranial Nerve VIII

The eighth cranial nerve, cranial nerve VIII or **vestibulocochlear nerve** (ves-tib-you-lo-**kok**-lere), serves as an afferent nerve for hearing and balance.

Ophthalmic Division of the Trigeminal Nerve (Ophthalmic Nerve)

The first and smallest division of the sensory root is the ophthalmic division of the trigeminal nerve, the **ophthalmic nerve** (of-**thal**-mik) or V_1 (Figure 8–4). The ophthalmic nerve serves as an afferent nerve for the conjunctiva, cornea, eyeball, orbit, forehead, and ethmoid and frontal sinuses, plus a portion of the dura mater.

The ophthalmic nerve carries this sensory information toward the brain by way of the superior orbital fissure of the sphenoid bone. Other nerves that traverse this fissure include cranial nerves III, IV, and VI. The ophthalmic nerve arises from three major nerves: the frontal, lacrimal, and nasociliary nerves.

FRONTAL NERVE

The **frontal nerve** (**frunt**-il) is an afferent nerve located in the orbit and is composed of a merger of the **supraorbital nerve** (soo-prah-**or**-bit-al) from the forehead and anterior scalp and the **supratrochlear nerve** (soo-prah-**trok**-lere) from the bridge of the nose and medial parts of the upper eyelid and forehead. The frontal nerve courses along the roof of the orbit toward the superior orbital fissure of the sphenoid bone where it is joined by the lacrimal and nasociliary nerves to form the ophthalmic nerve or V_1.

LACRIMAL NERVE

The **lacrimal nerve** (**lak**-ri-mal) serves as an afferent nerve for the lateral part of the upper eyelid, conjunctiva, and lacrimal gland. The nerve also delivers the postganglionic parasympathetic nerves to the lacrimal gland. These nerves are responsible for the production of lacrimal fluid or tears. The lacrimal nerve runs posteriorly along the lateral roof of the orbit and then joins the frontal and nasociliary nerves near the superior orbital fissure of the sphenoid bone to form the ophthalmic nerve or V_1.

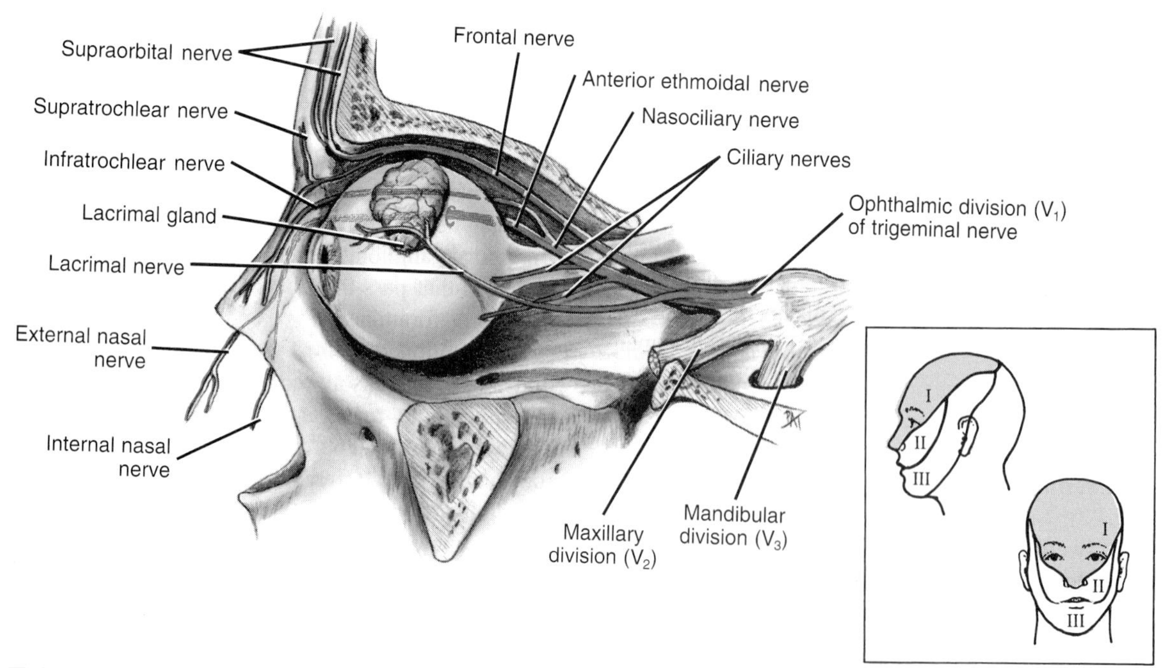

Figure 8–4
Lateral view of the cut-away orbit with the pathway of the ophthalmic division of the trigeminal nerve highlighted.

▼ NASOCILIARY NERVE

Several afferent nerve branches converge to form the **nasociliary nerve** (nay-zo-**sil**-eh-a-re). These branches include the **infratrochlear nerve** (in-frah-**trok**-lere) from the skin of the medial part of the eyelids and the side of the nose, **ciliary nerves** (**sil**-eh-a-re) to and from the eyeball, and **anterior ethmoidal nerve** (eth-**moy**-dal) from the nasal cavity and paranasal sinuses. The anterior ethmoidal nerve is formed by the **external nasal nerve** (**nay**-zil) from the skin of the ala and apex of the nose and the **internal nasal nerves** from the anterior part of the nasal septum and lateral wall of the nasal cavity.

The nasociliary nerve is an afferent nerve that runs within the orbit, superior to cranial nerve II, the optic nerve, to join the frontal and lacrimal nerves near the superior orbital fissure of the sphenoid bone to form the ophthalmic nerve or V_1.

▼ Maxillary Division of the Trigeminal Nerve (Maxillary Nerve)

The second division from the sensory root is the maxillary division of the trigeminal nerve, the **maxillary nerve** (**mak**-sil-ar-ee) or V_2 (Figure 8–5). The afferent nerve branches of the maxillary nerve carry sensory information for the maxilla and overlying skin, maxillary sinuses, nasal cavity, palate, and nasopharynx and a portion of the dura mater.

The maxillary nerve is a nerve trunk formed in the pterygopalatine fossa by the convergence of many nerves. The largest contributor is the infraorbital nerve. Tributaries of the infraorbital nerve or maxillary nerve trunk include the zygomatic, anterior, middle and posterior superior alveolar, greater and lesser palatine, and nasopalatine nerves.

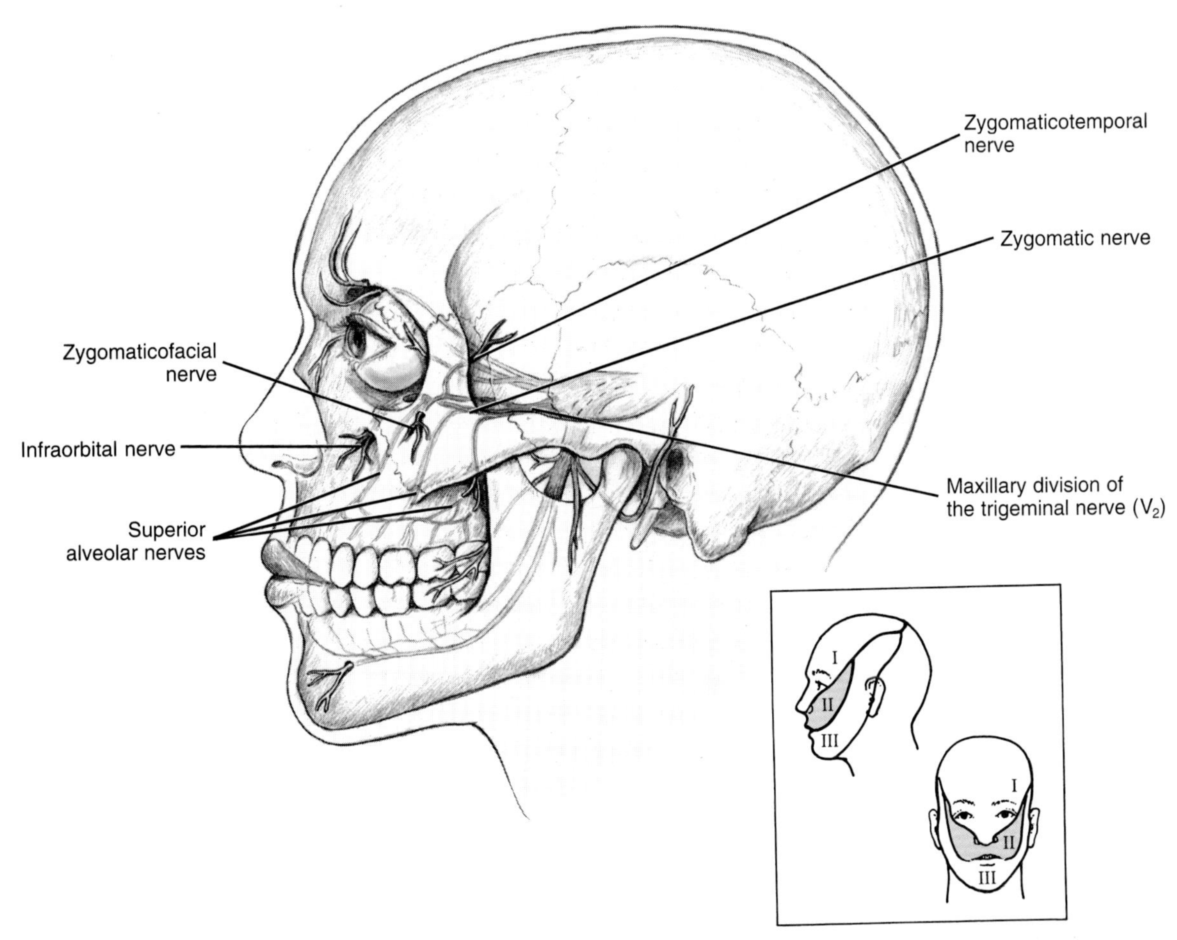

Figure 8–5
The pathway of the maxillary division of the trigeminal nerve highlighted.

After all these branches come together in the pterygopalatine fossa to form the maxillary nerve and thus V_2 of the trigeminal nerve, the maxillary nerve enters the skull through the foramen rotundum of the sphenoid bone. Small afferent meningeal branches (me-**nin**-je-al) from portions of the dura mater join the maxillary division as it enters the trigeminal ganglion.

Another ganglion, the **pterygopalatine ganglion** (teh-ri-go-**pal**-ah-tine), lies just below the maxillary nerve in the pterygopalatine fossa. This ganglion serves as a relay station for parasympathetic nerves that arise in the facial nerve (described later). Fibers from the ganglion (postganglionic) are then distributed to the various tissues, such as the minor salivary glands, by the nerves of the maxillary division.

Because the pterygopalatine ganglion lies between the maxillary nerve and its tributaries from the palate, the sensory fibers actually pass through the ganglion. However, unlike the parasympathetic fibers, the sensory fibers do not synapse in the ganglion.

▼ ZYGOMATIC NERVE

The **zygomatic nerve** (zy-go-**mat**-ik) is an afferent nerve composed of the merger of the zygomaticofacial nerve and the zygomaticotemporal nerve in the orbit (Figure 8–5). This nerve also conveys the postganglionic parasympathetic fibers for the lacrimal gland to the lacrimal nerve. The zygomatic nerve courses posteriorly along the lateral orbit floor, enters the pterygopalatine fossa through the inferior orbital fissure, between the sphenoid bone and maxilla, and finally joins the maxillary nerve.

The rather small **zygomaticofacial nerve** (zy-go-mat-i-ko-**fay**-shal) serves as an afferent nerve for the skin of the cheek. This nerve pierces the frontal process of the zygomatic bone and enters the orbit through its lateral wall. The zygomaticofacial nerve then turns posteriorly to join with the zygomaticotemporal nerve.

The other nerve, the **zygomaticotemporal nerve** (zy-go-mat-i-ko-**tem**-poh-ral) serves as an afferent nerve for the skin of the temporal region, pierces the temporal surface of the zygomatic bone, and traverses the lateral wall of the orbit to join the zygomaticofacial nerve, forming the zygomatic nerve.

▼ INFRAORBITAL NERVE

The **infraorbital nerve** (in-frah-**or**-bit-al) or **IO nerve** is an afferent nerve formed from the merger of cutaneous branches from the upper lip, medial portion of the cheek, lower eyelid, and side of the nose (Figure 8–6). The IO nerve then passes into the infraorbital foramen of the maxilla and travels posteriorly through the infraorbital canal, along with the infraorbital blood vessels, where it is joined by the anterior superior alveolar nerve.

From the infraorbital canal and groove, the IO nerve passes into the pterygopalatine fossa through the inferior orbital fissure. After it leaves the infraorbital groove and within the pterygopalatine fossa, the IO nerve receives the posterior superior alveolar (PSA) nerve.

▼ ANTERIOR SUPERIOR ALVEOLAR NERVE

The **anterior superior alveolar nerve** (al-**ve**-o-lar) or **ASA nerve** serves as an afferent nerve of sensation (including pain) for the maxillary central incisors, lateral incisors, and maxillary canines, as well as their associated tissues.

The anterior superior alveolar nerve originates from dental branches in the pulp tissue of these teeth that exit through the apical foramina (Figure 8–6). The nerve also receives interdental branches from the surrounding periodontium, forming a dental plexus or nerve network in the maxilla for the region. The ASA nerve also innervates the overlying facial gingiva. The ASA nerve then ascends along the anterior wall of the maxillary sinus to join the infraorbital nerve in the infraorbital canal.

Many times the anterior superior alveolar nerve crosses over the midline to the opposite side in a patient. This is important to consider when giving a local anesthetic injection for the maxillary anterior teeth or associated tissues (see Chapter 9).

▼ MIDDLE SUPERIOR ALVEOLAR NERVE

The **middle superior alveolar nerve** (al-**ve**-o-lar) or **MSA nerve** serves as an afferent nerve of sensation (including pain), typically for the maxillary premolar teeth and mesial buccal root of the maxillary first molar and their associated periodontium and overlying buccal gingiva.

The middle superior alveolar nerve originates from dental branches in the pulp tissue that exit the teeth through the apical foramina, as well as interdental and interradicular branches from the periodontium (Figure 8–6). The MSA nerve, like the PSA and

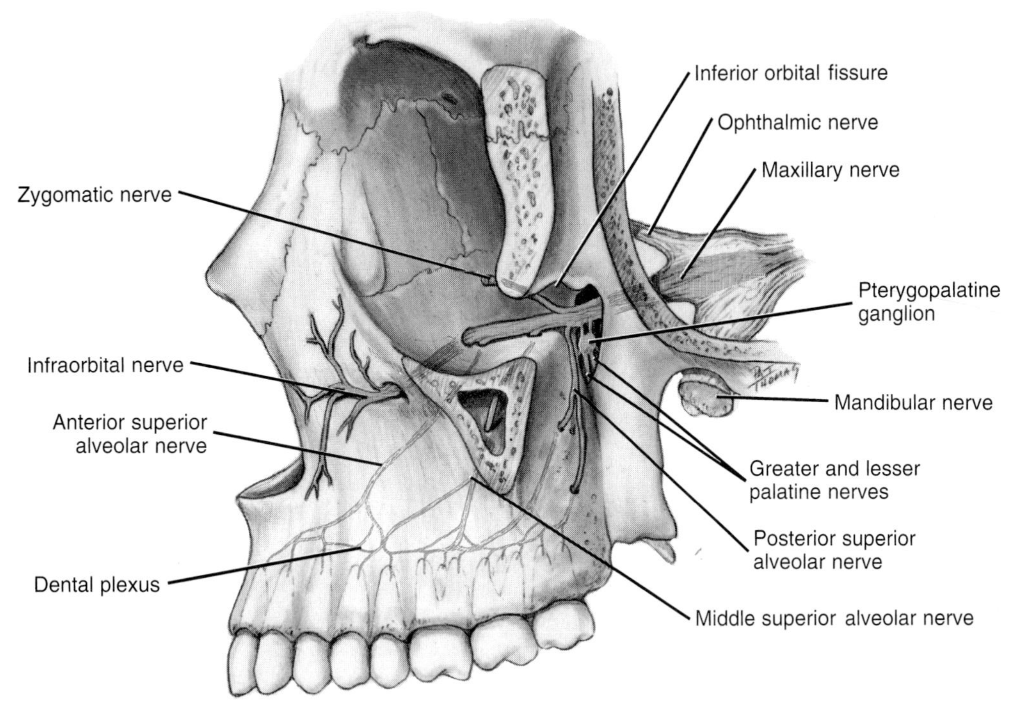

▼ **Figure 8–6**
Lateral view of the skull (a portion of the lateral wall of the orbit removed) with the branches of the maxillary nerve highlighted.

ASA nerves, forms the dental plexus or nerve network in the maxilla. The MSA nerve then ascends to join the infraorbital nerve by running in the lateral wall of the maxillary sinus.

The middle superior alveolar nerve is not present in all patients. If this nerve is not present, the area is innervated by both the ASA and PSA nerves, but mainly the ASA nerve. If the MSA nerve is present, there is communication between the MSA nerve and both the ASA and PSA nerves. These considerations are important when giving local anesthetic injections to the maxillary teeth (see Chapter 9).

▼ POSTERIOR SUPERIOR ALVEOLAR NERVE

The **posterior superior alveolar nerve** (al-**ve**-o-lar) or **PSA nerve** joins the infraorbital nerve (or the maxillary nerve directly) in the pterygopalatine fossa (Figure 8–6). The PSA nerve serves as an afferent nerve of sensation (including pain) for most portions of the maxillary molar teeth and their periodontium and buccal gingiva, as well as the maxillary sinus.

Some branches of the posterior superior alveolar nerve remain external to the posterior surface of the maxilla. These branches provide afferent innervation for the buccal gingiva overlying the maxillary molars.

Other afferent nerve branches of the posterior superior alveolar nerve originate from dental branches in the pulp tissue of each of the maxillary molar teeth that exit the teeth by way of the apical foramina. These dental branches are joined by interdental and interradicular branches from the periodontium, forming a dental plexus or nerve network in the maxilla for the region.

All these internal branches of the PSA nerve exit from several posterior superior alveolar foramina on the maxillary tuberosity of the maxilla. The posterior superior alveolar arteries (from the maxillary artery) enter the maxillary tuberosity through these same foramina.

Both the external and internal branches of the posterior superior alveolar nerve then ascend together along the maxillary tuberosity, which forms the posterolateral wall of the maxillary sinus, to join the IO nerve or maxillary nerve. The PSA nerve typically provides afferent innervation for the maxillary second and third molars and the palatal and distal buccal root of the maxillary first molar, as well as the mucous membranes of the maxillary sinus.

▼ GREATER AND LESSER PALATINE NERVES

Both palatine nerves join with the maxillary nerve from the palate (Figure 8–7). The **greater palatine nerve** (**pal**-ah-tine) or anterior palatine nerve is located between the mucoperiosteum and bone of the anterior hard palate. This nerve serves as an afferent nerve for the posterior hard palate and posterior lingual gingiva. There is communication with the terminal fibers of the nasopalatine nerve in the hard palate area, lingual to the maxillary canines.

Posteriorly, the greater palatine nerve enters the greater palatine foramen in the palatine bone near the maxillary second or third molar to travel in the pterygopalatine canal, along with the greater palatine blood vessels.

The **lesser palatine nerve** (**pal**-ah-tine) or posterior palatine nerve serves as an afferent nerve for the soft palate and palatine tonsillar tissues. The lesser palatine nerve enters the lesser palatine foramen in the palatine bone near its junction with the pterygoid process of the sphenoid bone, along with the lesser palatine blood vessels (Figure 8–7). The lesser palatine nerve then joins the greater palatine nerve in the pterygopalatine canal.

Both palatine nerves ascend through the pterygopalatine canal, toward the maxillary nerve in the pterygopalatine fossa. On the way, the palatine nerves are joined by lateral nasal branches (**nay**-zil), which are afferent nerves from the posterior nasal cavity.

▼ NASOPALATINE NERVE

The **nasopalatine nerve** (nay-zo-**pal**-ah-tine) originates in the mucosa of the anterior hard palate, lingual to the maxillary anterior teeth (Figure 8–7). The right and left nasopalatine nerves enter the incisive canal by way of the incisive foramen, thus exiting the oral cavity. The nerve then travels along the nasal septum. The nasopalatine nerve serves as an afferent nerve for the anterior hard palate and the lingual gingiva of the maxillary anterior teeth, as well as the nasal septal tissues. There is also communication with the greater palatine nerve in the area lingual to the maxillary canines.

▼ Mandibular Division of the Trigeminal Nerve (Mandibular Nerve)

The mandibular division of the trigeminal nerve, the **mandibular nerve** (man-**dib**-you-lar) or V_3, is a short

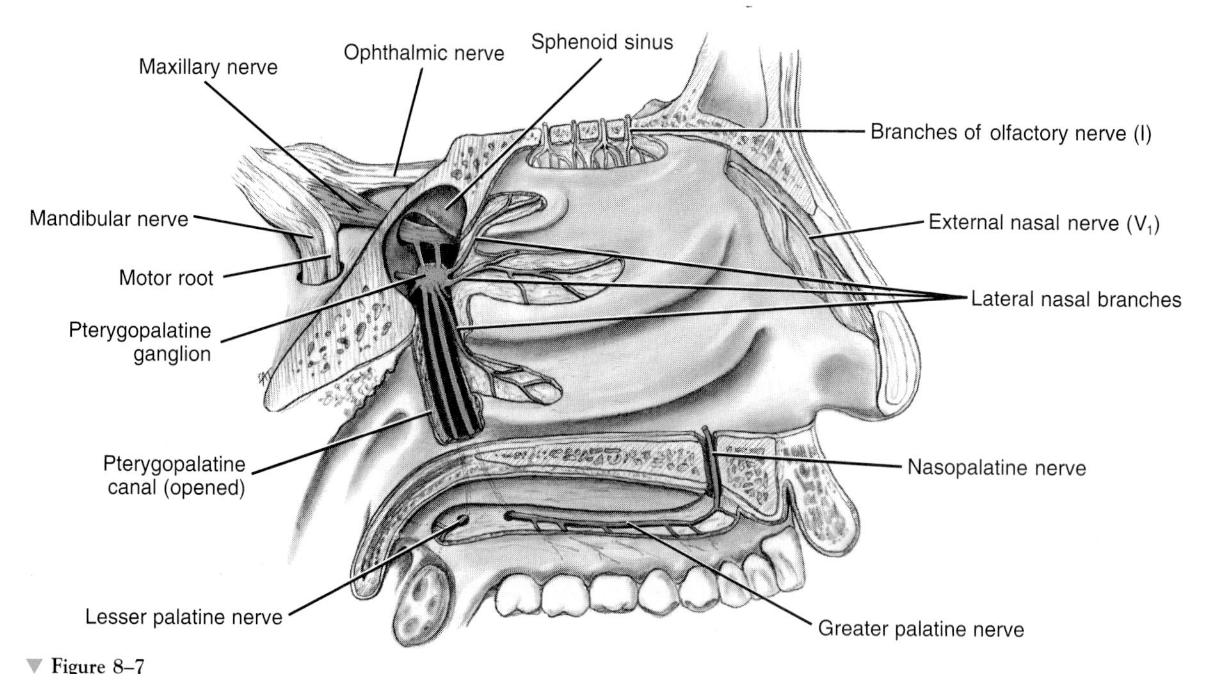

▼ **Figure 8–7**
Medial view of the lateral nasal wall and opened pterygopalatine canal highlighting the maxillary nerve and its palatine branches. The nasal septum has been removed, thus severing the nasopalatine nerve.

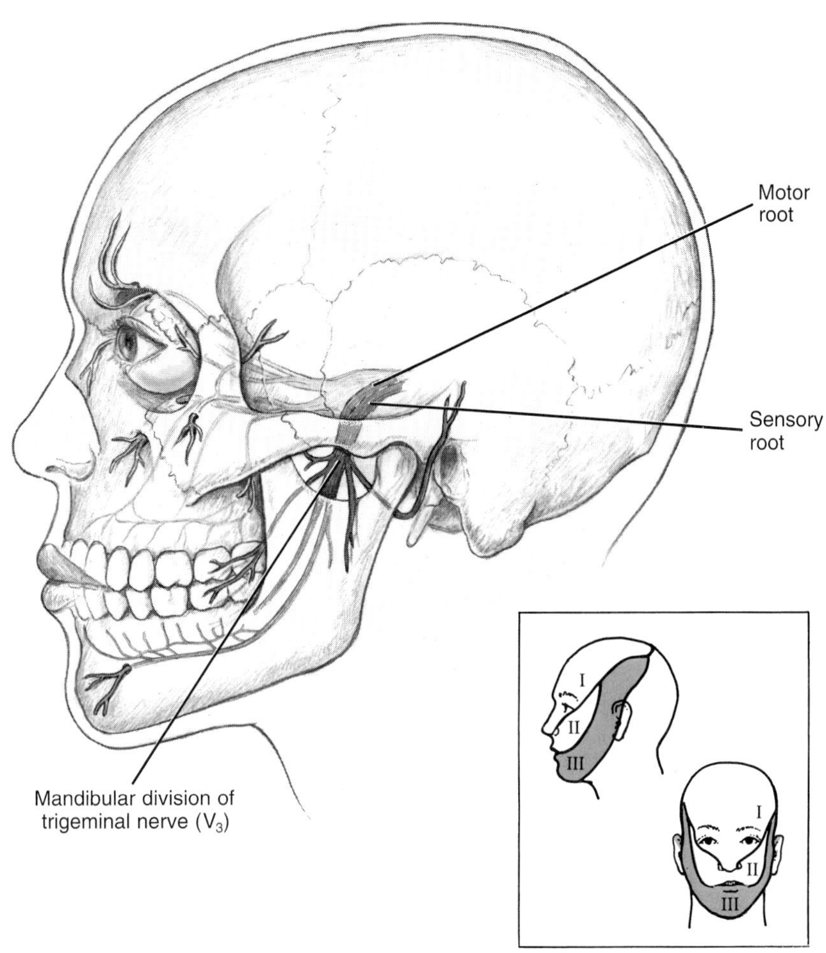

▼ **Figure 8–8**
The pathway of the mandibular division of the trigeminal nerve highlighted.

main trunk formed by the merger of a smaller anterior trunk and a larger posterior trunk in the infratemporal fossa before the nerve passes through the foramen ovale of the sphenoid bone (Figure 8–8). The mandibular nerve then joins with the ophthalmic and maxillary nerves to form the trigeminal ganglion of the trigeminal nerve. The mandibular nerve is the largest of the three divisions that form the trigeminal nerve.

A few small branches arise from the V_3 trunk before its separation into the anterior and posterior trunks (Figure 8–11). These branches from the undivided mandibular division include the meningeal branches (me-**nin**-je-al), which are afferent nerves for portions of the dura mater. Also from the undivided mandibular division are muscular branches, which are efferent nerves for the medial pterygoid, tensor tympani, and tensor veli palatini muscles.

The anterior trunk of the mandibular division is formed by the merger of the buccal nerve and additional muscular nerve branches (Figure 8–9). The posterior trunk of the mandibular division is formed by the merger of the auriculotemporal, lingual, and inferior alveolar nerves (Figure 8–10).

▼ BUCCAL NERVE

The **buccal nerve** or long buccal nerve serves as an afferent nerve for the skin of the cheek, buccal mucous membranes, and buccal gingiva of the mandibular posterior teeth. The nerve is located on the surface of the buccinator muscle (Figure 8–9). The buccal nerve then travels posteriorly in the cheek, deep to the masseter muscle.

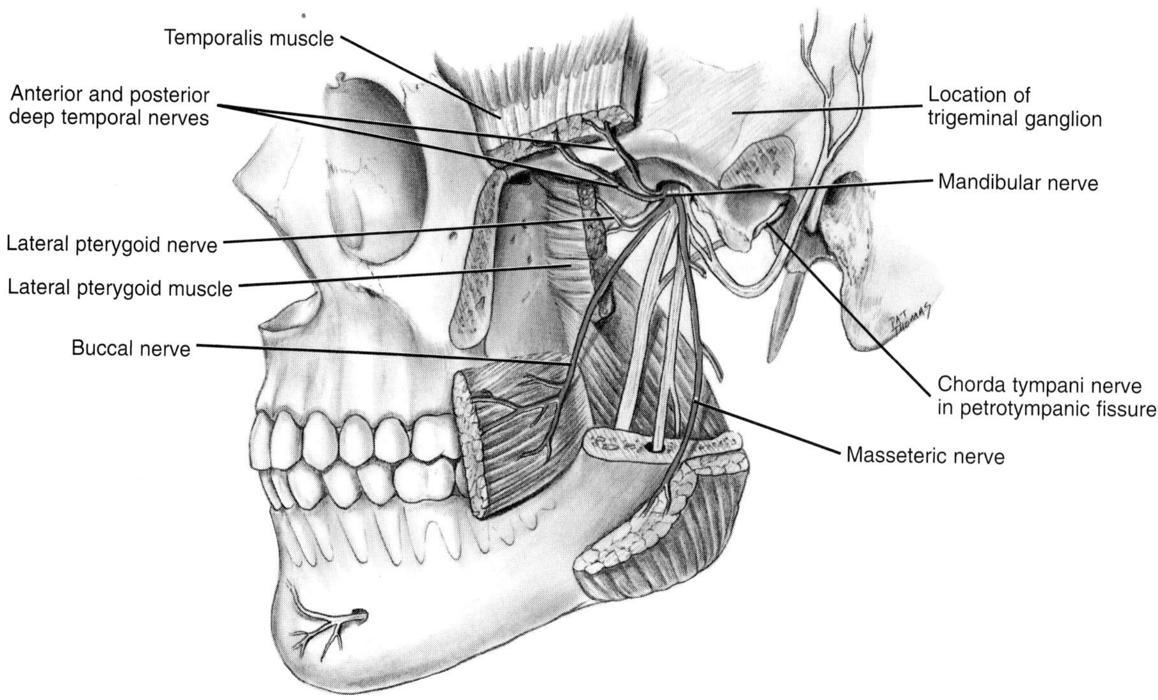

Figure 8–9
The pathway of the anterior trunk of the mandibular division of the trigeminal nerve highlighted.

At the level of the occlusal plane of the last mandibular molar, the nerve crosses in front of the anterior border of the ramus and goes between the two heads of the lateral pterygoid muscle to join the anterior trunk of V_3. It is important not to confuse this nerve with the nerve to the buccinator muscle, an efferent nerve branch from the facial nerve.

▼ MUSCULAR BRANCHES

There are several muscular branches that are part of the anterior trunk of the V_3 (Figure 8–9). They arise from the motor root of the trigeminal nerve. The **deep temporal nerves** (**tem**-poh-ral), usually two in number, anterior and posterior, are efferent nerves that pass between the sphenoid bone and the superior border of the lateral pterygoid muscle and turn around the infratemporal crest of the sphenoid bone to end in the deep surface of the temporal muscle that they innervate. The posterior temporal nerve may arise in common with the masseteric nerve, and the anterior temporal nerve may be associated at its origin with the buccal nerve.

The **masseteric nerve** (mass-et-**tehr**-ik) is also an efferent nerve that passes between the sphenoid bone and the superior border of the lateral pterygoid muscle. The nerve then accompanies the masseteric blood vessels through the mandibular notch to innervate the masseter muscle. A small sensory branch also goes to the temporomandibular joint. The **lateral pterygoid nerve** (**teh**-ri-goid), after a short course, enters the deep surface of the lateral pterygoid muscle between the muscle's two heads of origin and serves as an efferent nerve for that muscle.

▼ AURICULOTEMPORAL NERVE

The nerve known as the **auriculotemporal nerve** (aw-rik-yule-lo-**tem**-poh-ral) travels with the superficial temporal artery and vein and serves as an afferent nerve for the external ear and scalp (Figures 8–10 and 8–11). The nerve also carries postganglionic parasympathetic nerve fibers to the parotid salivary gland. It is important to note that these parasympathetic fibers arise from the lesser petrosal branch of the glossopharyngeal nerve or ninth cranial nerve, joining the

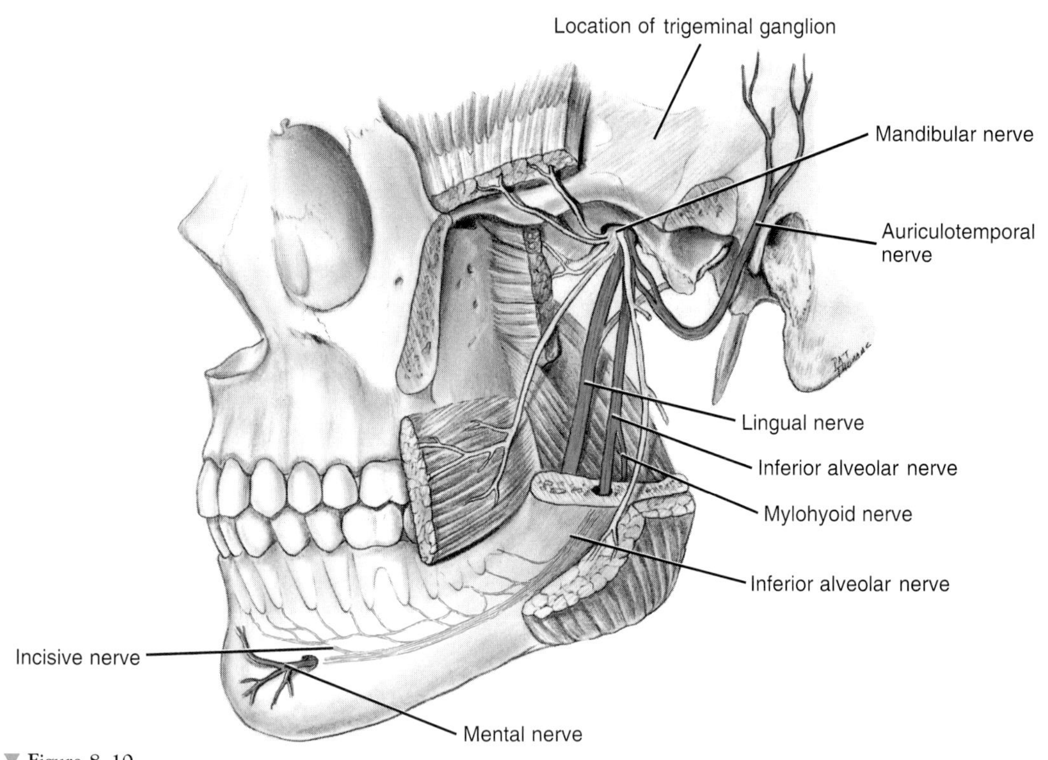

▼ **Figure 8–10**
The pathway of the posterior trunk of the mandibular division of the trigeminal nerve highlighted.

auriculotemporal nerve only after relaying in the otic ganglion near the foramen ovale.

There is also communication of the auriculotemporal nerve with the facial nerve near the ear. The nerve courses deep to the lateral pterygoid muscle and neck of the mandible, then splits to encircle the middle meningeal artery, and finally joins the posterior trunk of V_3.

▼ LINGUAL NERVE

The **lingual nerve** is formed from afferent branches from the body of the tongue that travel along the lateral surface of the tongue (Figures 8–10 and 8–11). The nerve then passes posteriorly, passing from the medial to the lateral side of the duct of the submandibular salivary gland by going under the duct.

The lingual nerve communicates with the **submandibular ganglion** (sub-man-**dib**-you-lar) located superior to the deep lobe of the submandibular salivary gland (Figure 8–12). The submandibular ganglion is part of the parasympathetic system. Parasympathetic efferent innervation for the sublingual and submandibular salivary glands arises from the facial nerve (specifically, a branch of the facial nerve, the chorda tympani, which is discussed later in this chapter) but travels along with the lingual nerve.

At the base of the tongue, the lingual nerve ascends and runs between the medial pterygoid muscle and the mandible, anterior and slightly medial to the inferior alveolar nerve. Thus the lingual nerve is anesthetized when giving an inferior alveolar anesthetic nerve block. When the nerve is only a short distance posterior to the roots of the last mandibular molar tooth, it is covered only by a thin layer of oral mucous membrane and thus is sometimes visible clinically. Thus the lingual nerve is often endangered by dental procedures in this region, such as the extraction of mandibular third molars.

The lingual nerve then continues to travel upward to join the posterior trunk of V_3. Thus the lingual nerve serves as an afferent nerve for general sensation for the body of the tongue, floor of the mouth, and lingual gingiva of the mandibular teeth.

▼ INFERIOR ALVEOLAR NERVE

The **inferior alveolar nerve** (al-**ve**-o-lar) or **IA nerve** is an afferent nerve formed from the merger of the

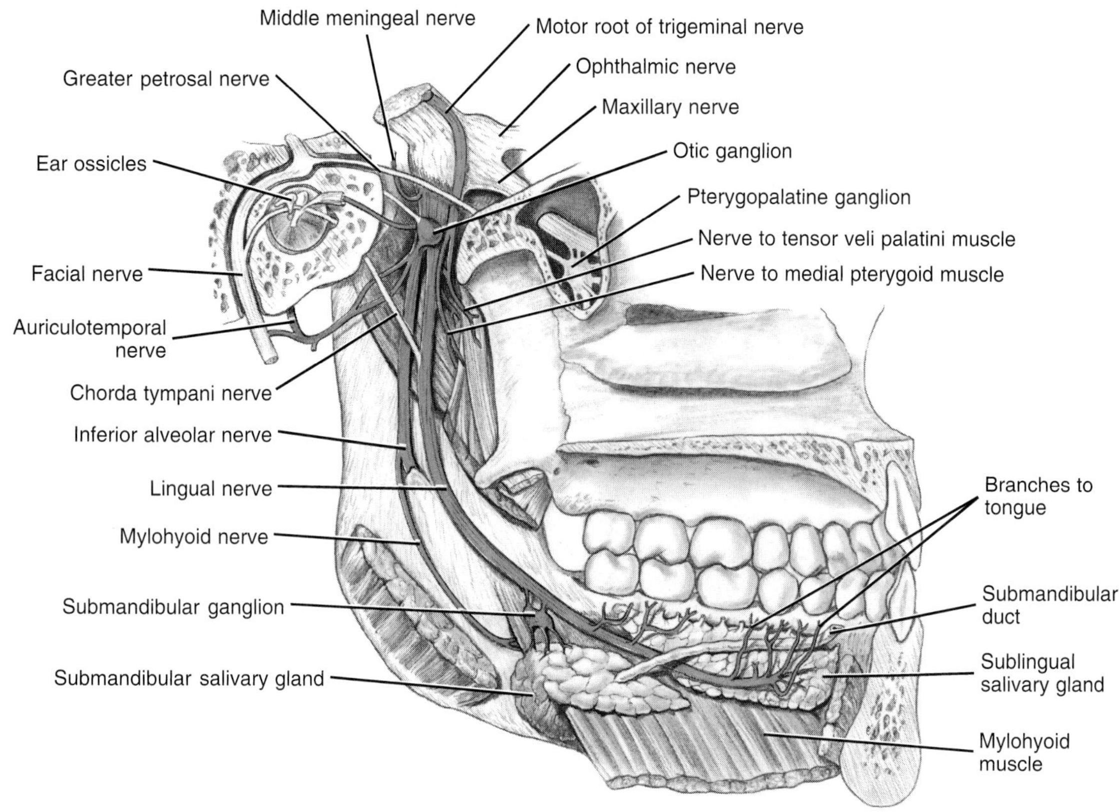

▼ **Figure 8–11**
Medial view of the mandible with the motor and sensory branches of the mandibular nerve highlighted.

mental and incisive nerves (Figure 8–10). The mental and incisive nerves are discussed later in this section.

After forming, the inferior alveolar nerve continues to travel posteriorly through the mandibular canal, along with the inferior alveolar artery and vein. The nerve is joined by dental branches and interdental and interradicular branches from the mandibular posterior teeth, forming a dental plexus or nerve network in the region. The inferior alveolar nerve then exits the mandible through the mandibular foramen, where it is joined by the mylohyoid nerve (discussed later). The mandibular foramen is a central opening on the internal surface of the ramus.

The inferior alveolar nerve then travels lateral to the medial pterygoid muscle, between the sphenomandibular ligament and ramus of the mandible within the pterygomandibular space. This is posterior and slightly lateral to the lingual nerve. The inferior alveolar nerve then joins the posterior trunk of V_3 (Figures 8–10 and 8–11). The inferior alveolar nerve carries afferent innervation for the mandibular teeth.

In some cases, there are two nerves present on the same side, creating bifid inferior alveolar nerves. This situation can occur unilaterally or bilaterally and can be detected on a radiograph by a double mandibular canal. As discussed in Chapter 3, there can be more than one mandibular foramen, usually inferiorly placed, either unilaterally or bilaterally, along with the bifid inferior alveolar nerves. These considerations must be kept in mind when administering a local anesthetic for the mandibular arch (see Chapter 9).

▼ MENTAL NERVE

The **mental nerve** (**ment**-il) is composed of external branches that serve as an afferent nerve for the chin, lower lip, and labial mucosa near the mandibular anterior teeth (Figure 8–10). The mental nerve then enters the mental foramen on the anterior-lateral surface of the mandible, usually between the apices of the first and second mandibular premolars, and merges with the incisive nerve to form the inferior alveolar nerve in the mandibular canal.

▼ INCISIVE NERVE

The **incisive nerve** (in-**sy**-ziv) is an afferent nerve composed of dental branches from the anterior mandibular teeth that originate in the pulp tissue, exit the teeth through the apical foramina, and join with interdental branches from the surrounding periodontium, forming a dental plexus in the region (Figure 8–10). The incisive nerve then merges with the mental nerve, just posterior to the mental foramen, to form the inferior alveolar nerve in the mandibular canal. The incisive nerve serves as an afferent nerve for the anterior mandibular teeth.

There is sometimes crossover from the opposite incisive nerve, which is an important consideration when using a local anesthetic in the area of the mandibular anterior teeth (see Chapter 9).

▼ MYLOHYOID NERVE

The **mylohyoid nerve** (my-lo-**hi**-oid) is a small branch of the inferior alveolar nerve after the IA nerve exits the mandibular foramen (Figures 8–10 and 8–11). This nerve pierces the sphenomandibular ligament and runs inferiorly and anteriorly in the mylohyoid groove and then onto the lower surface of the mylohyoid muscle. The mylohyoid nerve serves as an efferent nerve to the mylohyoid muscle and anterior belly of the digastric muscle (the posterior belly of the digastric muscle is innervated by a branch from the facial nerve).

▼ FACIAL NERVE

The dental professional must also have an understanding of the **facial nerve** or cranial nerve VII. The facial nerve carries both efferent and afferent nerves. The facial nerve emerges from the brain and enters the internal acoustic meatus in the petrous portion of the temporal bone. Within the bone, the nerve gives off a small efferent branch to the muscle in the middle ear (stapedius) and two larger branches, the greater petrosal and chorda tympani nerves, both of which carry parasympathetic fibers (Figure 8–12).

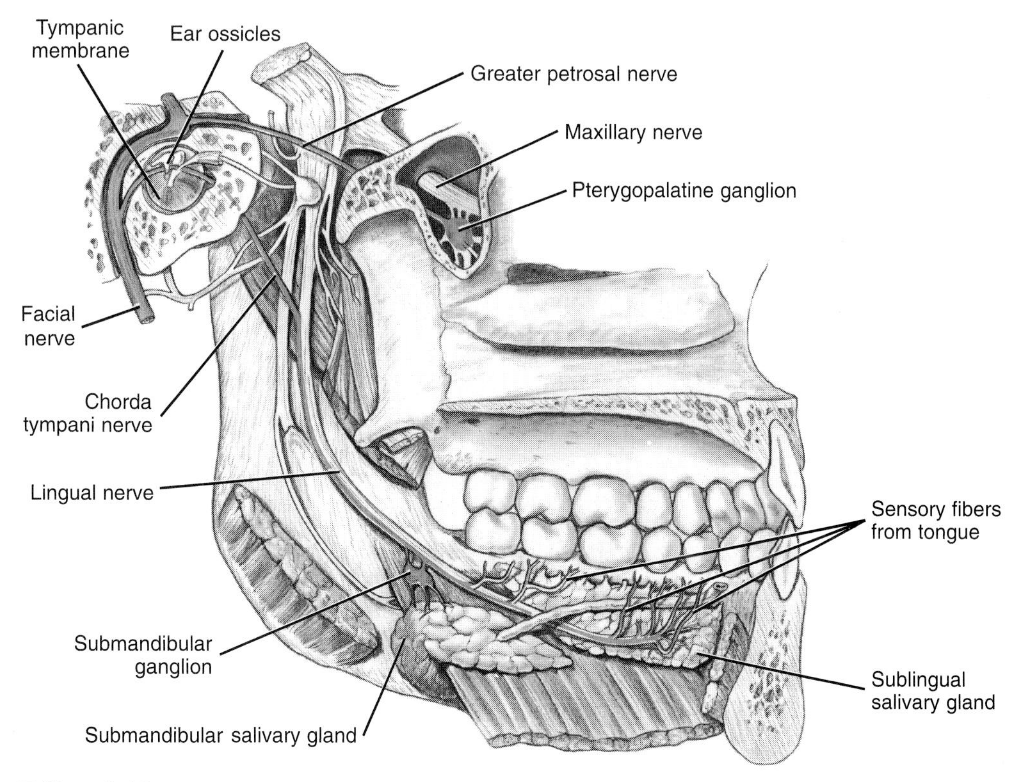

▼ **Figure 8–12**
The pathway of the trunk of the facial nerve, greater petrosal nerve, and chorda tympani nerve (note relationship with lingual nerve) highlighted.

The main trunk of the nerve emerges from the skull through the stylomastoid foramen of the temporal bone and gives off two branches, the posterior auricular nerve and a branch to the posterior belly of the digastric and stylohyoid muscles (Figure 8–13). The facial nerve then passes into the parotid salivary gland and divides into numerous branches to supply the muscles of facial expression, but not the parotid gland itself.

Greater Petrosal Nerve

The **greater petrosal nerve** (peh-**troh**-sil) is a branch off the facial nerve before it exits the skull (Figure 8–12). The greater petrosal nerve carries efferent nerve fibers, preganglionic parasympathetic fibers to the **pterygopalatine ganglion** (teh-ri-go-**pal**-ah-tine) in the pterygopalatine fossa.

The postganglionic fibers arising in the pterygopalatine ganglion then join with branches of the maxillary division of the trigeminal nerve to be carried to the lacrimal gland (via the zygomatic and lacrimal nerves), nasal cavity, and minor salivary glands of the hard and soft palate. The greater petrosal nerve also carries afferent nerve fibers for taste sensation in the palate.

Chorda Tympani Nerve

This small branch of the facial nerve, the **chorda tympani nerve** (**kor**-dah **tim**-pan-ee), is a parasympathetic efferent nerve for the submandibular and sublingual salivary glands and also serves as an afferent nerve for taste sensation for the body of the tongue.

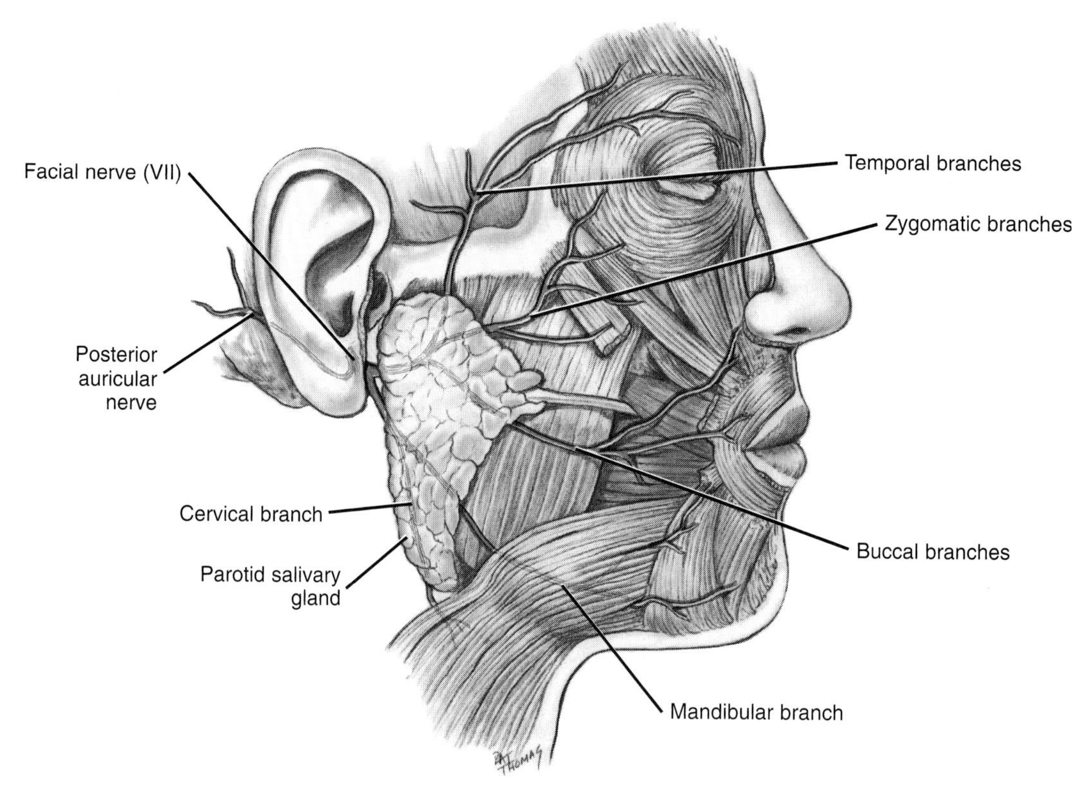

Figure 8–13
The pathway of the branches of the facial nerve to the muscles of facial expression highlighted.

After branching off the facial nerve within the petrous portion of the temporal bone, the chorda tympani nerve crosses the medial surface of the tympanic membrane (eardrum) (Figure 8–12), which thereby exits the skull by the **petrotympanic fissure** (pe-troh-tim-**pan**-ik), located immediately posterior to the temporomandibular joint (Figures 8–10 and 8–11). The chorda tympani nerve then travels with the lingual nerve along the floor of the mouth in the same nerve bundle.

In the submandibular triangle, the chorda tympani nerve, appearing as part of the lingual nerve, has communication with the submandibular ganglion. The **submandibular ganglion** (sub-man-**dib**-you-lar) is located superior to the deep lobe of the submandibular salivary gland, for which it supplies parasympathetic efferent innervation (see Figure 8–12).

Posterior Auricular, Stylohyoid, and Posterior Digastric Nerves

The **posterior auricular nerve** (aw-**rik**-you-lar), **stylohyoid nerve** (sty-lo-**hi**-oid), and **posterior digastric nerve** (di-**gas**-trik) are branches of the facial nerve after it exits the stylomastoid foramen. All are efferent nerves. The posterior auricular nerve supplies the occipital belly of the epicranial muscle (Figure 8–13). The other two nerves supply the stylohyoid muscle and the posterior belly of the digastric muscle, respectively.

Branches to the Muscles of Facial Expression

Additional efferent nerve branches of the facial nerve originate in the parotid salivary gland and pass to the muscles they innervate (Figure 8–13). These branches to the muscles of facial expression are the temporal, zygomatic, buccal, (marginal) mandibular, and cervical branches. However, these branches are rarely seen as five independent nerves. They may vary in number and connect irregularly. For convenience, they are described as five simple branches.

The temporal branch(es) (**tem**-poh-ral) supplies the muscles anterior to the ear, frontal belly of the epicranial muscle, superior portion of the orbicularis oculi muscle, and corrugator supercilii muscle.

The zygomatic branch(es) (zy-go-**mat**-ik) supplies the inferior portion of the orbicularis oculi muscle and zygomatic major and minor muscles. The buccal branch(es) supplies the muscles of the upper lip and nose and buccinator, risorius, and orbicularis oris muscles. The zygomatic and buccal branches are usually closely associated, exchanging many fibers.

The (marginal) mandibular branch (man-**dib**-you-lar) supplies the muscles of the lower lip and mentalis muscle. The mandibular branch should not be confused with the mandibular nerve or V_3. The cervical branch (**ser**-vi-kal) runs below the mandible to supply the platysma muscle.

Nerve Lesions of the Head and Neck

A dental professional needs to have a background on certain lesions of the head and neck that pertain to the nervous system before treating patients. These lesions are facial paralysis, Bell's palsy, and trigeminal neuralgia.

Facial paralysis (pah-**ral**-i-sis) is a loss of muscular action of the muscles of facial expression (see Chapter 4 and Figure 4–6). This lesion can occur secondarily to a brain injury by way of a stroke (cerebrovascular accident), with other muscles of the head and neck also affected. The lesion can also occur by directly injuring the nerve that supplies the efferent nerves to the muscles of facial expression, the seventh cranial or facial nerve. Facial paralysis can be unilateral or bilateral depending on the nature of nerve damage. These injuries are common because the facial nerve branches are superficially located and vulnerable to trauma.

The injury to the facial nerve can occur secondarily through injury to the parotid salivary gland or surrounding region. If cancer (malignant neoplasm) occurs in the parotid salivary gland, the facial nerve can be injured since it travels through the gland. During facial surgery or due to a laceration (deep wound) to the parotid gland region, the facial nerve within the tissues can be affected. Transient facial paralysis can occur due to injection into the gland during an incorrectly administered inferior alveolar local anesthetic block (see Chapter 9).

The affected facial muscles lose tone on the involved side (see Chapter 4). Clinically, the patient with facial paralysis has a drooping eyebrow, eyelid, and labial commissure, with a dribbling of saliva. There is also an inability to show normal expression, close the eye, or whistle. As a result, there can be infection in the involved eye. Speech and eating are also difficult.

Bell's palsy (**pawl**-ze) involves unilateral facial paralysis with no known cause, except that there is a loss of excitability of the involved facial nerve. All or just some of the branches of the facial nerve are affected. The onset of this paralysis is abrupt. One theory of its cause is that the facial nerve becomes inflamed within the temporal bone.

Bell's palsy may undergo remission or may become chronic depending on the amount of loss of facial nerve excitability. There is no specific treatment, but injections of anti-inflammatory medications or physical therapy may be helpful.

Trigeminal neuralgia (try-**jem**-i-nal noor-**al**-je-ah) also has no known cause but involves the afferent nerves of the fifth cranial or trigeminal nerve. One theory is that this lesion is caused by pressure on the sensory root of the trigeminal ganglion by area blood vessels.

Clinically, the patient feels excruciating short-term pain when facial trigger zones are touched. These trigger zones vary per patient but can include areas around the eyes or the ala of the nose. The right side of the face is affected more than the left.

Treatment for trigeminal neuralgia can include peripheral neurectomy by surgical or ultrasonic means where the sensory root of the trigeminal nerve is sectioned, cutting off innervation to the tissues. Alcohol injection of the trigeminal nerve has also been used to cause necrosis of the nerve, as well as systemic anticonvulsant drugs, with varying degrees of success.

Identification Exercise

Identify the structures on the following diagrams by filling in the blanks with the correct anatomical term. You can check your answers by looking back at the figure indicated in parentheses for each identification diagram.

1. (Figure 8–2)

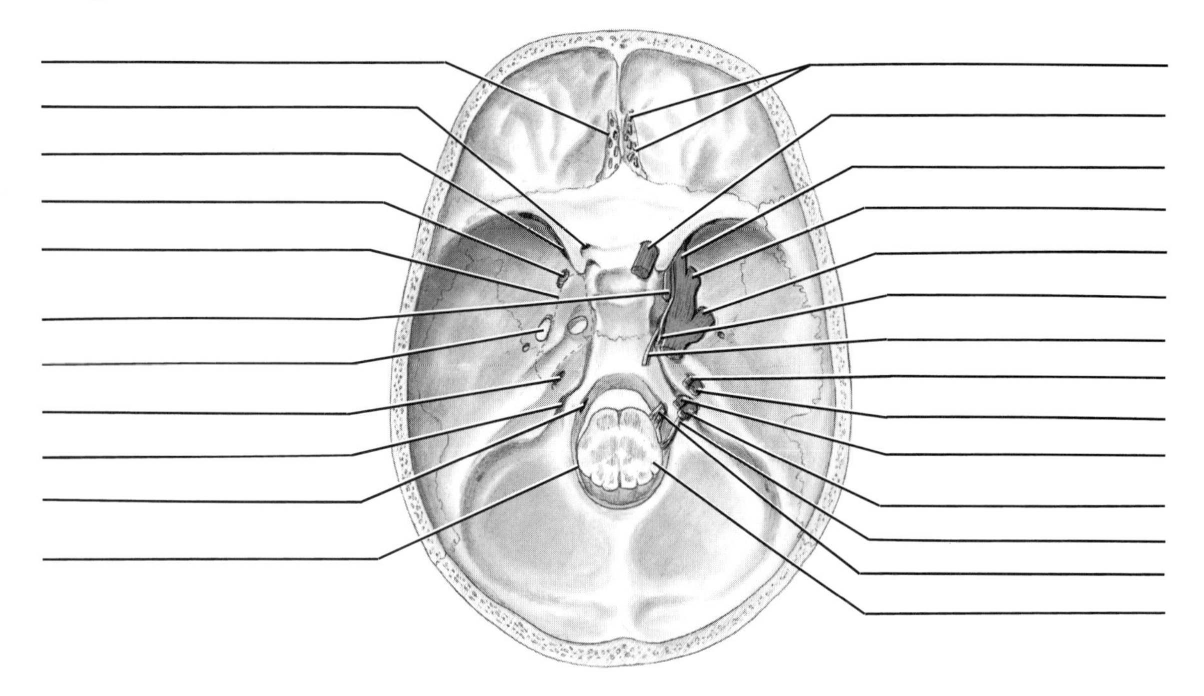

2. (Figure 8–3)

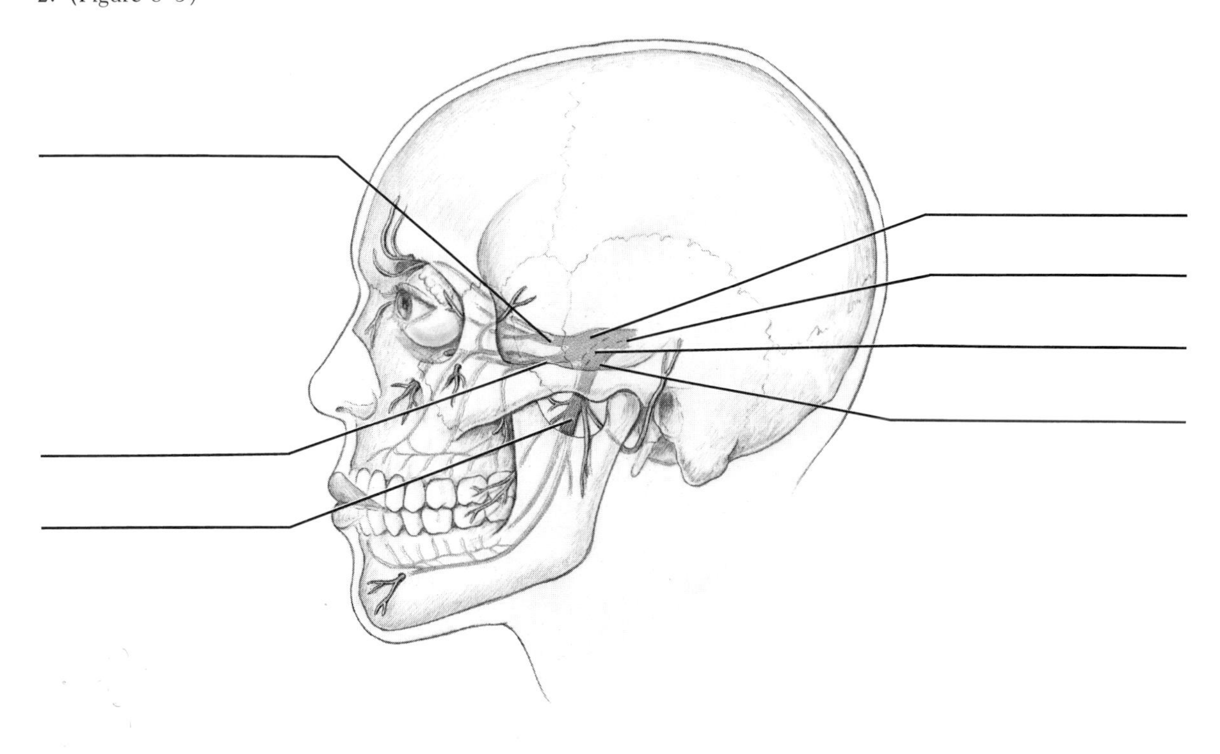

Identification Exercise *Continued*

3. (Figure 8–6)

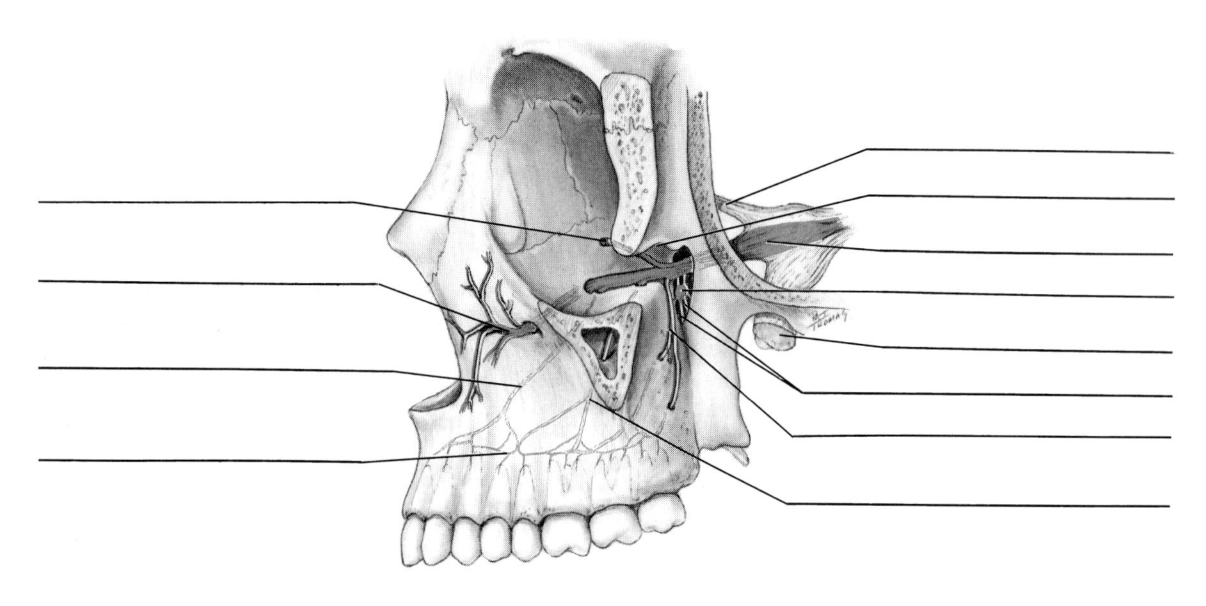

4. (Figure 8–10)

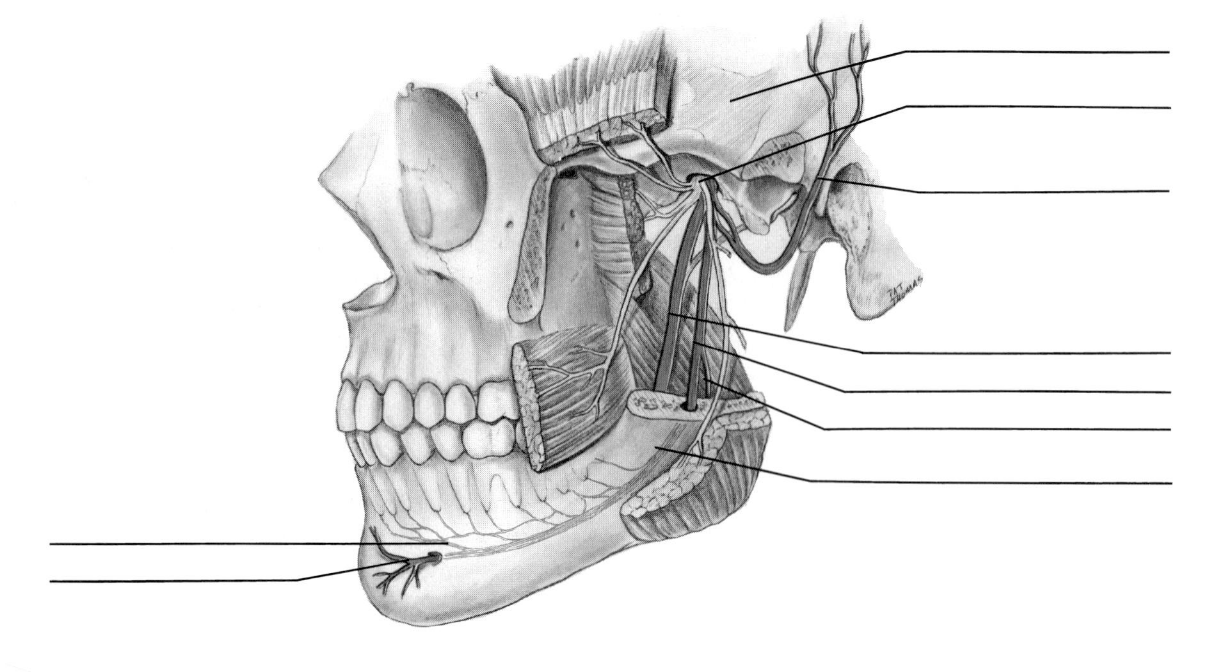

Identification Exercise *Continued*

5. (Figure 8–11)

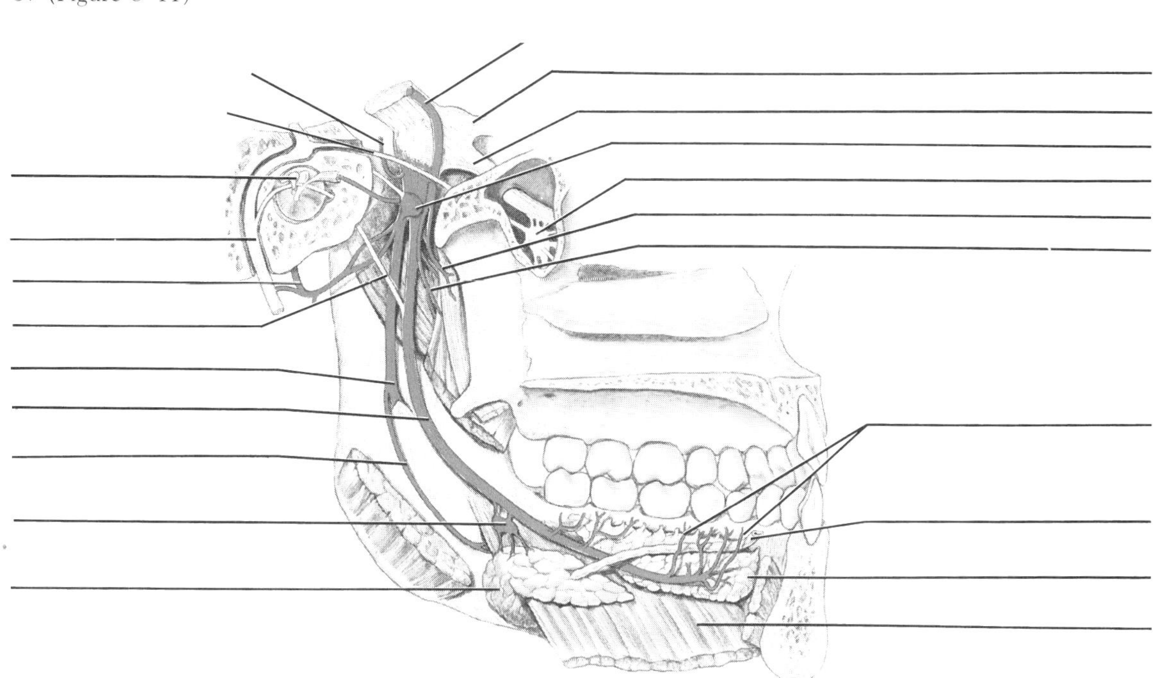

REVIEW QUESTIONS

1. The posterior superior alveolar nerve and its branches supply the:
   A. frontal sinus.
   B. maxillary posterior teeth.
   C. parotid salivary gland.
   D. temporalis muscle.

2. Through which of the following foramina does the facial nerve pass through the skull?
   A. Foramen rotundum
   B. Foramen ovale
   C. Jugular foramen
   D. Stylomastoid foramen

3. Which of the following cranial nerves is involved in Bell's palsy?
   A. Trigeminal nerve
   B. Facial nerve
   C. Glossopharyngeal nerve
   D. Vagus nerve

4. Which of the following nerve and muscle pairs is correctly matched?
   A. Long buccal nerve, buccinator muscle
   B. Accessory nerve, platysma muscle
   C. Hypoglossal nerve, intrinsic tongue muscles
   D. Auriculotemporal nerve, temporalis muscle

5. Which of the following nerve and tissue pairs is correctly matched?
   A. Facial nerve, parotid salivary gland
   B. Chorda tympani, sublingual salivary gland
   C. Vagus nerve, temporomandibular joint
   D. Lingual nerve, base of tongue

6. Which of the following cranial nerves has fibers that cross to the opposite side in the skull before continuing into the brain?
   A. Facial nerve
   B. Optic nerve
   C. Trochlear nerve
   D. Vestibulocochlear nerve

7. Which of the following cranial nerves carries taste sensation for the base of the tongue?
   A. Trigeminal nerve
   B. Facial nerve
   C. Vagus nerve
   D. Glossopharyngeal nerve

8. In which of the following areas is the trigeminal ganglion located?
   A. Superior to the deep lobe of the submandibular salivary gland

REVIEW QUESTIONS *Continued*

B. Anterior surface of the petrous portion of the temporal bone
C. Posterior surface of the maxillary tuberosity of the maxilla
D. Anterior to the infraorbital foramen of the maxilla

9. Sensory information is supplied for the soft palate by the:
A. greater palatine nerve.
B. lesser palatine nerve.
C. nasopalatine nerve.
D. posterior alveolar nerve.

10. The posterior belly of the digastric muscle is innervated by branches from the:
A. facial nerve.
B. mylohyoid nerve.
C. buccal nerve.
D. maxillary nerve.

11. Which of the following nerves may show crossover from the opposite side in a patient?
A. Posterior superior alveolar nerve
B. Anterior superior alveolar nerve
C. Posterior auricular nerve
D. Buccal nerve

12. Which of the following nerves is part of the ophthalmic division of the trigeminal nerve?
A. Nasociliary nerve
B. Maxillary nerve
C. Zygomaticotemporal nerve
D. Zygomaticofacial nerve

13. Which of the following anatomical names is also used for cranial nerve X?
A. Hypoglossal nerve
B. Vagus nerve
C. Glossopharyngeal nerve
D. Accessory nerve

14. Which of the following nerves is located in the mandibular canal?
A. Lingual nerve
B. Mylohyoid nerve
C. Inferior alveolar nerve
D. Masseteric nerve

15. Which of the following nerves exits the foramen ovale of the sphenoid bone?
A. Chorda tympani of the facial nerve
B. Greater petrosal nerve of the facial nerve
C. Ophthalmic division of the trigeminal nerve
D. Motor root of the trigeminal nerve

▼ *This chapter discusses the following topics:*

I. Anatomical Considerations for Local Anesthesia

II. Maxillary Nerve Anesthesia
   - A. Posterior Superior Alveolar Block
   - B. Middle Superior Alveolar Block
   - C. Infraorbital Block
   - D. Greater Palatine Block
   - E. Nasopalatine Block

III. Mandibular Nerve Anesthesia
   - A. Inferior Alveolar Block
   - B. Buccal Block
   - C. Mental Block
   - D. Incisive Block

▼ *Key words*

Local infiltration (in-fil-**tray**-shun) Type of injection that anesthetizes a small area, one or two teeth and associated structures when the anesthetic solution is deposited near terminal nerve endings.

Nerve block Type of injection that anesthetizes a larger area than the local infiltration because the anesthetic solution is deposited near large nerve trunks.

Chapter 9

Anatomy of Local Anesthesia

▼ ***After studying this chapter, the reader should be able to:***

1. Define and pronounce all the key words and anatomical terms in this chapter.
2. List the tissues anesthetized by each type of injection and describe the target areas.
3. Locate and identify the anatomical structures used to determine the local anesthetic needle's penetration site for each type of injection on a skull and a patient.
4. Demonstrate the correct placement of the local anesthetic needle for each type of injection on a skull and a patient.
5. Identify the correct depth and tissues penetrated by the local anesthetic needle for each type of injection.
6. Discuss the complications of local anesthesia of the oral cavity associated with anatomical considerations for each type of injection.
7. Correctly complete the review questions for this chapter.
8. Integrate the knowledge about the nerves into the administration of local anesthesia and control of pain during dental procedures.

Anatomical Considerations for Local Anesthesia

The management of pain through local anesthesia by dental professionals requires a thorough knowledge of the anatomy of the skull, fifth cranial nerve, and related tissues. This text discusses the anatomical considerations for local anesthesia. Dental professionals will also want to refer to a current textbook on local anesthesia for more information on this topic.

The skull bones involved in the local anesthetic administration are the maxilla, palatine bone, and mandible (see Chapter 3 for review). Soft tissues of the face and oral cavity may serve the dental professional as initial landmarks to visualize and palpate for local anesthesia (see Chapter 2). However, there are many variations in soft tissue topographical anatomy among patients. Thus to increase the reliability of anesthesia, the dental professional must learn to rely mainly on the visualization and palpation of hard tissues for landmarks while injecting patients.

The dental professional must also know the location of certain adjacent soft tissue structures, such as major blood vessels and glandular tissue, so as to avoid inadvertently injecting these structures (see Chapters 6 and 7 for review). If certain soft tissue structures are accidentally injected with anesthetic solution, complications may occur. Infections also may be spread to deeper tissues by needle-tract contamination (see Chapter 12 for more information).

The fifth cranial nerve or trigeminal nerve provides the sensory information for the teeth and associated tissues (see Chapter 8 for review). Thus branches of the trigeminal nerve are those anesthetized prior to most dental procedures. Knowledge of the location of these nerve branches in relationship to the facial skull bones, as well as soft tissues, again increases the reliability of each injection.

There are two types of local anesthetic injections commonly used in dentistry: the local infiltration and nerve block. The type of injection used for a given dental procedure is determined by the type and length of the procedure.

The ***local infiltration*** (in-fil-**tray**-shun) anesthetizes a small area, one or two teeth and associated tissues by injection near their apices. For this type of local anesthetic injection, the anesthetic solution is deposited near terminal nerve endings. This localized deposition has varying degrees of success depending on the anatomy of the region. This text does not discuss local infiltration in detail.

The ***nerve block*** affects a larger area than the local infiltration. With a nerve block, the anesthetic solution is deposited near large nerve trunks. All the injections, such as the posterior alveolar block, discussed in this chapter are nerve block injections.

Dental procedures can usually commence within 3-5 minutes after the correct administration of local anesthesia. If failure or incomplete anesthesia occurs, reassess the injection technique and consider the need for reinjection using an alternate technique or other injections. Never reinject with the same technique or in the same area unless careful consideration is given as to why failure or incomplete anesthesia has occurred.

The dental professional also needs to keep in mind the total number of injections and dosage of anesthetic solution needed to complete the planned dental procedures. This will help prevent systemic complications of local anesthesia that may occur when a patient receives more than the maximum safe dosage (see other texts for specific recommendations). It is also important never to inject through an area with abscess, cellulitis, or osteomyelitis so as to prevent the spread of dental infection (see Chapter 12). The effectiveness of local anesthetic agents is also greatly reduced when administered in areas of infection.

Maxillary Nerve Anesthesia

The maxillary nerve and its branches can be anesthetized in a number of ways depending on the tissues requiring anesthesia (Figure 9–1, Chart 9–1). Most anesthesia of the maxilla is more successful than that of the mandible because the bone over the facial surface of the maxillary teeth is less dense than that of the mandible over similar teeth (see Chapter 3). There is also less variation in the anatomy of the maxillary and palatine bones and associated nerves with respect to local anesthetic landmarks as compared with similar mandibular structures.

Pulpal anesthesia is achieved through anesthesia of each nerve's dental branches as they extend into the pulp tissue by way of each tooth's apical foramen. The hard and soft tissues of the periodontium are anesthetized by way of each nerve's interdental and interradicular branches.

The posterior superior alveolar block is generally recommended for anesthesia of the maxillary molar teeth and associated buccal tissues in one quadrant. The middle superior alveolar block is generally recommended for anesthesia of the maxillary premo-

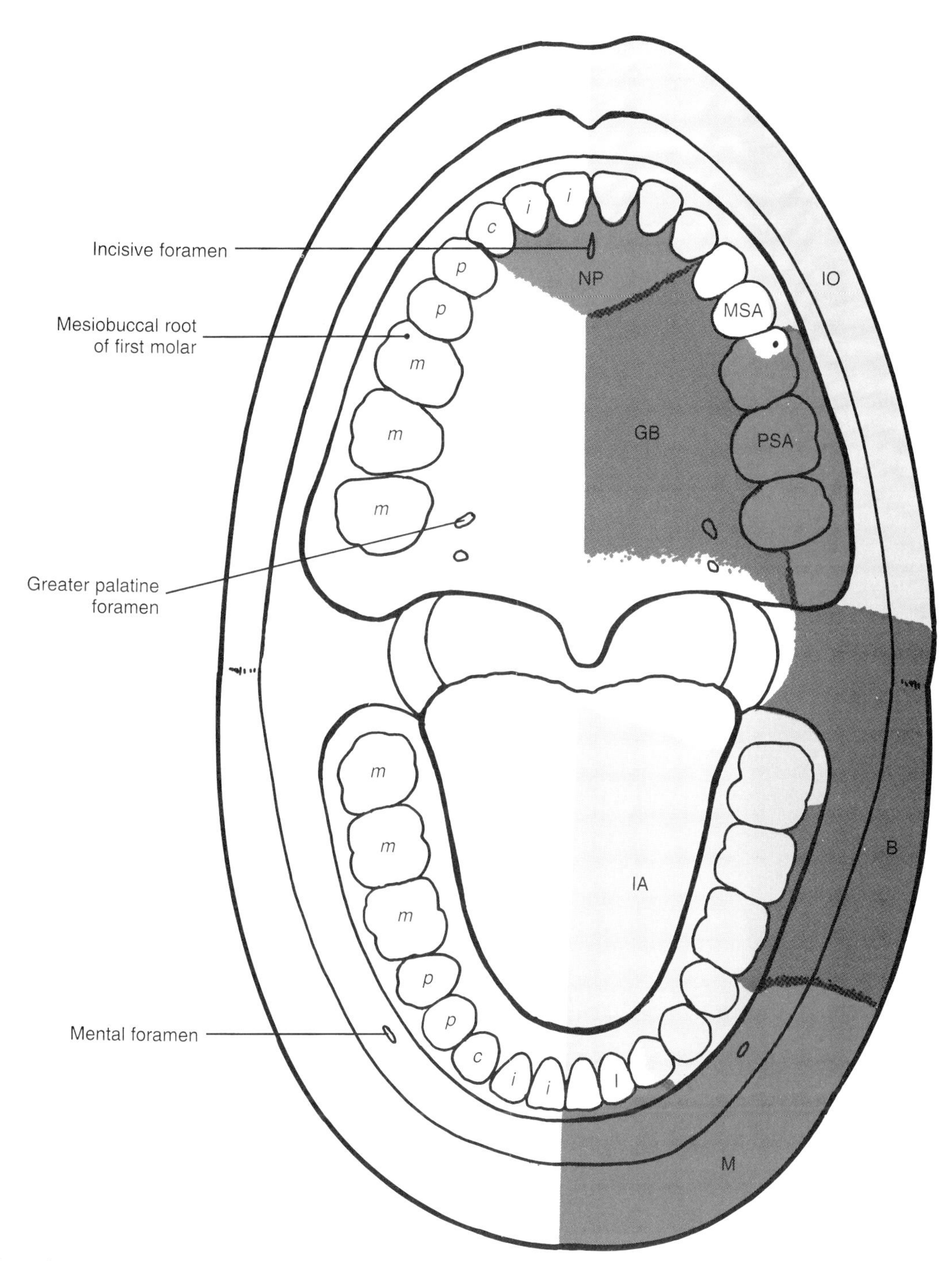

Figure 9–1
A diagram of the types of local anesthetic blocks of oral tissues innervated by branches of the maxillary nerve and mandibular nerve. PSA = posterior superior alveolar block, MSA = middle superior alveolar block, IO = infraorbital block, GP = greater palatine block, NP = nasopalatine block, IA = inferior alveolar block, B = buccal block, M = mental block, I = incisive block.

Chart 9–1

SUMMARY OF MAXILLARY LOCAL ANESTHESIA OF THE TEETH AND ASSOCIATED TISSUES					
Maxillary Tooth and Tissues	PSA Block	MSA Block	IO Block	NP Block	GP Block
Central Incisor					
Facial/pulpal			X		
Lingual				X	
Lateral Incisor					
Facial/pulpal			X		
Lingual				X	
Canine					
Facial/pulpal			X		
Lingual				X	
First Premolar					
Buccal/pulpal		X	X		
Lingual					X
Second Premolar					
Buccal/pulpal		X	X		
Lingual					X
First Molar					
Buccal/pulpal	X				
Lingual					X
Second Molar					
Buccal/pulpal	X				
Lingual					X
Third Molar					
Buccal/pulpal	X				
Lingual					X

The anesthesia is not included for the mesiobuccal root of the first molar or for anatomical variants. PSA = posterior superior alveolar, MSA = middle superior alveolar, IO = infraorbital, NP = nasopalatine, GP = greater palatine.

lars and associated buccal tissues in one quadrant. The infraorbital block is generally recommended for anesthesia of the maxillary anterior and premolar teeth and associated facial tissues in one quadrant.

Palatal anesthesia involves anesthesia of the soft and hard tissues of the periodontium of the palatal area, such as the gingiva, periodontal ligament, and alveolar bone. Palatal anesthesia does not provide any pulpal anesthesia to the maxillary teeth or associated facial or buccal tissues. The greater palatine block is generally recommended for anesthesia of the palatal tissues distal to the maxillary canine in one quadrant. The nasopalatine block is generally recommended for anesthesia of the palatal tissues between the right and left maxillary canines.

POSTERIOR SUPERIOR ALVEOLAR BLOCK

The **posterior superior alveolar block** (al-**ve**-o-lar) or **PSA block** is used to achieve pulpal anesthesia in the maxillary third, second, and first molars in most patients. Thus the PSA block is indicated when the dental procedure involves two or more maxillary molars or their associated buccal tissues. This block also is useful for restorative procedures on these teeth.

In some patients, the mesiobuccal root of the maxillary first molar is not innervated by the posterior superior alveolar nerve, but by the middle superior alveolar nerve. Therefore a second injection to anesthetize the middle superior alveolar nerve may be needed to achieve pulpal anesthesia of all the roots of the maxillary first molar.

The PSA block also anesthetizes the buccal periodontium overlying the maxillary third, second, and first molars, including the associated gingiva, periodontal ligament, and alveolar bone. This buccal tissue anesthesia is useful for periodontal treatment of these teeth. If anesthesia of the lingual tissues is desired, such as with extraction, the greater palatine block also may be needed.

Target Area and Injection Site for PSA Block

The target area for the PSA block is the PSA nerve as it enters the maxilla through the posterior superior alveolar foramina on the maxilla's infratemporal surface (Figure 9–2). This area is posterosuperior and medial on the maxillary tuberosity.

The injection site for the PSA block is into the tissues at the height of the mucobuccal fold at the apex of the maxillary second molar, distal to the zygomatic process of the maxilla (Figure 9–3). The needle is inserted into the mucobuccal fold tissues to the desired depth in a distal and medial direction to the tooth and maxilla without touching the maxillary bone in order to reduce trauma (Figures 9–4 and 9–5). The angulation of the needle should be upward or superiorly at a 45-degree angle to the occlusal plane, inward or medially at a 45-degree angle to the occlusal plane,

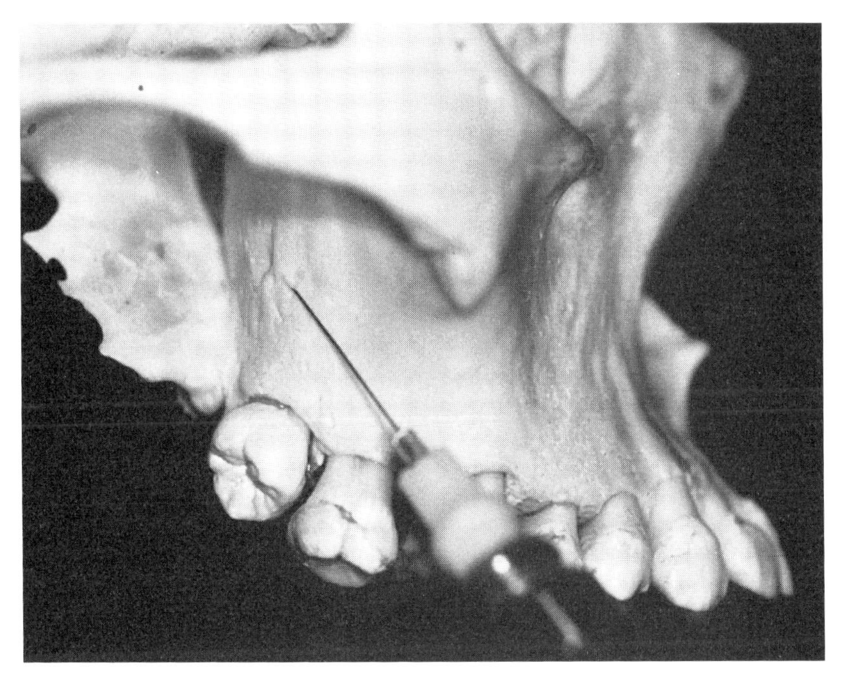

Figure 9–2
Target area for the PSA block is the PSA nerve located at the posterior superior alveolar foramina on the infratemporal surface of the maxilla, posterosuperior and medial to the maxillary tuberosity.

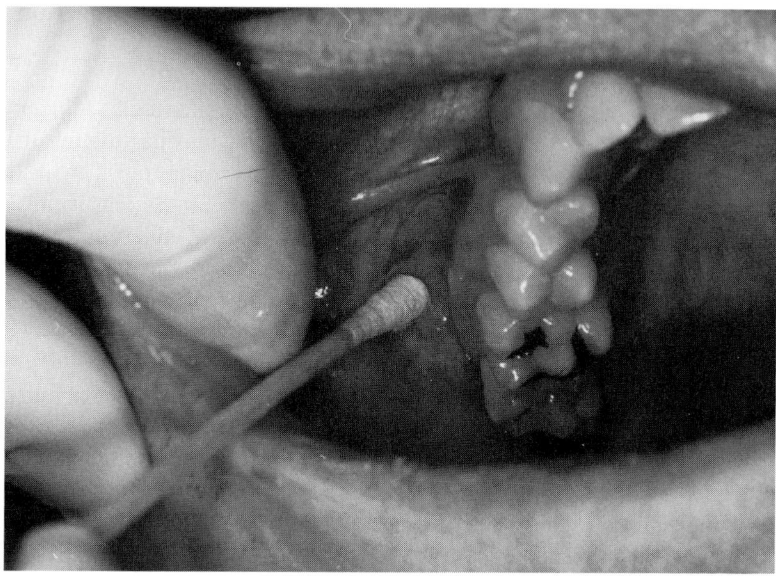

Figure 9–3
The injection site for the PSA block is probed at the height of the mucobuccal fold at the apex of the maxillary second molar.

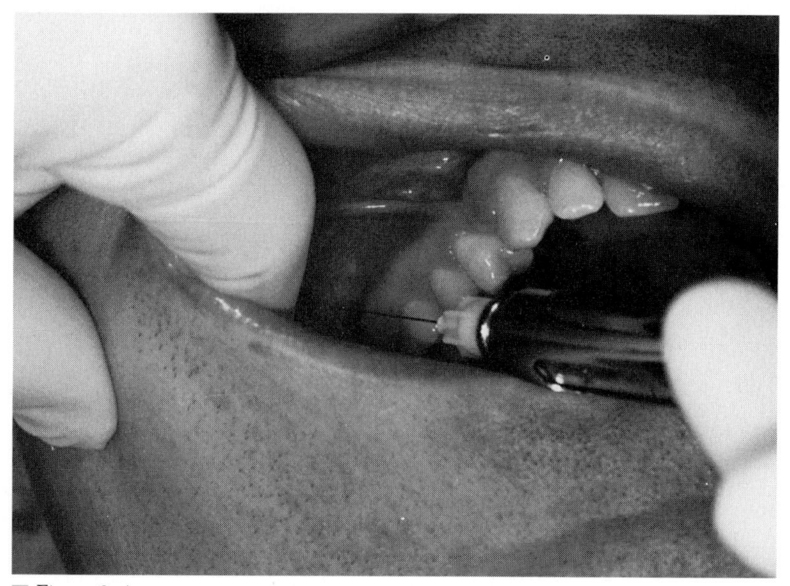

Figure 9–4
Needle penetration of the mucobuccal fold tissues for the PSA block is demonstrated at the apex of the maxillary second molar.

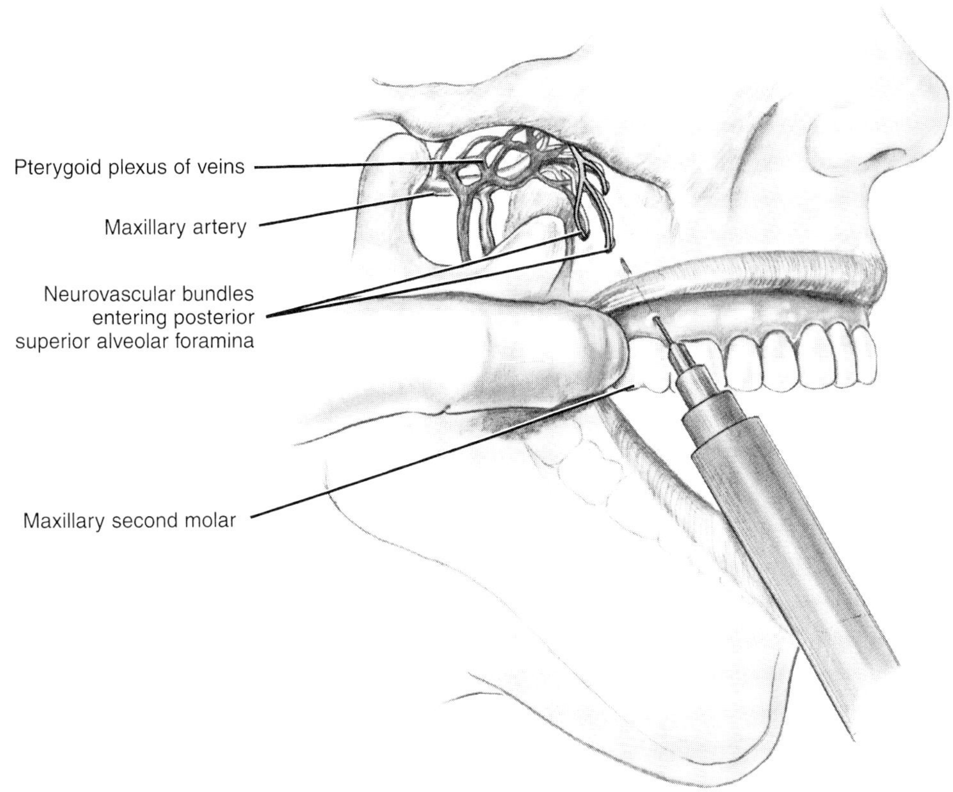

Figure 9–5
Correct insertion of the needle during a PSA block. If the needle is overinserted, it can penetrate the pterygoid plexus of the veins and maxillary artery, which may lead to complications such as a hematoma.

and backward or posteriorly at a 45-degree angle to the long axis of the second maxillary molar.

The desired depth of needle insertion for the PSA block depends on the size of the maxilla. With an adult of average size, the penetration is 16 mm. For small adults and children, it is important not to advance the needle more than a depth of 10–14 mm. This conservative insertion technique may also be used in all adults so as to reduce possible complications (discussed later in this chapter).

Symptoms and Possible Complications of PSA Block

There are usually no symptoms of soft tissue anesthesia with a PSA block. Thus the patient frequently has difficulty determining the extent of anesthesia since a lip or the tongue is not anesthetized. Instead, the patient will state that the teeth in the area feel dull when gently tapped, and there will be an absence of discomfort during dental procedures.

Complications can occur if the needle is advanced too far distally into the tissues during a PSA block (Figure 9–5). The needle may penetrate the pterygoid plexus of veins and the maxillary artery if overinserted, possibly causing a hematoma in the infratemporal fossa (see Chapter 6 for more information and Figure 6–15). This results in a bluish-reddish extraoral swelling of hemorrhaging blood in the tissues on the side of the face a few minutes after the injection, progressing over time inferiorly and anteriorly toward the lower anterior region of the cheek. If the needle is contaminated, there may be a spread of infection to the cavernous venous sinus (see Chapter 12). Aspiration should always be attempted in all injections in order to avoid injection into blood vessels, as well as strict aseptic protocol.

Inadvertent and harmless anesthesia of branches of the mandibular nerve with a PSA block may occur since they are located lateral to the PSA nerve. This may result in varying degrees of lingual anesthesia and anesthesia of the lower lip.

MIDDLE SUPERIOR ALVEOLAR BLOCK

The **middle superior alveolar block** (al-**ve**-o-lar) or **MSA block** has a limited clinical usefulness since the middle superior alveolar nerve is not present in all patients. However, when the more useful infraorbital block fails to achieve pulpal anesthesia distal to the maxillary canine, the MSA block is indicated for dental procedures on the maxillary premolars and mesiobuccal root of the maxillary first molar, such as with restorative procedures.

The MSA block is also used when the dental procedure only involves the maxillary premolars or their associated buccal tissues. Where the MSA nerve is absent, the area is innervated by the anterior superior alveolar nerve and can be anesthetized using the same MSA field block technique.

Thus the MSA block anesthetizes the pulp tissue of the maxillary first and second premolars, possibly the mesiobuccal root of the maxillary first molar and the associated buccal periodontal tissues, including the gingiva, periodontal ligament, and alveolar bone. The buccal tissue anesthesia is useful for periodontal treatment of these teeth. If lingual tissue anesthesia of these teeth is desired, as with an extraction or periodontal treatment, the greater palatine block may also be needed.

Target Area and Injection Site for MSA Block

The target area for the MSA block is the MSA nerve at the apex of the maxillary second premolar (Figure 9–6). Thus the injection site is the tissues at the height of the mucobuccal fold at the apex of the maxillary second premolar (Figure 9–7). The needle is inserted into the mucobuccal fold tissues until its tip is located well above the apex of the maxillary second premolar without touching the bone in order to reduce trauma (Figure 9–8).

Symptoms and Possible Complications of MSA Block

Symptoms of anesthesia with the MSA block include harmless tingling or numbness of the upper lip and

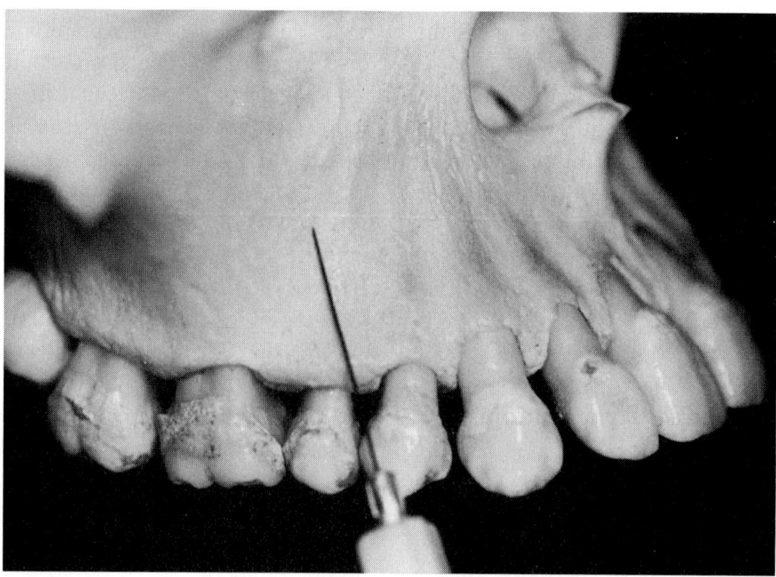

Figure 9–6
Target area for the MSA block is the MSA nerve located at the apex of the maxillary second premolar.

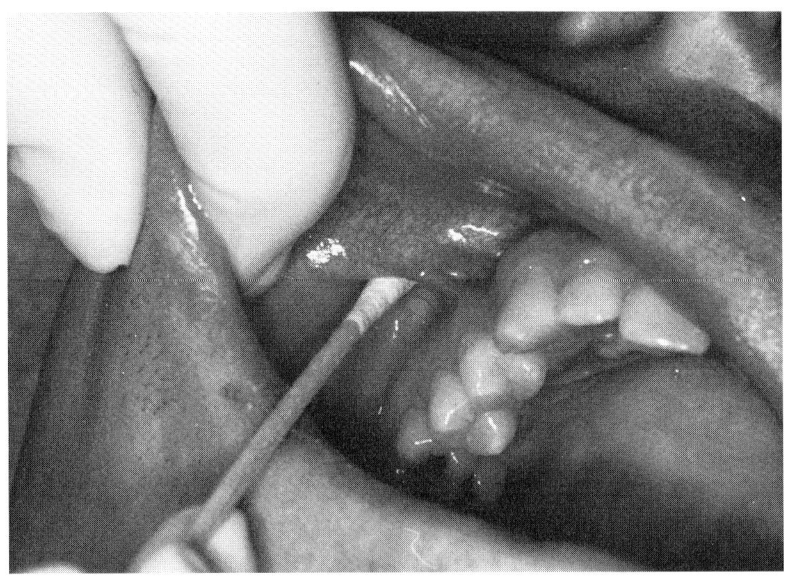

▼ Figure 9–7
The injection site for the MSA block is probed at the height of the mucobuccal fold at the apex of the maxillary second premolar.

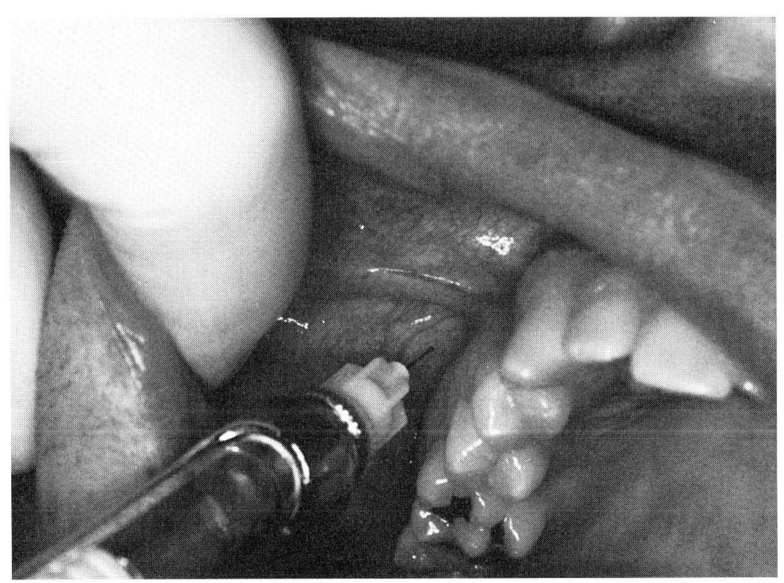

▼ Figure 9–8
Needle penetration of the mucobuccal fold tissues for the MSA block is demonstrated at the apex of the maxillary second premolar.

absence of discomfort during dental procedures. Overinsertion with complications such as a hematoma is rare with the MSA block.

INFRAORBITAL BLOCK

The **infraorbital block** (in-frah-**or**-bit-al) or **IO block** is a useful nerve block since it anesthetizes both the middle and anterior superior alveolar nerves. The IO block is used for anesthesia of the maxillary premolars, maxillary canine, and maxillary incisors. The IO block is indicated when the dental procedures involve more than two maxillary premolars or anterior teeth and the overlying facial periodontium, including the gingiva, periodontal ligament, and alveolar bone.

Thus the IO nerve injection is useful for restorative procedures since it gives pulpal anesthesia of these teeth and also for periodontal treatment since it gives facial tissue anesthesia. If lingual tissue anesthesia is needed, such as with an extraction or periodontal treatment, a nasopalatine block may also be needed.

In many patients, the ASA nerve crosses over the midline from the opposite side, so bilateral injections of the IO block or local infiltration over the opposite maxillary central incisor may be indicated. The maxillary central incisor may be additionally innervated by branches from the palatally located nasopalatine nerve, which may need to be anesthetized as well (discussed later).

Some dental professionals use multiple local infiltrations at the apex of each involved tooth in the area of the ASA nerve instead of the IO block to achieve anesthesia. This multiple local infiltration technique can be successful, but possibly traumatic for the patient. This text only discusses the use of the IO block to achieve anesthesia of these teeth.

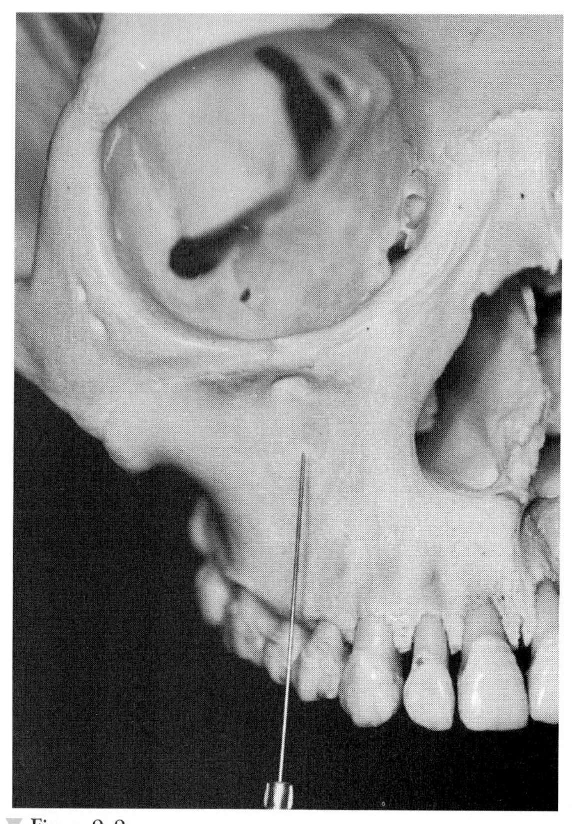

Figure 9–9
Target area for the IO block is the ASA and MSA nerve located at the infraorbital foramen. Note the position of the infraorbital rim.

Target Area and Injection Site for IO Block

The target area for the IO block is the ASA and MSA nerves as they ascend to join the infraorbital nerve after the infraorbital nerve enters the infraorbital foramen (Figure 9–9). Branches of the infraorbital nerve to the lower eyelid, side of the nose, and upper lip are also inadvertently anesthetized.

To locate the infraorbital foramen, palpate extraorally the patient's infraorbital rim and then move slightly downward about 10 mm applying pressure until a depression created by the infraorbital foramen is located (Figure 9–10). The patient may feel a dull aching sensation when pressure is applied to the nerves in the region. The infraorbital foramen is about 1–4 mm medial to the pupil of the eye if the patient looks straight forward.

The injection site for the IO block is the tissues at the height of the muccobuccal fold at the apex of the maxillary first premolar. A preinjection approximation of the depth of needle penetration for the IO block can be made by placing one finger on the infraorbital foramen and the other one on the injection site and estimating the distance between them (Figure 9–11). The approximate depth of needle penetration for the IO block will be 16 mm for an adult of average height, although the depth of penetration may vary. On a patient with a high or deep mucobuccal fold or low infraorbital foramen, less tissue penetration will be

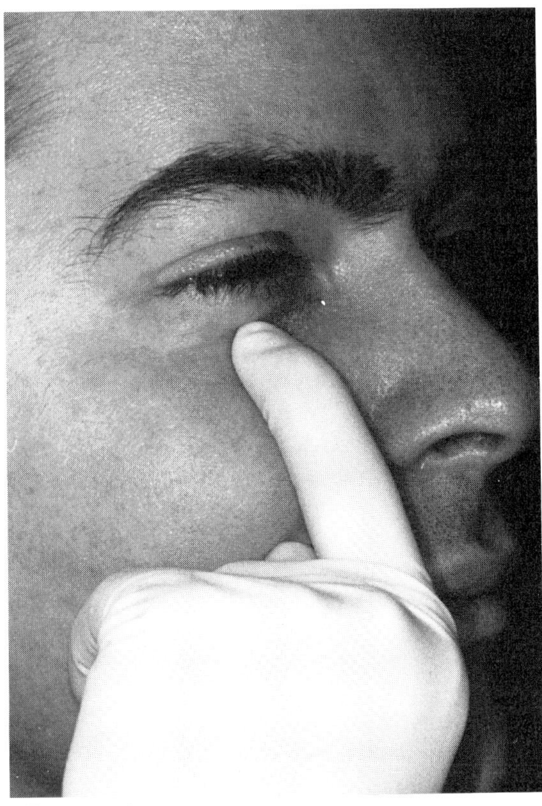

▼ **Figure 9–10**
Palpation of the depression created by the infraorbital foramen for the IO block is demonstrated by moving slightly downward on the face from the infraorbital rim.

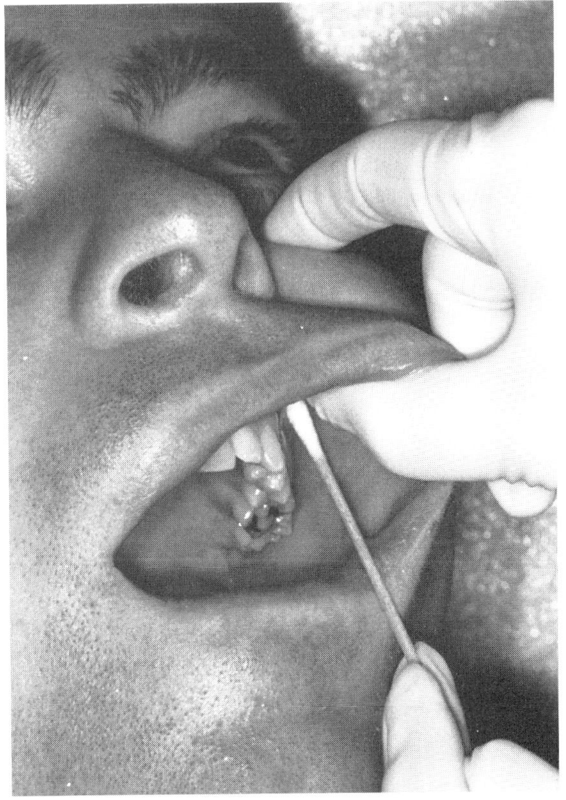

▼ **Figure 9–11**
The injection site for the IO block is probed at the height of the mucobuccal fold at the apex of the maxillary first premolar. The palpation over the infraorbital foramen is maintained.

required than in a patient with a shallow mucobuccal fold or high infraorbital foramen.

The needle is inserted for the IO block into the mucobuccal fold tissues while keeping the finger of the other hand on the infraorbital foramen during the injection to help keep the syringe toward the foramen (Figure 9–12). The needle is advanced while keeping it parallel with the long axis of the tooth to avoid premature contact with the maxillary bone. The point of contact of the needle with the maxillary bone should be the upper rim of the infraorbital foramen. Keeping the needle in contact with the bone at the roof of the infraorbital foramen prevents overinsertion and possible puncture of the orbit.

Symptoms and Possible Complications of IO Block

Symptoms of the IO block include harmless tingling and numbness of the eyelid, side of the nose, and upper lip since there is inadvertent anesthesia of the branches of the infraorbital nerve. Additionally, there is numbness in the teeth and associated tissue along the distribution of the ASA and MSA nerves and absence of discomfort during dental procedures. Rarely the complication of a hematoma may develop across the lower eyelid and the tissues between it and the infraorbital foramen (see Chapter 6 for more information and Figure 6–15).

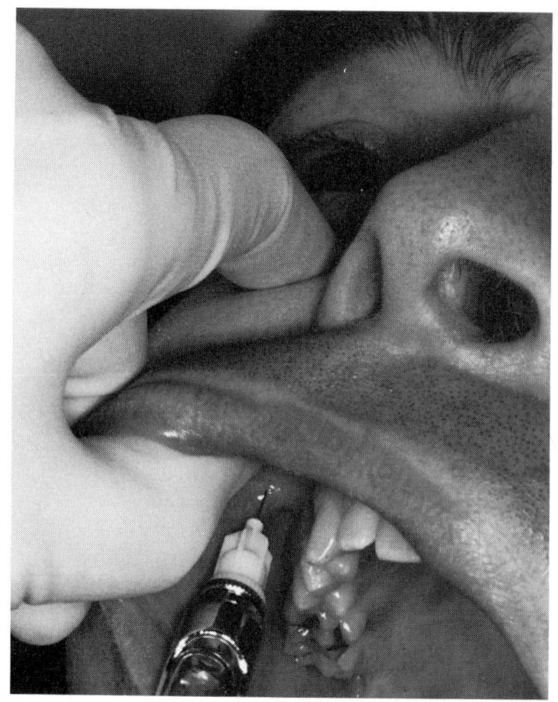

Figure 9–12
Needle penetration of the mucobuccal fold tissues for the IO block is demonstrated at the apex of the maxillary first premolar. The palpation over the infraorbital foramen is maintained.

GREATER PALATINE BLOCK

The **greater palatine block** (**pal**-ah-tine) is used during dental procedures that involve more than two maxillary posterior teeth or palatal soft tissues distal to the maxillary canine. This maxillary block anesthetizes the posterior portion of the hard palate, anteriorly as far as the maxillary first premolar and medially to the midline.

Since the greater palatine block does not provide pulpal anesthesia of the area teeth, the use of the PSA, MSA, or IO block may also be indicated during restorative or extraction procedures. The greater palatine block is useful for periodontal treatment of the involved teeth due to its lingual tissue anesthesia. However, soft tissue anesthesia in the palatal area of the maxillary first premolar may prove inadequate because of overlapping nerve fibers from the nasopalatine nerve. This lack of anesthesia may be corrected by additional administration of the nasopalatine block.

Because the overlying palatal tissues are very dense and adhere firmly to the underlying palatal bone, the use of pressure anesthesia by an applicator stick to the site before and during the injection to blanch the tissues will reduce patient discomfort. This pressure anesthesia on the tissues produces a dull ache to block pain impulses that arise from needle penetration. The very slow deposition of anesthetic solution will also reduce patient discomfort.

Target Area and Injection Site for Greater Palatine Block

The target area for the greater palatine block is the greater palatine nerve as it enters the greater palatine foramen from its location between the mucoperiosteum and bone of the hard palate (Figure 9–13). The greater palatine foramen is located at the junction of the maxillary alveolar process and posterior hard palate, at the apex of the maxillary second (for

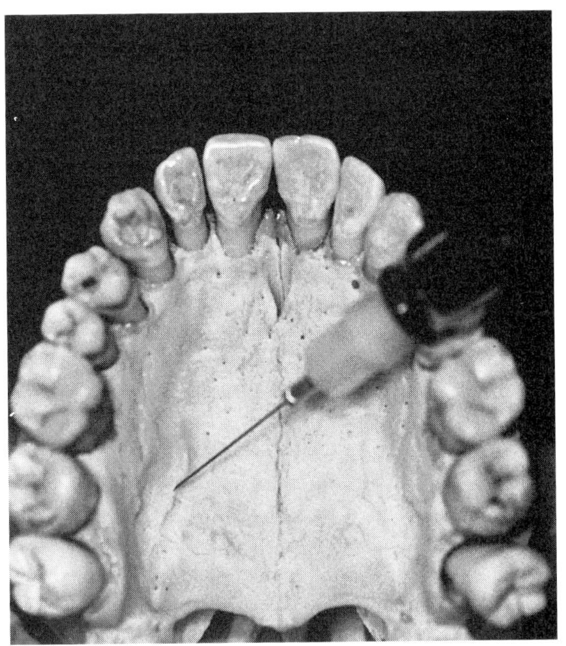

Figure 9–13
Target area for the greater palatine block is the greater palatine nerve located at the greater palatine foramen, which is at the junction of the maxillary alveolar process and the hard palate, at the apex of the maxillary second or third molar.

children) or third molar, about 10 mm medial and directly above the lingual gingival margin.

The site of injection is the palatal tissues at the depression created by the greater palatine foramen (Figure 9–14). This depression can be palpated about midway between the median palatine raphe and lingual gingival margin of the molar tooth.

The needle is inserted for the greater palatine block into the previously blanched palatal tissues at a right or 90-degree angle to the palate (Figure 9–15). The needle is advanced during the greater palatine block, applying enough pressure to bow the needle slightly, until the palatine bone is contacted. The depth of penetration of the needle into the palatal tissue will usually be less than 5 mm. There is no need to enter the greater palatine canal. Although this is not potentially hazardous, it is not necessary for adequate anesthesia.

Symptoms and Possible Complications of Greater Palatine Block

Symptoms of the greater palatine block are numbness in the posterior portion of the palate and absence of

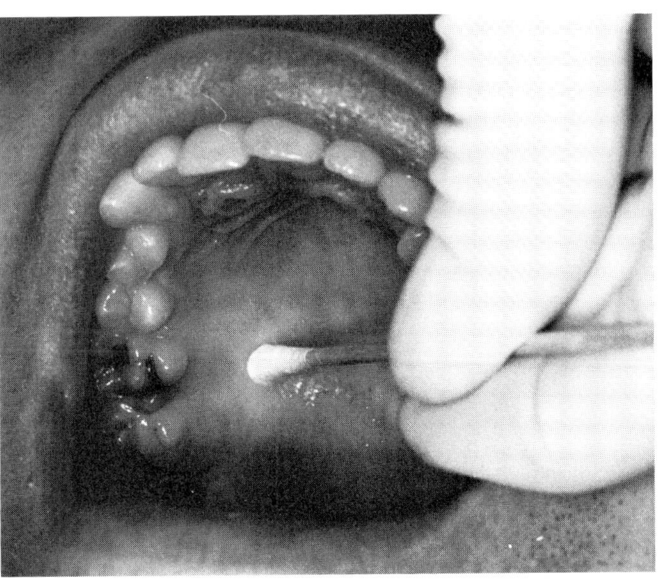

Figure 9–14
The injection site for the greater palatine block is probed at the depression caused by the greater palatine foramen. The palatal tissues around the greater palatine foramen are blanched to cause pressure anesthesia.

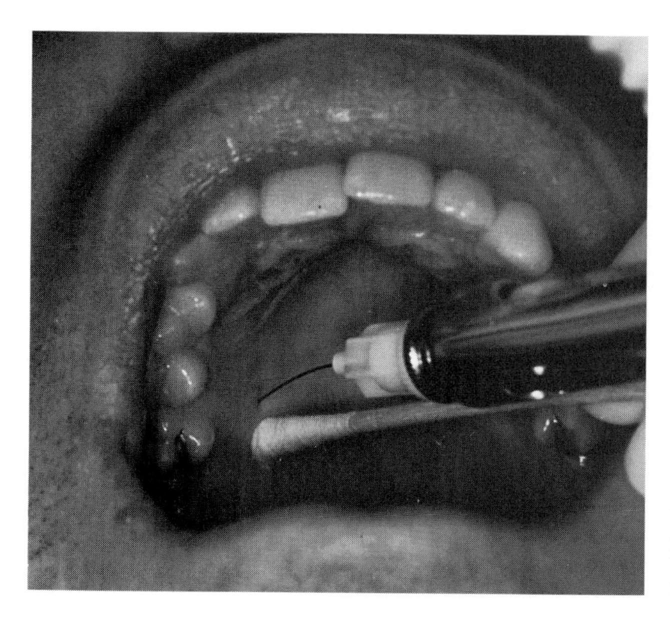

▼ Figure 9–15
Needle penetration of the palatal tissues for the greater palatine block is demonstrated at the depression created by the greater palatine foramen. The needle may bow slightly, and the tissue blanching is maintained during the injection.

discomfort during dental procedures. Some patients may become uncomfortable and may gag if their soft palate becomes inadvertently and harmlessly anesthetized, which is a distinct possibility given the proximity of the lesser palatine nerve.

▼ NASOPALATINE BLOCK

The **nasopalatine block** (nay-zo-**pal**-ah-tine) is useful for anesthesia of the bilateral anterior portion of the hard palate from the mesial of the right maxillary first premolar to the mesial of the left maxillary first premolar. Both the right and left nasopalatine nerves are anesthetized by this block. The nasopalatine block is used when palatal soft tissue anesthesia is required for two or more maxillary anterior teeth, such as with periodontal treatment.

If restorative procedures or extraction are to be performed on the anterior maxillary teeth, additional anesthesia may be indicated, such as the MSA or IO blocks, since the nasopalatine block does not provide pulpal anesthesia of these teeth.

Because the dense overlying palatal tissues adhere firmly to the underlying maxillary bone, the use of pressure anesthesia by an applicator stick to blanch the tissues will reduce patient discomfort. This pressure anesthesia to the tissues produces a dull ache to block pain impulses that arise from needle penetration. Slow deposition of the anesthetic solution will also help reduce patient discomfort.

In addition, the labial soft tissues between the maxillary central incisors may be anesthetized prior to the palatal injection. This initial injection technique allows some anesthesia of the nasopalatine nerve in the area of the palatal injection before needle insertion in the area.

Some dental professionals follow this initial injection for the nasopalatine block by directing the needle from the labial aspect through the interdental gingiva or papilla between the maxillary central incisors toward the incisive papilla to anesthetize the nasopalatine nerve. This may be followed by injection into the palatal area of the nerve if needed. This text discusses only a single-needle injection on the palate.

▼ Target Area and Injection Site for Nasopalatine Block

The target area for the nasopalatine block is both the right and left nasopalatine nerves as they enter the incisive foramen from the mucosa of the anterior hard palate, beneath the incisive papilla (Figure 9–16).

The injection site is the palatal tissues lateral to the incisive papilla, which is located at the midline, about 10 mm lingual to the maxillary central incisor teeth (Figure 9–17). The needle is inserted for the nasopalatine block into the previously blanched palatal tissues at a 45-degree angle to the palate (Figure 9–18). The initial depth of penetration of the needle into the palatal tissues will be 6–10 mm. The needle is

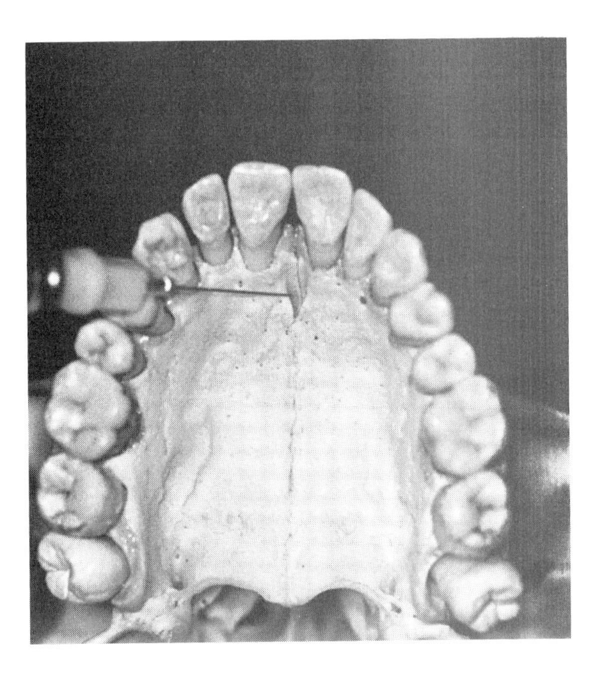

Figure 9–16
Target area for the nasopalatine block is the right and left nasopalatine nerves located at the incisive foramen on the anterior hard palate of the maxilla, lingual to the maxillary central incisors.

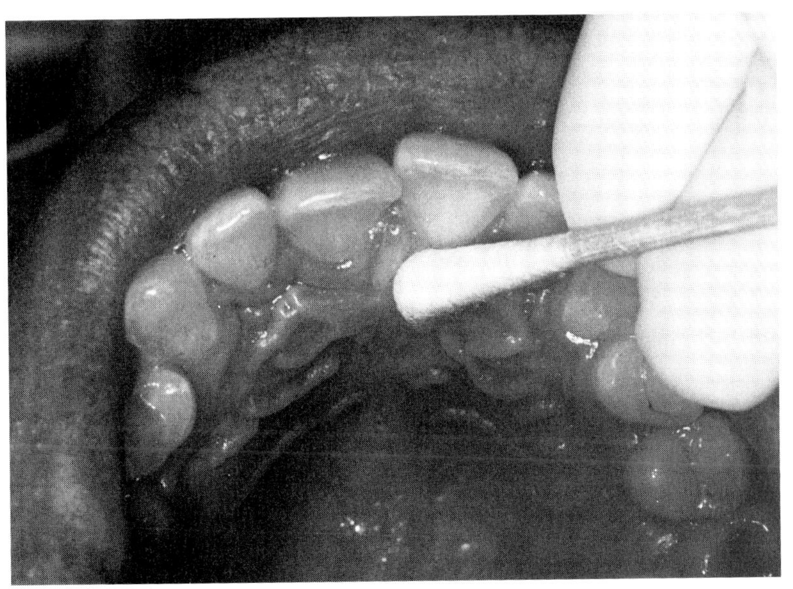

Figure 9–17
The injection site for the nasopalatine block is probed at the palatal tissues lateral to the incisive papilla on the anterior portion of the palate. The palatal tissues lateral to the incisive papilla have blanched to cause pressure anesthesia.

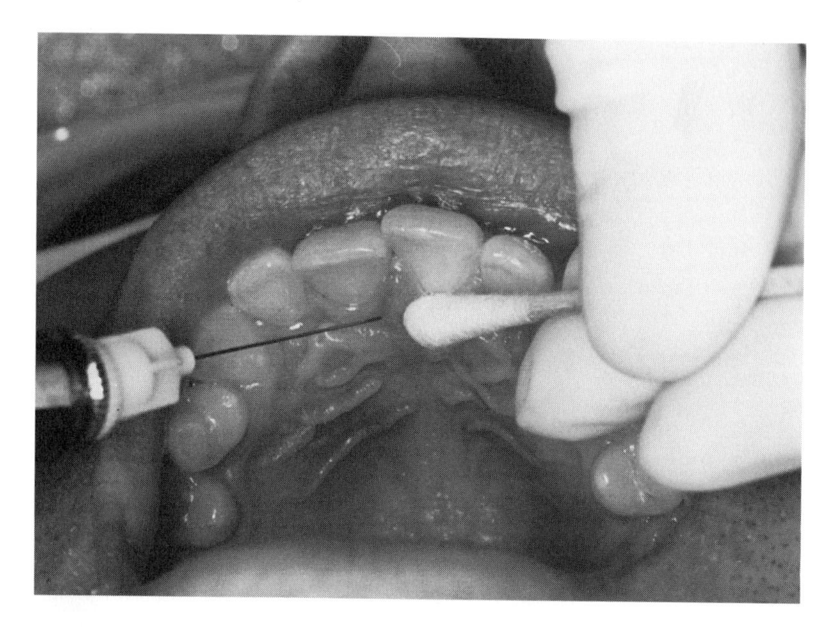

▼ Figure 9–18
Needle penetration of the palatal tissues for the nasopalatine block is demonstrated lateral to the incisive papilla. Blanching of the palatal tissues is continued during the injection.

advanced into the tissues until the maxillary bone is contacted. The needle is withdrawn 1 mm from the palatal tissue until the needle tip is directly over the opening of the incisive foramen before depositing the anesthetic solution.

Never insert the needle directly into the incisive papilla since this can be extremely painful to the patient. Do not enter deeply into the incisive canal, so as to reduce possible complications (discussed later). There is no reason for the needle to enter the incisive canal to achieve adequate palatal anesthesia.

▼ Symptoms and Possible Complications of Nasopalatine Block

Symptoms of the nasopalatine block include numbness in the anterior portion of the palate and absence of discomfort during dental procedures. It is important not to enter the incisive canal during the block. If the needle is advanced more than 5 mm into the incisive canal, the floor of the nose may be entered and infection of the nasal cavity may result. Other complications such as hematoma are extremely rare.

Mandibular Nerve Anesthesia

The mandibular nerve and its branches can be anesthetized in a number of ways depending on the tissues requiring anesthesia (Figure 9–1, Chart 9–2). Infiltration anesthesia of the mandible is not as successful as that of the maxilla since overall the mandible is more dense than the maxilla over similar teeth, especially in the area of the posterior teeth (see Chapter 3). For this reason, nerve blocks are preferred to local infiltrations in most parts of the mandible.

There is also substantial variation in the anatomy of local anesthetic landmarks of the mandibular bone and nerves compared with similar structures in the maxilla, complicating mandibular anesthesia. This text covers some of the most common mandibular variations.

Pulpal anesthesia is achieved through anesthesia of each nerve's dental branches as they extend into the pulp tissue by way of each tooth's apical foramen. The hard and soft tissues of the periodontium are anesthetized by way of each nerve's interdental and interradicular branches.

The inferior alveolar block is used for anesthesia of the mandibular teeth and their associated lingual tissues to the midline, as well as the facial tissues anterior to the mandibular first molar. The buccal block is useful for anesthesia of the tissues buccal to the mandibular molars. The mental block is used for anesthesia of the facial tissues anterior to the mental foramen (usually the mandibular premolars and anterior teeth). The incisive block is used for anesthesia of the teeth and associated facial tissues anterior to the mental foramen (usually the mandibular premolars and anterior teeth).

Chart 9–2

SUMMARY OF MANDIBULAR LOCAL ANESTHESIA OF THE TEETH AND ASSOCIATED TISSUES

Mandibular Tooth and Tissues	IA Block	Buccal Block	Mental Block	Incisive Block
Central Incisor				
Pulpal	X			X
Facial	X		X	X
Lingual	X			
Lateral Incisor				
Pulpal	X			X
Facial	X		X	X
Lingual	X			
Canine				
Pulpal	X			X
Facial	X		X	X
Lingual	X			
First Premolar				
Pulpal	X			X
Facial	X		X	X
Lingual				
Second Premolar				
Pulpal	X			X
Facial	X		X	X
Lingual	X			
First Molar				
Lingual/pulpal	X			
Buccal		X		
Second Molar				
Lingual/pulpal	X			
Buccal		X		
Third Molar				
Lingual/pulpal	X			
Buccal		X		

The anesthesia for anatomical variants is not included. IA = inferior alveolar.

INFERIOR ALVEOLAR BLOCK

The **inferior alveolar block** (al-**ve**-o-lar) or **IA block** or also mandibular block is the most commonly used injection in dentistry. It is used when dental procedures are performed on the mandibular teeth and pulpal anesthesia is needed, such as in the case of restorative procedures and extraction. This block also gives anesthesia of the lingual periodontium of all the mandibular teeth, as well as anesthesia of the facial periodontium of the mandibular anterior and premolar teeth, both of which are needed during periodontal treatment.

Additional use of the buccal nerve block may be considered if anesthesia of the buccal periodontium of the mandibular molars is also needed, such as with the periodontal treatment of these teeth.

Sometimes there is overlap of the left and right incisive nerves. The incisive nerve is a branch of the mandibular nerve that serves the pulp tissue of the mandibular anterior teeth.

If this is the case, a bilateral IA block can be used, but it is not recommended. More often, the use of an incisive block or local infiltration at the apices of the mandibular teeth that fail to achieve initial pulpal anesthesia may be indicated.

Local infiltrations on the facial surface of the anterior mandible are more successful than more posterior injections but less successful than injections over the maxilla in similar locations. Again, these differences in success rates are due to differences in the density of the facial bones.

Bilateral inferior alveolar injections are usually avoided unless absolutely needed. Bilateral mandibular injections produce complete anesthesia of the body of the tongue and floor of the mouth, which can cause difficulty with swallowing and speech, especially in patients with full or partial removable mandibular dentures, until the effects of the anesthetic drug wear off. Dental treatment planning can usually prevent the need for bilateral inferior alveolar injections.

Even though the IA block is the most commonly used dental injection, it is not always initially successful. This may mean that the patient must be reinjected to achieve the necessary anesthesia of the tissues. This lack of consistent success is due in part to anatomical variation in the height of the mandibular foramen on the medial side of the ramus and the great depth of soft tissue penetration required to achieve pulpal anesthesia. Other techniques to achieve mandibular anesthesia, such as the Gow-Gates nerve block, may also be employed but are not discussed in this text.

Target Area and Injection Site for IA Block

The target area for the IA block is slightly superior to the entry point of the inferior alveolar nerve into the mandibular foramen, overhung anteriorly by the lingula (Figure 9–19). The solution must be accurately deposited within 1 mm of the target area to achieve anesthesia. The adjacent anteriorly placed lingual nerve will also be anesthetized as the anesthetic solution diffuses.

The injection site for the IA block is the mandibular tissues on the medial border of the mandibular ramus at the correct height and anteroposterior direction for the injection (Figure 9–20). To locate the injection site, use mainly hard tissues as landmarks, such as the coronoid notch and occlusal plane of the mandibular molars, to reduce errors caused by patient soft tissue variance.

The correct height of the injection for the IA block is determined by palpating the coronoid notch, the greatest depression on the anterior border of the ramus.

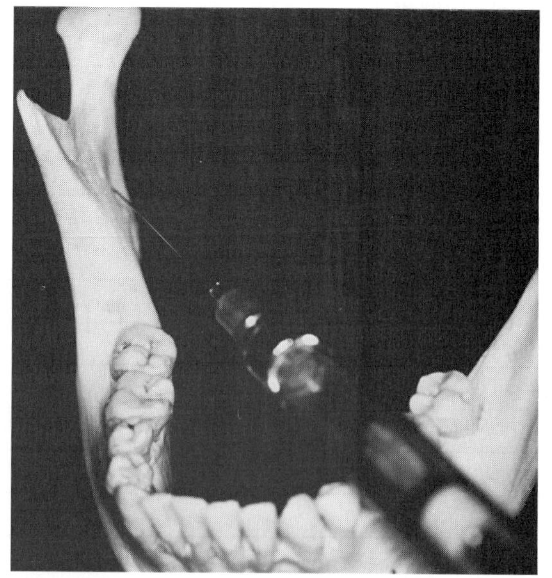

Figure 9–19
Target area for the IA block is the inferior alveolar nerve located at the mandibular foramen on the medial surface of the mandibular ramus, inferior to the lingula.

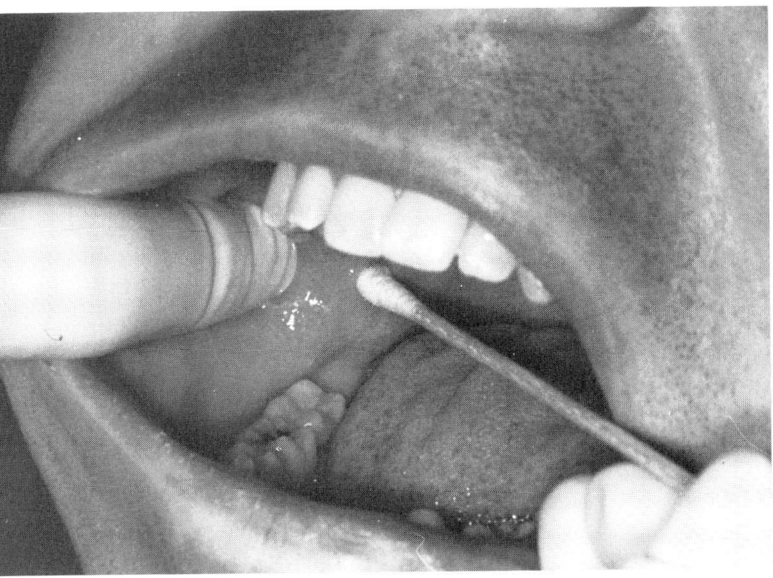

Figure 9–20
The injection site for the IA block is probed at the depth of the pterygomandibular space on the medial surface of the ramus.

To determine this level of height, it helps to visualize an imaginary horizontal line that extends posteriorly from the coronoid notch to the pterygomandibular fold as it turns upward toward the palate (Figure 9–21A and B). The pterygomandibular fold covers the deeper pterygomandibular raphe between the buccinator and superior pharyngeal constrictor muscles. This fold is accentuated as the patient opens the mouth wider.

This imaginary horizontal line showing the height of inferior alveolar nerve injection is also parallel to and 6–10 mm above the occlusal plane of the mandibular molar teeth in the majority of adults. For children or small adults, this imaginary horizontal line for the height of the injection should be at the occlusal plane of the mandibular molars. For edentulous patients or when the mandibular molars are absent, the mandibular foramen may appear to be higher than when the dentition is present because the occlusal plane of molars is not present as a guide.

The correct anteroposterior direction of the IA block injection is achieved at the same time as determining the correct height of the injection. To determine this direction, it helps to visualize an imaginary vertical line, two thirds to three fourths of the distance between the coronoid notch and the posterior border of the ramus (Figure 9–21 A and B). This area will be at the deepest or most posterior portion of the pterygomandibular space, lateral to the pterygomandibular fold and the sphenomandibular ligament (see Chapter 11 for review). The syringe barrel is usually over the contralateral mandibular second premolar.

Thus the injection site on the ramus for the IA block is the mandibular tissues at the intersection of the imaginary horizontal line for the height of injection with the imaginary vertical line for the anteroposterior direction of the injection (Figure 9–21 A and B). The needle is inserted into the tissues of the pterygomandibular space until the mandible is contacted (Figures 9–22 and 9–23). The needle is then withdrawn 1 mm from the tissues before injecting the anesthetic solution. The average depth of penetration by the needle into the soft tissues is 20–25 mm.

It is not necessary to deposit small amounts of the anesthetic solution as the needle enters the tissue for the IA block to anesthetize the adjacent anteriorly placed lingual nerve since anesthesia of the lingual nerve will occur through diffusion of the anesthetic solution placed near the inferior alveolar nerve. These small amounts injected early will not reduce tissue discomfort to the patient.

If bone is contacted too soon when trying to administer an IA block, the needle tip is located too far anterior on the ramus. Correction is made by withdrawing the needle slightly and bringing the syringe

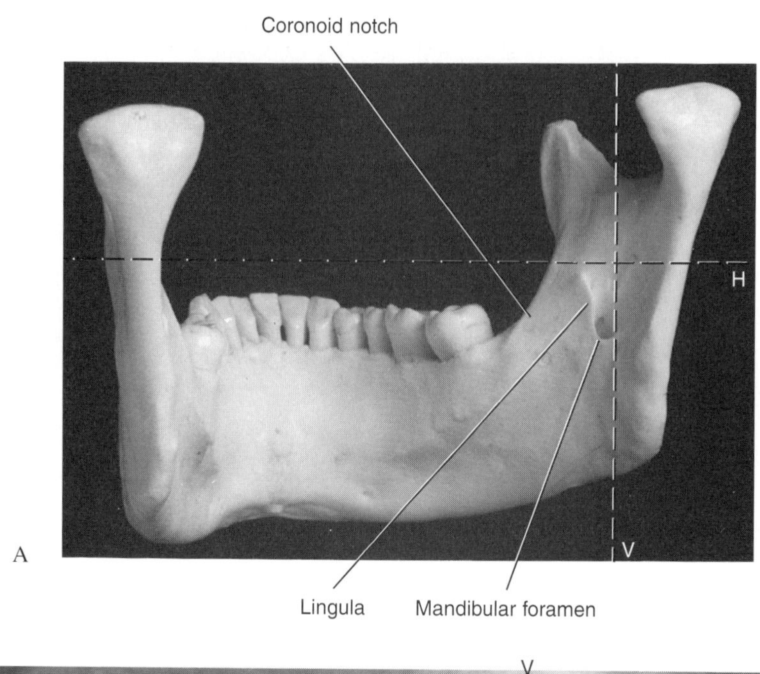

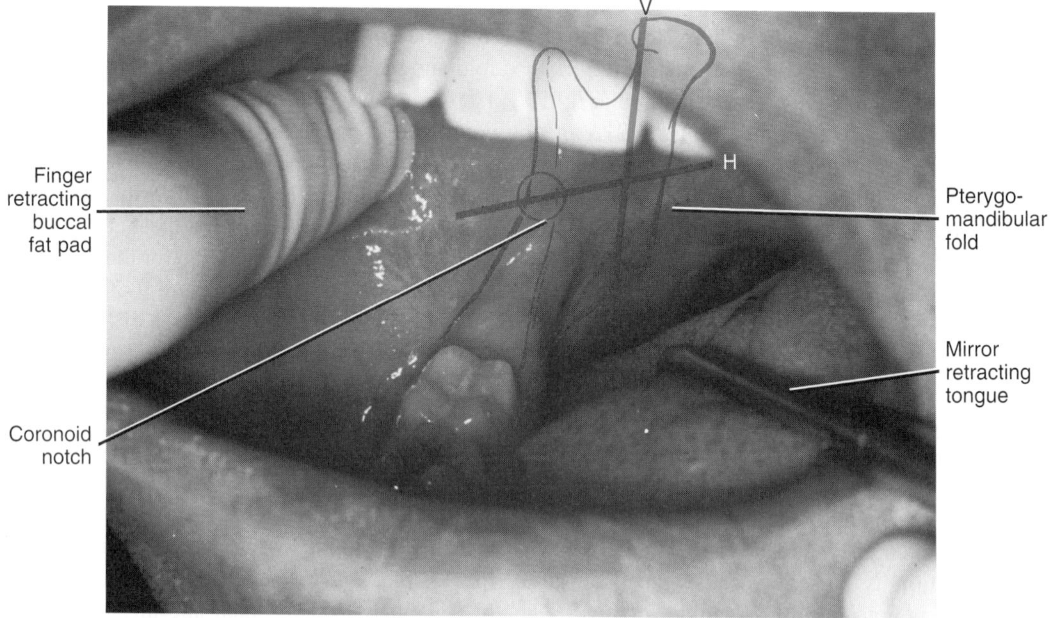

Figure 9–21

Imaginary horizontal line showing the correct height and imaginary vertical line showing the correct anteroposterior direction of the injection for the IA block. The intersection of these two lines is the correct injection site for this block. **A:** Posteromedial surface of the mandible. **B:** Oral view of the pterygomandibular space.

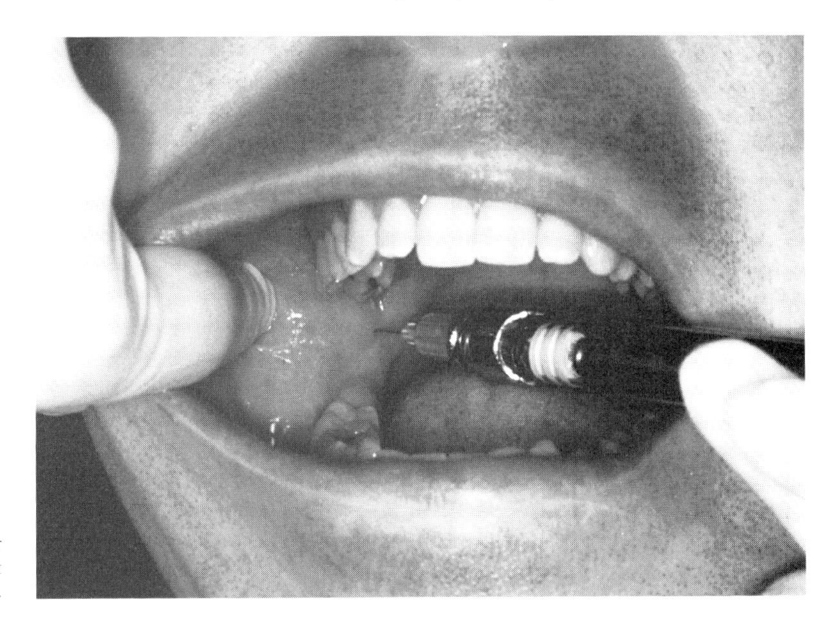

▼ Figure 9–22
Needle penetration of the mandibular tissues for the IA block is demonstrated at the depth of the pterygomandibular space.

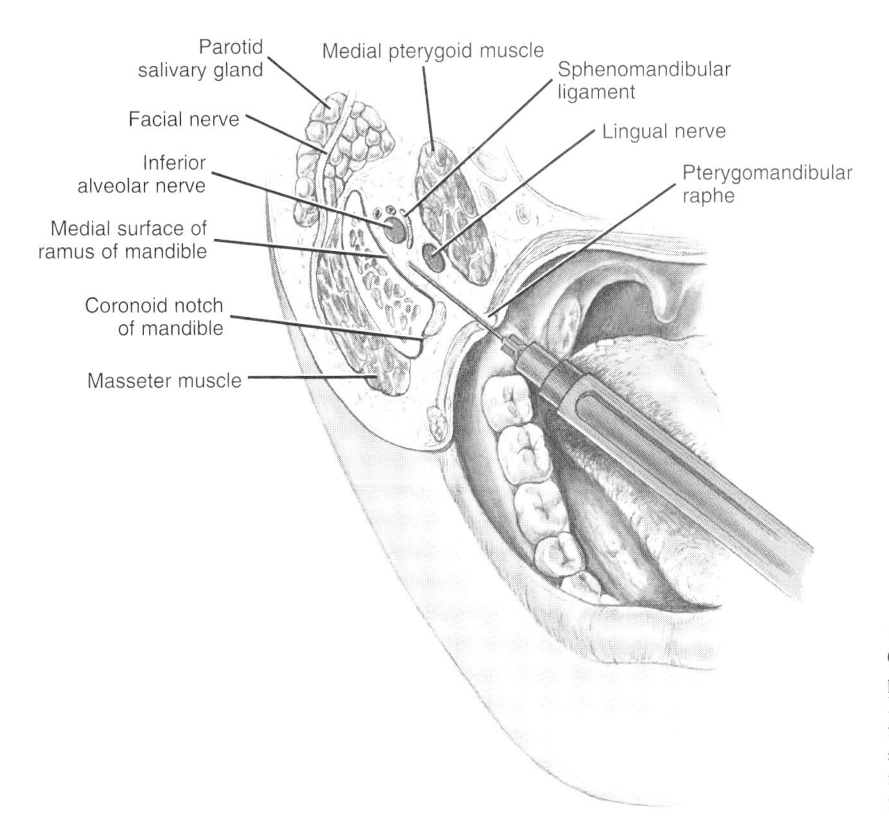

▼ Figure 9–23
Correct needle penetration into the pterygomandibular space during an IA block. If the needle is inserted too far posterior, it may enter the parotid salivary gland containing the facial nerve, causing a complication such as transient facial paralysis.

barrel more over the mandibular anterior teeth. This correction brings the needle tip more posteriorly when reinserted.

If bone is not contacted when trying to administer an IA block, the needle tip is located too far posterior on the ramus. Correction is made by withdrawing the needle slightly and bringing the syringe barrel more over the mandibular molars. This correction brings the needle tip more anteriorly when reinserted.

Do not deposit the anesthetic solution if bone is not contacted upon initial insertion of the needle for an IA block. The needle tip may be too posterior and thus resting within the parotid salivary gland near the facial nerve, resulting in complications (discussed later, Figure 9–23). If the insertion and deposition are too shallow and bone is not contacted, the medially located sphenomandibular ligament can become a physical barrier that stops the important diffusion of the anesthetic solution to the mandibular foramen and inferior alveolar nerve.

If there is failure of anesthesia, there may be accessory innervation of the mandibular teeth. Current thinking supports the mylohyoid nerve as the nerve that may be involved in this accessory mandibular innervation. To correct this problem, anesthesia of the mylohyoid nerve on the lingual border of the mandible is indicated. This additional anesthetic technique is discussed in most current local anesthesia textbooks.

Whenever a bifid inferior alveolar nerve is detected by noting a doubled mandibular canal on a radiograph (see Chapters 3 and 8 for more information), incomplete anesthesia of the mandible may follow an IA block. In many such cases, a second mandibular foramen, more inferiorly placed, exists. To correct this, deposit more anesthetic solution inferior to the normal anatomical landmarks.

Symptoms and Possible Complications of IA Block

Symptoms of the IA block include harmless numbness or tingling of the lower lip since the mental nerve, a branch of the inferior alveolar nerve, is anesthetized. This is a good indication that the inferior alveolar nerve is anesthetized but is not a reliable indicator of the depth of anesthesia, especially concerning pulpal anesthesia.

Another symptom is harmless numbness or tingling of the body of the tongue and floor of the mouth, which indicates that the lingual nerve, a branch of the mandibular nerve, is anesthetized. It is important to note that this anesthesia of the tongue may occur without anesthesia of the inferior alveolar nerve. Possibly the needle was not advanced deeply enough into the tissues to anesthetize the deeper inferior alveolar nerve. Finally, there is absence of discomfort during dental procedures with a successful IA nerve block.

Another symptom that sometimes can occur is "lingual shock" as the needle passes by the lingual nerve. The patient may make an involuntary movement, varying from a slight opening of the eyes to jumping in the chair. This symptom is only momentary, and anesthesia will quickly occur.

One of the complications with an IA block is transient facial paralysis if the facial nerve is mistakenly anesthetized. This can occur because of an incorrect administration of anesthetic into the deeper parotid salivary gland (Figure 9–23). Symptoms of this paralysis include an inability to close the eyelid and drooping of the lips on the affected side (see Chapter 4 for more information and Figure 4–6). Other complications such as hematoma can occur (see Chapter 6 for more information and Figure 6–16). Muscle soreness or limited movement of the mandible is rarely seen with this injection.

BUCCAL BLOCK

The buccal block or long buccal block is useful for anesthesia of the buccal periodontium of the mandibular molars, including the gingiva, periodontal ligament, and alveolar bone. This type of anesthesia is needed during dental procedures, such as certain restorative procedures or periodontal treatment of these teeth. Many times this block is administered when it is not needed. This is a very successful dental injection since the buccal nerve is readily located on the surface of the tissue and not within bone.

Target Area and Injection Site for Buccal Block

The target area for the buccal block is the buccal nerve (or long buccal nerve) as it passes over the anterior border of the ramus and through the buccinator muscle

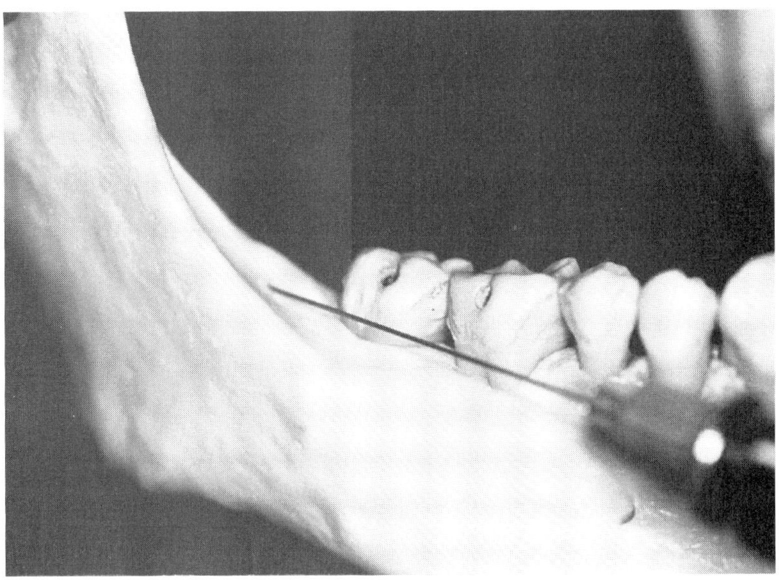

Figure 9–24
Target area for the buccal block is the buccal nerve located on the anterior border of the mandibular ramus.

before it enters the cheek region (Figure 9–24). Thus the injection site is the buccal tissues distal and buccal to the most distal molar tooth in the arch, on the anterior border of the ramus (Figure 9–25).

The needle is advanced until it contacts the mandible (Figure 9–26). The usual depth of penetration of the needle into the buccal tissues is only 1–2 mm. If the overlying buccal tissue balloons, deposition of the solution should be stopped. If the solution runs out of the puncture site, the deposition should be stopped and then the needle advanced deeper into the tissue and the injection continued.

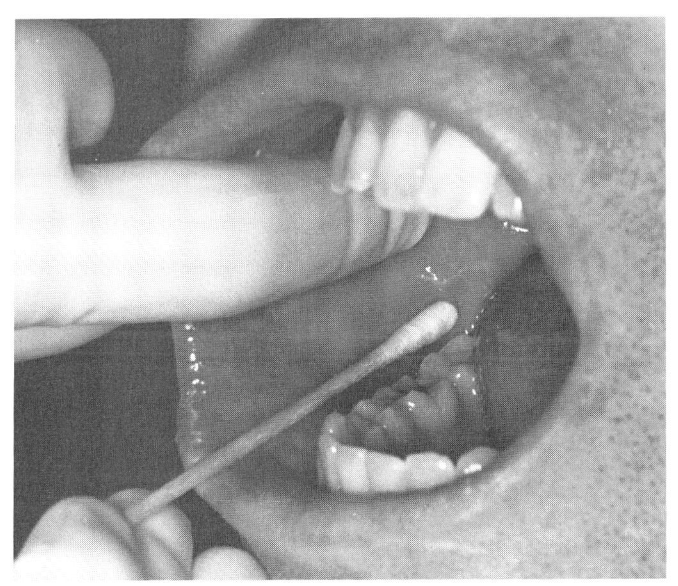

Figure 9–25
The injection site for the buccal block is probed in the buccal tissues that are distal and buccal to the most distal molar tooth in the arch.

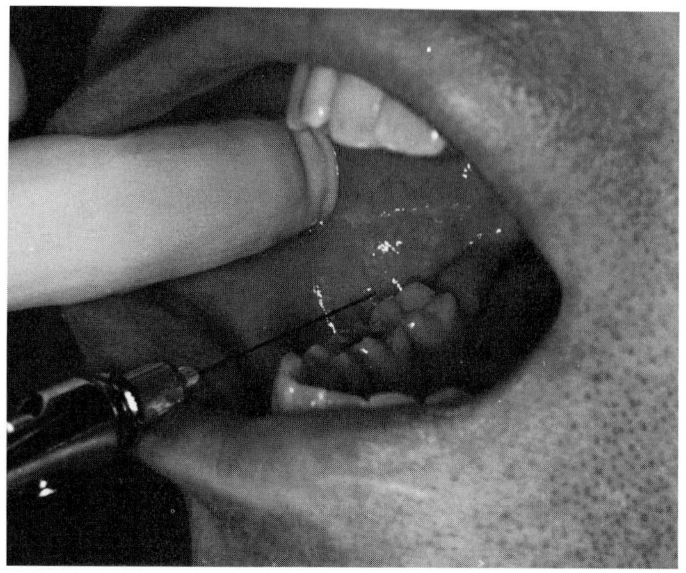

Figure 9–26
Needle penetration of the buccal tissues for the buccal block is demonstrated, distal and buccal to the most distal molar tooth in the arch.

Symptoms and Possible Complications of Buccal Block

The patient rarely feels any symptoms of the buccal nerve block because of the location and small size of the anesthetized area. There is usually only absence of discomfort with dental procedures. The complication of a hematoma rarely occurs.

MENTAL BLOCK

The **mental block** (**ment**-il) is used to anesthetize the facial periodontium of the mandibular premolars and anterior teeth on one side, including the gingiva, periodontal ligament, and other alveolar tissues. This is useful before dental procedures, such as periodontal treatment of these teeth. It is a very successful injection because of the ease of access to the nerve.

If restorative procedures are also to be performed on these lower anterior teeth, local infiltrations at the facial surface of apices of the involved teeth may be additionally performed, or administration of an incisive block may be considered instead (discussed later in this chapter).

Target Area and Injection Site of Mental Block

The target area for the mental block is the mental nerve before it enters the mental foramen to merge with the incisive nerve to form the inferior alveolar nerve (Figure 9–27).

The mental foramen is usually located on the surface of the mandible between the apices of the first and second mandibular premolars, yet it may be anterior or posterior to this site. The mental foramen in adults faces posterosuperiorly. The mental foramen can be located on a radiograph before the injection to allow a better determination of its position.

To locate the mental foramen for the mental block, palpate intraorally the depth of the mucobuccal fold between the apices of the mandibular first and second premolars or at a site indicated by a radiograph until a depression is felt, surrounded by smoother bone (Figure 9–28). The patient will comment that pressure in this area produces soreness as the mental nerve is compressed against the mandible near the foramen.

The insertion site for a mental block is the mucobuccal fold tissues directly over or slightly anterior to the depression created by the mental

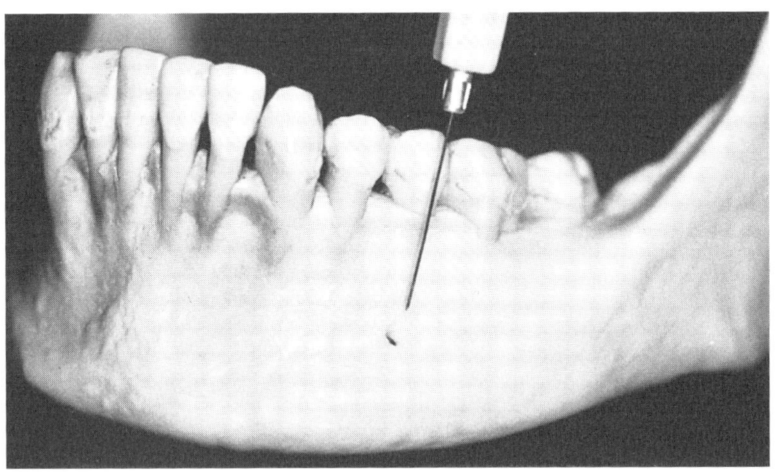

▼ Figure 9–27
Target area for the mental block is the mental nerve located at the mental foramen on the surface of the mandible, usually between the apices of the mandibular first and second premolar. This is the same target area for the incisive block.

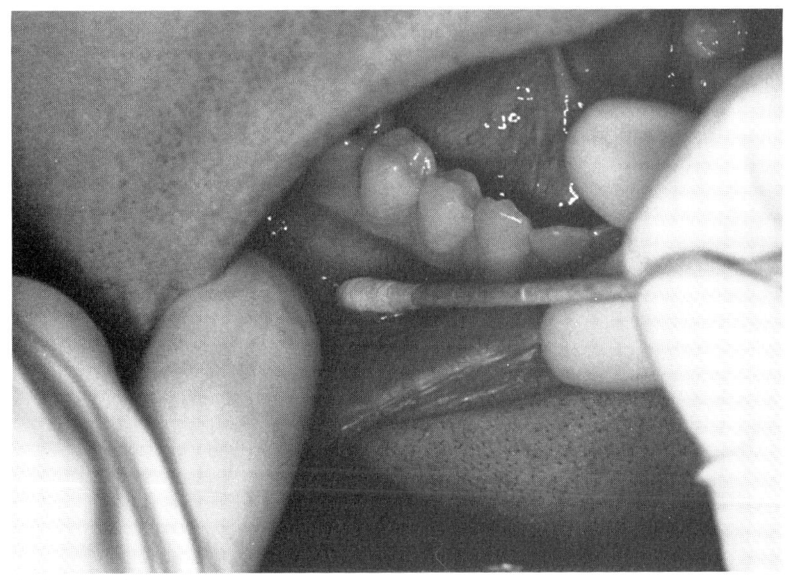

▼ Figure 9–28
The injection site for the mental block is probed at the depression caused by the mental foramen, usually located in the depth of the mucobuccal fold between the apices of the mandibular first and second premolar. This is the same injection site for the incisive block.

foramen (Figure 9–29). The needle is advanced until the level of the foramen is reached, avoiding contact with the mandible. The depth of penetration into the tissues will be 5–6 mm. There is no need to enter the mental foramen to achieve facial anesthesia.

Symptoms and Possible Complications of Mental Block

The symptoms of a mental block are harmless tingling or numbness of the lower lip and absence of discomfort during dental procedures. The complication of a hematoma rarely occurs.

INCISIVE BLOCK

The **incisive block** (in-**sy**-ziv) anesthetizes the pulp tissue and facial tissues of the mandibular teeth anterior to the mental foramen (usually the first premolar and anterior teeth). This block is useful for restorative and periodontal treatment procedures done on these teeth. If extraction procedures are to be performed, an inferior alveolar block would be administered instead since the incisive block does not provide lingual anesthesia. This block has a high success rate since the incisive nerve is readily accessible.

Target Area and Injection Site for Incisive Block

The target area for the incisive block is the mental foramen, the opening of the mandibular canal where the incisive nerve is located. The mental foramen is also where the mental nerve enters to join with the incisive nerve to form the inferior alveolar nerve (Figure 9–30).

The mental foramen is usually located on the surface of the mandible between the apices of the first and second mandibular premolars, yet it may be anterior or posterior to this site. The mental foramen in adults faces posteriorly. The mental foramen can be located on a radiograph before the injection to allow a better determination of its position.

The injection site for an incisive block is the same as for a mental nerve block: directly over or slightly anterior to the depression created by the mental

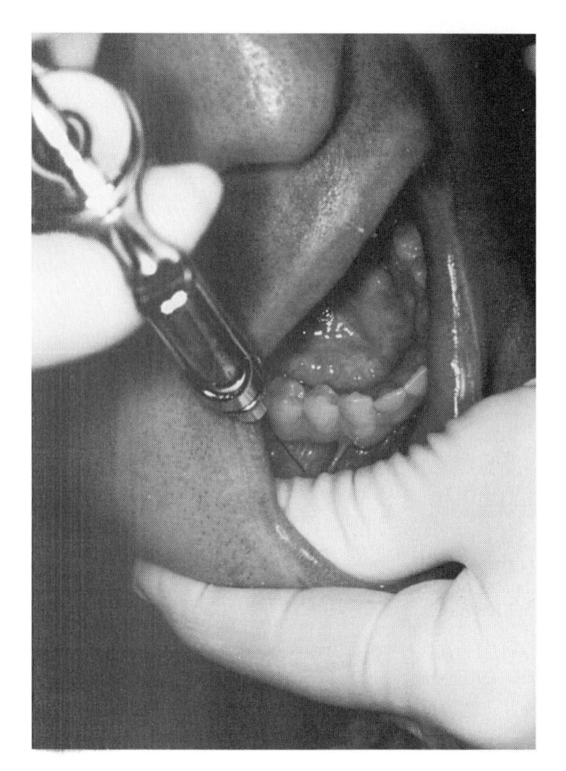

Figure 9–29
Needle penetration of the mucobuccal fold tissues for the mental block is demonstrated at the depression created by the mental foramen.

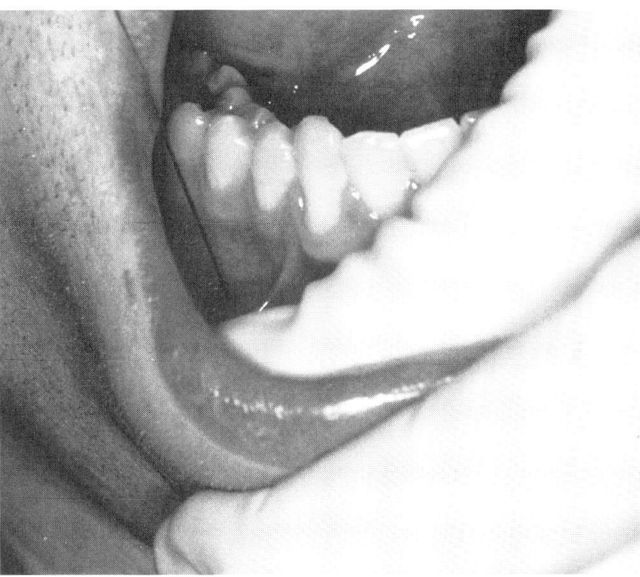

▼ Figure 9–30
Needle penetration of the mucobuccal fold tissues for the incisive block is demonstrated at the depression created by the mental foramen. Pressure near the mental foramen is maintained.

foramen, in the depth of the mucobuccal fold between the apices of the first and second mandibular premolars. More anesthetic solution is deposited within the tissue for the incisive block while applying pressure during the injection. This pressure forces more anesthetic solution into the mental foramen, thus anesthetizing first the shallow mental nerve and then the deeper incisive nerve. It is important to note that it is not necessary to have the needle enter the mental foramen for a successful incisive block.

While injecting these increased amounts of anesthetic solution during an incisive block, the facial tissue in the area may balloon slightly without complications. Anesthesia of the tissues innervated by the mental nerve will precede that of the deeper incisive nerve's tissues. If an inadequate amount of anesthetic solution is placed initially in the mental foramen, with subsequent lack of pulpal anesthesia, correction can be made by injecting more solution into the proper region and applying pressure again to the site.

▼ Symptoms and Possible Complications of Incisive Block

The symptoms of an incisive block are the same symptoms of a mental block, except that there is pulpal anesthesia of the involved teeth. There also is an absence of discomfort during dental procedures. As with a mental block, a hematoma rarely occurs.

REVIEW QUESTIONS

1. A hematoma can result from an incorrectly administered PSA block because the needle was overinserted and penetrated the:
   A. parotid salivary gland.
   B. pterygoid plexus of veins.
   C. floor of the nose.
   D. facial or cranial nerve VII.

2. Which of the following local anesthetic blocks has the same target area as the incisive block?
   A. Nasopalatine block
   B. Greater palatine block
   C. Inferior alveolar block
   D. Buccal block
   E. Mental block

3. Which of the following nerves is not anesthetized during an inferior alveolar block?
   A. Buccal nerve
   B. Lingual nerve
   C. Mental nerve
   D. Incisive nerve

4. Which of the following anesthetic blocks uses pressure anesthesia of the tissue with an applicator stick to reduce patient discomfort?
   A. PSA block
   B. IO block
   C. Greater palatine block
   D. Inferior alveolar block
   E. Buccal block

5. Which of the following tissues are anesthetized during an IO block?
   A. Bilateral anterior hard palate
   B. Buccal periodontium of maxillary molars
   C. Upper lip, side of nose, and lower eyelid
   D. Lingual periodontium of maxillary anteriors
   E. Side of face, upper eyelid, and bridge of nose

6. If the mesiobuccal root of the maxillary first molar is not anesthetized by a PSA block, the dental professional should:
   A. reinject at the same site as for the PSA block.
   B. perform a buccal block injection.
   C. administer an MSA block injection.
   D. perform a nasopalatine block injection.

7. Which of the following is an important landmark to palpate before an inferior alveolar block?
   A. Coronoid notch
   B. Tongue
   C. Buccal fat pad
   D. Mental foramen

8. The injection site for the greater palatine block is located on the palate near the:
   A. maxillary first premolar.
   B. maxillary second or third molar.
   C. incisive papilla.
   D. midline portion.

9. If an extraction of a maxillary lateral incisor is scheduled, which of the following anesthetic blocks should be administered besides the IO block?
   A. PSA block
   B. MSA block
   C. Nasopalatine block
   D. Greater palatine block

10. Transient facial paralysis can occur with an incorrectly administered:
    A. PSA block.
    B. MSA block.
    C. nasopalatine block.
    D. inferior alveolar block.
    E. mental block.

Chapter 10

Lymphatics

▼ *This chapter discusses the following topics:*

I. Lymphatic System
   A. Lymphatic Vessels
   B. Lymph Nodes
   C. Tonsillar Tissue
   D. Lymphatic Ducts
II. Lymph Nodes of the Head and Neck
   A. Lymph Nodes of the Head
   B. Cervical Lymph Nodes
III. Tonsils
   A. Palatine and Lingual Tonsils
   B. Pharyngeal and Tubal Tonsils
IV. Lymphadenopathy
V. Metastasis and Cancer

▼ *Key words*

Afferent vessels (**af**-er-int) Type of lymphatic vessel in which lymph flows into the lymph node.

Efferent vessel (**ef**-er-ent) Type of lymphatic vessel in which lymph flows out of the lymph node in the area of the node's hilus.

Hilus (**hi**-lus) Depression on one side of the lymph node where lymph flows out by way of an efferent lymphatic vessel.

Lymph (limf) Tissue fluid that drains from the surrounding region and into the lymphatic vessels.

Lymphadenopathy (lim-fad-in-**op**-ah-thee) Process in which there is an increase in the size and a change in the consistency of lymphoid tissue.

Lymphatic ducts (lim-**fat**-ik dukts) Larger lymphatic vessels that drain smaller vessels and then empty into the venous system.

Lymphatic vessels (lim-**fat**-ik **ves**-els) System of channels that drain tissue fluid from the surrounding regions.

Lymph nodes (limf nodes) Organized bean-shaped lymphoid tissue that filters the lymph by way of lymphocytes to fight disease and is grouped into clusters along the connecting lymphatic vessels.

Metastasis (meh-**tas**-tah-sis) Spread of cancer from the original or primary site to another or secondary site.

Primary node Lymph node that drains lymph from a particular region.

Secondary node Lymph node that drains lymph from a primary node.

Tonsillar tissue (**ton**-sil-lar) Masses of lymphoid tissue located in the oral cavity and pharynx to protect the body against disease processes.

Chapter 10

Lymphatics

▼ ***After studying this chapter, the reader should be able to:***

1. Define and pronounce all the key words and anatomical terms in this chapter.
2. List and discuss the lymphatic system and its components.
3. Locate and identify all the major groups of lymph nodes of the head and neck on a diagram and a patient.
4. Locate and identify all the tonsillar tissues of the head and neck on a diagram and intraorally on a patient.
5. Identify the patterns of lymph drainage for each head and neck tissue or region.
6. Describe and discuss lymphadenopathy of lymphoid tissue.
7. Discuss the spread of cancer in the head and neck region and its relationship to lymph nodes.
8. Correctly complete the identification exercise and review questions for this chapter.
9. Integrate the knowledge about head and neck lymphatics into the clinical practice of patient examination and any related lymphatic diseases of the region.

Lymphatic System

The **lymphatics** (lim-**fat**-iks) are part of the immune system and help fight disease processes, as well as serve other functions in the body. The lymphatic system consists of a network of lymphatic vessels linking lymph nodes throughout most of the body. Tonsillar tissue located in the oral cavity and pharynx is also part of the lymphatic system. Although not part of the lymphatic system, the thymus gland also works as part of the immune system and is discussed in Chapter 7.

▼ LYMPHATIC VESSELS

The ***lymphatic vessels*** (lim-**fat**-ik **ves**-els) are a system of channels that mainly parallel the venous blood vessels in location yet are more numerous (Figure 10–1). Tissue fluid drains from the surrounding region

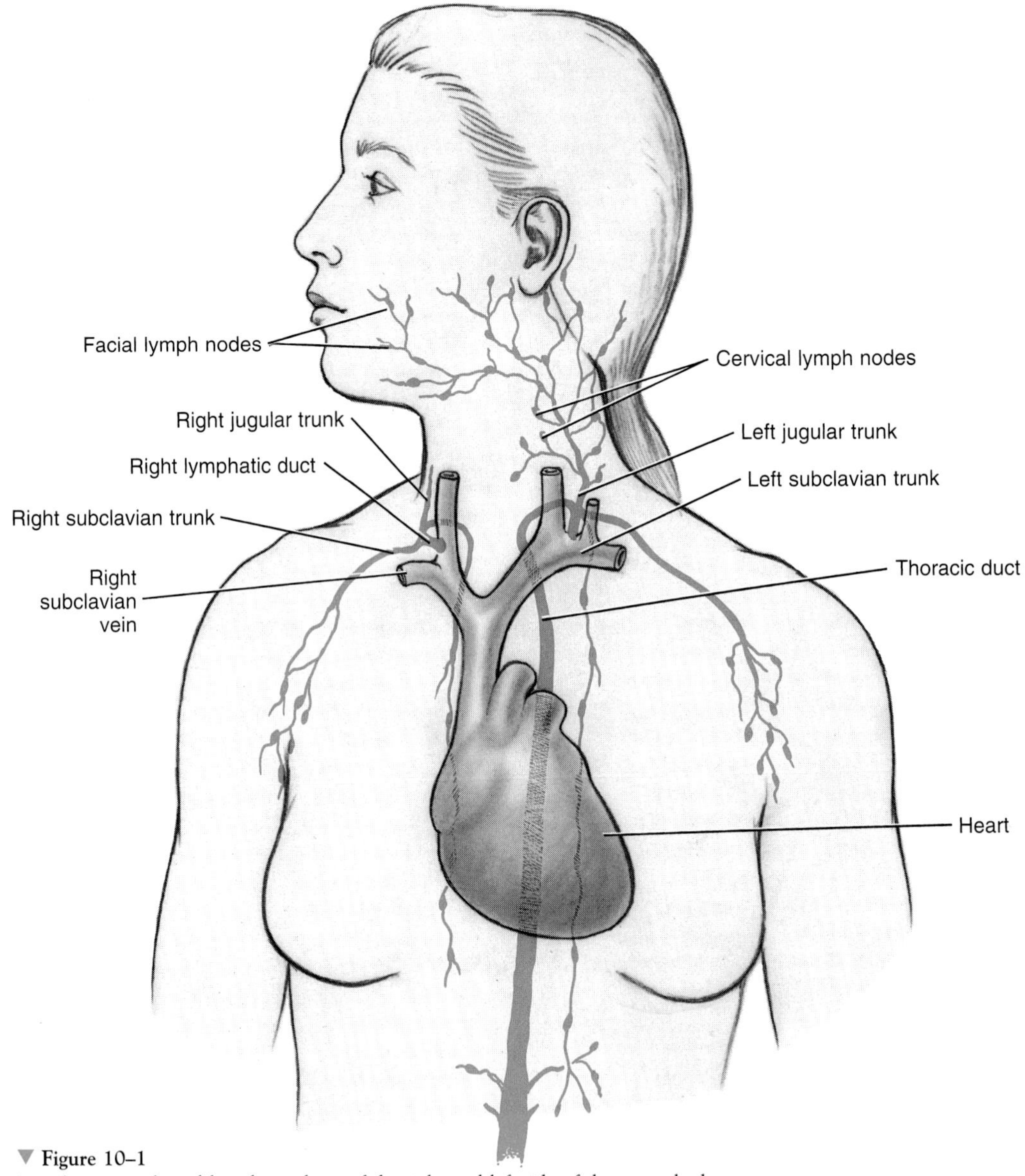

▼ **Figure 10–1**
Lymphatic vessels and lymphatic ducts of the right and left side of the upper body.

Table 10–2

LYMPH NODE DRAINAGE PATTERNS FOR TISSUES OF THE SCALP AND FACE

Tissues	Primary Nodes	Secondary Nodes
Scalp	Occipital, retroauricular, anterior auricular, superficial parotid, and accessory	Superior deep cervical, inferior deep cervical, and supraclavicular
Lacrimal gland	Superficial parotid	Superior deep cervical
External ear	Retroauricular, anterior auricular, and superficial parotid	Superior deep cervical
Middle ear	Deep parotid	Superior deep cervical
Pharyngeal tonsil and tubal tonsil	Superior deep cervical	Inferior deep cervical
Paranasal sinuses	Retropharyngeal	Superior deep cervical
Infraorbital region and nasal cavity	Malar, nasolabial, retropharyngeal, and superior deep cervical	Submandibular and deep cervical
Cheek	Buccal, malar, mandibular, and submandibular	Superior deep cervical
Parotid gland	Deep parotid	Superior deep cervical
Upper lip	Submandibular	Superior deep cervical
Lower lip	Submental	Submandibular and deep cervical
Chin	Submental	Submandibular and deep cervical
Sublingual gland	Submandibular	Superior deep cervical
Submandibular gland	Submandibular	Superior deep cervical

may be present in the patient. It is therefore important that a dental professional understand the relationship between the node location and node drainage patterns for the tissues of the oral cavity (Table 10–1), face and scalp (Table 10–2), and neck (Table 10–3).

The dental professional also needs to keep in mind that these lymph nodes drain not only intraoral structures, such as the teeth, but also the eyes, ears, nasal cavity, and deeper areas of the throat. Many times a patient needs a referral to a physician when lymph nodes are palpable due to a disease process in these other regions.

LYMPH NODES OF THE HEAD

The lymph nodes of the head are in either a superficial or deep position relative to the surrounding tissues. All the nodes of the head drain either the right or left tissues in the area depending on their location.

Table 10–3

LYMPH NODE DRAINAGE PATTERNS FOR TISSUES OF THE NECK

Tissues	Primary Nodes	Secondary Nodes
Superficial anterior cervical triangle	Anterior jugular	Inferior deep cervical
Superficial lateral and posterior cervical triangles	External jugular and accessory	Deep cervical and supraclavicular
Deep posterior cervical triangle	Inferior deep cervical	
Pharynx	Retropharyngeal	Superior deep cervical
Thyroid gland	Superior deep cervical	Inferior deep cervical
Larynx	Laryngeal	Inferior deep cervical
Esophagus	Superior deep cervical	Inferior deep cervical
Trachea	Superior deep cervical	Inferior deep cervical

Superficial Lymph Nodes of the Head

There are five groups of superficial lymph nodes in the head: the occipital, retroauricular, anterior auricular, superficial parotid, and facial nodes (Figure 10–3).

OCCIPITAL LYMPH NODES

The **occipital lymph nodes** (ok-**sip**-it-al) (1-3 in number) are located bilaterally on the posterior base of the head in the occipital region and drain this portion of the scalp. Having the patient lean the head forward allows for effective palpation for these nodes (Figure 10–4). The occipital nodes empty into the inferior deep cervical nodes of the neck.

RETROAURICULAR, AURICULAR, AND SUPERFICIAL PAROTID LYMPH NODES

The nodes known as **retroauricular lymph nodes** (ret-ro-aw-**rik**-you-lar) or mastoid or posterior auricular lymph nodes (1-3 in number) are located posterior to each ear, where the sternocleidomastoid muscle inserts on the mastoid process. The **anterior auricular lymph nodes** (aw-**rik**-you-lar) (1-3 in number) are located anterior to each ear.

The **superficial parotid lymph nodes** (pah-**rot**-id) or paraparotid lymph nodes (up to 10 in number along with the deep parotid group) are just superficial to each parotid salivary gland. Some anatomists group the anterior auricular and superficial parotid lymph nodes together.

The retroauricular, anterior auricular, and superficial parotid nodes drain the external ear, lacrimal gland, and adjacent regions of the scalp and face. All of these nodes empty into the superior deep cervical lymph nodes. Gentle pressure on the face and scalp around each ear will lead to effective palpation for these nodes.

FACIAL LYMPH NODES

The final group of superficial nodes of the head is the **facial lymph nodes** (up to 12 in number), which are positioned along the length of the facial vein. These nodes are typically small and variable in number. The

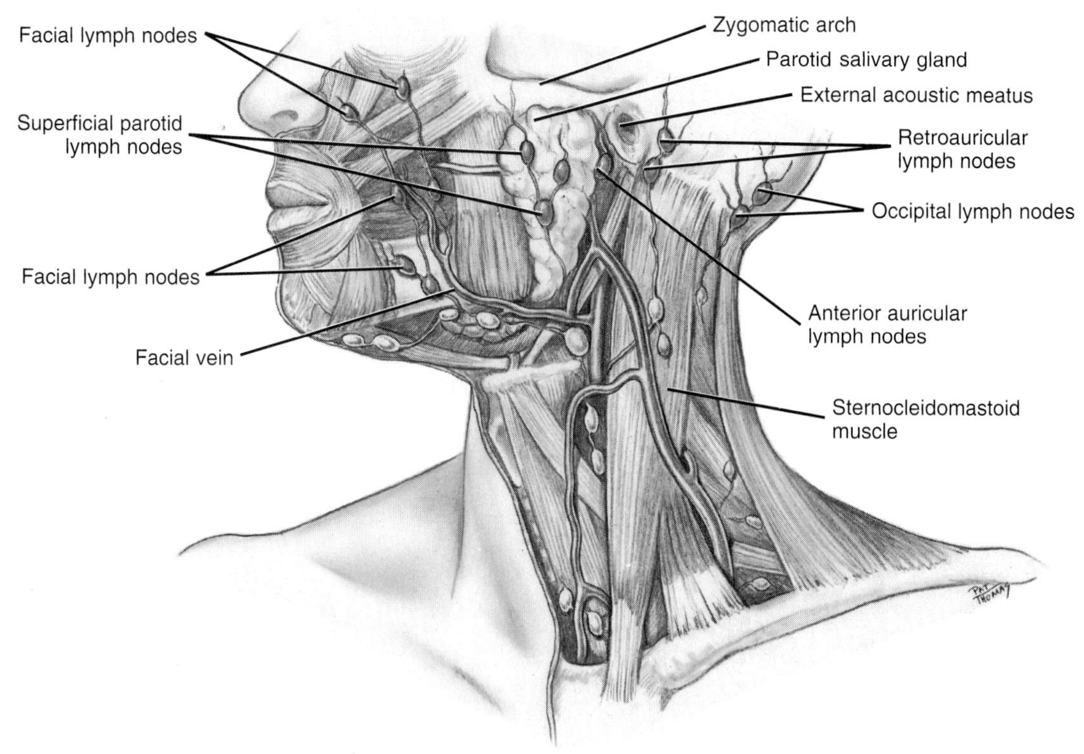

Figure 10–3
Superficial lymph nodes of the head and associated structures.

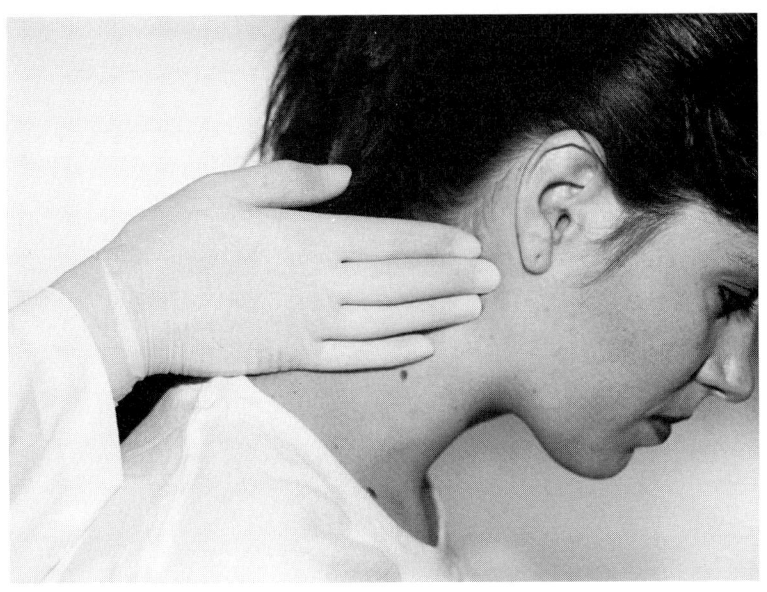

▼ **Figure 10–4**
Palpating the occipital lymph nodes by having the patient's head forward.

facial nodes are further categorized into four subgroups: malar, nasolabial, buccal, and mandibular nodes.

Nodes in the infraorbital region are the **malar lymph nodes** (**may**-lar) or infraorbital lymph nodes. Nodes near the nose are the **nasolabial lymph nodes** (nay-zo-**lay**-be-al). Nodes at the labial commissure and just superficial to the buccinator muscle are the **buccal lymph nodes.** Nodes over the surface of the mandible, anterior to the masseter muscle, are the **mandibular lymph nodes** (man-**dib**-you-lar).

Each facial node group drains the skin and mucous membranes where the nodes are located. The facial nodes also drain from one to the other, superior to inferior, and then finally drain into the submandibular nodes. To palpate effectively for the facial nodes, apply gentle pressure on each side of the face, moving from the infraorbital region to the labial commissure and then to the surface of the mandible.

▼ Deep Lymph Nodes of the Head

There are deep lymph nodes in the head region that can never be palpated during an extraoral examination due to their depth in the tissues. The deep nodes of the face include the deep parotid and retropharyngeal nodes (Figure 10–5).

▼ DEEP PAROTID LYMPH NODES

The **deep parotid lymph nodes** (pah-**rot**-id) (up to 10 in number along with the superficial parotid nodes) are located deep in the parotid salivary gland and drain the middle ear, auditory tube, and parotid salivary gland.

▼ RETROPHARYNGEAL LYMPH NODES

Also located near the deep parotid nodes and at the level of the atlas, the first cervical vertebra, are the **retropharyngeal lymph nodes** (ret-ro-far-**rin**-je-al) (up to 3 in number), which drain the pharynx, palate, paranasal sinuses, and nasal cavity. All of these deep nodes drain into the superior deep cervical lymph nodes of the neck.

▼ CERVICAL LYMPH NODES

Lymph nodes of the neck or cervical lymph nodes can be in either a superficial or deep position in the tissues. All the cervical lymph nodes drain either the right or left portions of the tissues where they are located, except the midline submental nodes that drain the tissues in the region bilaterally.

Many clinicians record all the cervical lymph nodes (except those directly inferior to the chin) in relationship to the sternocleidomastoid muscle. Thus, the cer-

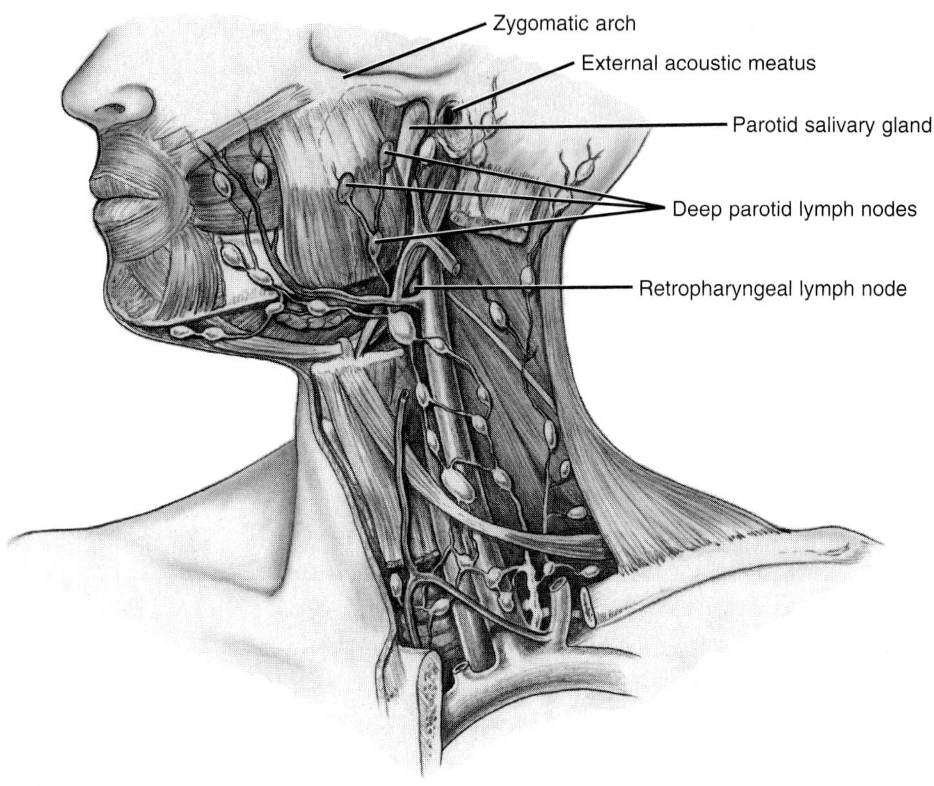

▼ Figure 10–5
Deep lymph nodes of the head and associated structures.

vical nodes are placed in three overlapping categories: superior or inferior, anterior or posterior, superficial or deep. This text is more specific but recognizes this other method as workable.

▼ Superficial Cervical Lymph Nodes

There are four groups of superficial cervical lymph nodes: the submental, submandibular, external jugular, and anterior jugular nodes (Figure 10–6).

▼ SUBMENTAL LYMPH NODES

The **submental lymph nodes** (sub-**men**-tal) (2-3 in number) are located inferior to the chin in the submental fascial space. The submental nodes are also just superficial to the mylohyoid muscle, near the midline between the mandible's symphysis and hyoid bone. The submental nodes drain both sides of the chin and the lower lip, floor of the mouth, apex of the tongue, and mandibular incisors and associated tissues. The submental nodes then empty into the submandibular nodes or directly into the deep cervical nodes.

▼ SUBMANDIBULAR LYMPH NODES

Submandibular lymph nodes (sub-man-**dib**-you-lar) (3-6 in number) are located at the inferior border of the ramus of the mandible, just superficial to the submandibular salivary gland, and within the submandibular fascial space. The submandibular nodes drain the cheeks, upper lip, body of the tongue, anterior portion of the hard palate, and teeth and associated tissues, except the mandibular incisors and maxillary third molars.

These nodes may be secondary nodes for the submental nodes and facial regions. Lymphatics from both the sublingual and submandibular salivary glands also drain into the submandibular nodes. The submandibular nodes then empty into the superior deep cervical nodes.

For the submental and submandibular nodes directly inferior to the chin, having the patient lower

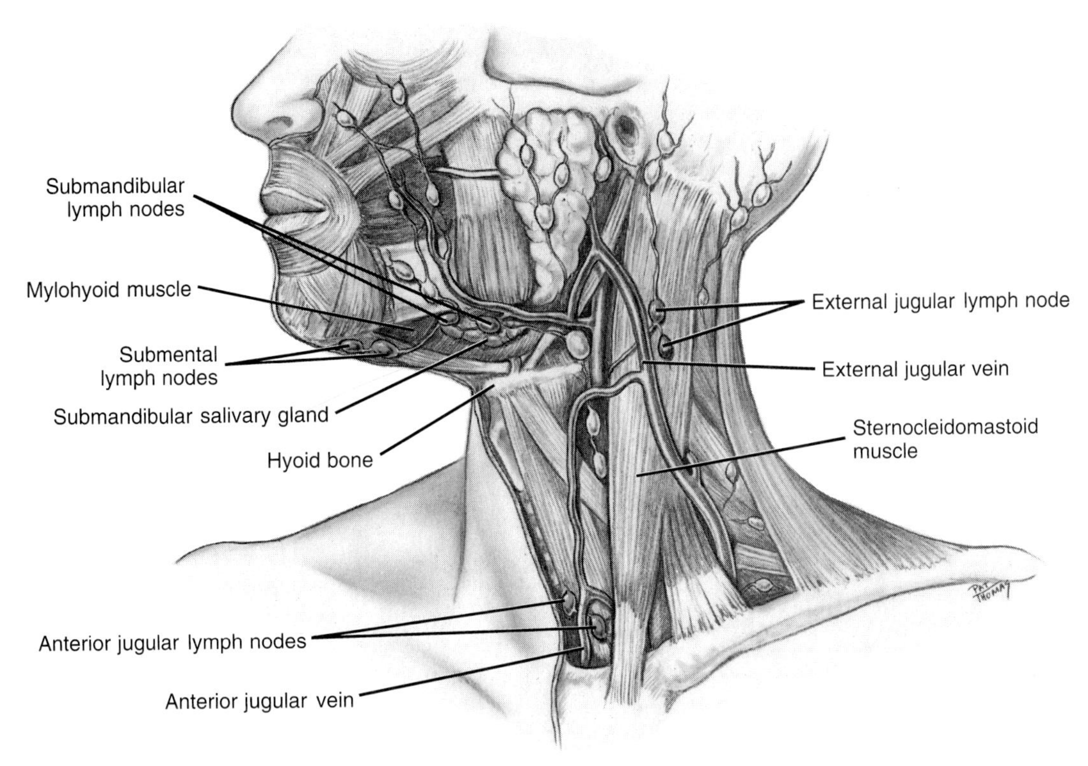

▼ **Figure 10–6**
Superficial cervical lymph nodes and associated structures.

the chin will allow more effective palpation (Figure 10–7). Pushing the tissue in the area from the patient's one side over to the opposite side where it is being grasped and rolled over the angle of the mandible allows any palpable nodes to be located. The opposite side is examined in the same way. Some dental professionals recommend that the patient put the chin up with the mouth slightly open and the tip of the tongue on the hard palate to palpate the nodes directly inferior to the chin.

▼ EXTERNAL JUGULAR LYMPH NODES

The **external jugular lymph nodes** (**jug**-you-lar) or superficial lateral cervical lymph nodes are located on each side of the neck along the external jugular vein, superficial to the sternocleidomastoid muscle. The external jugular nodes may be secondary nodes for the occipital, retroauricular, anterior auricular, and superficial parotid nodes. These nodes may then empty into the superior or inferior deep cervical nodes.

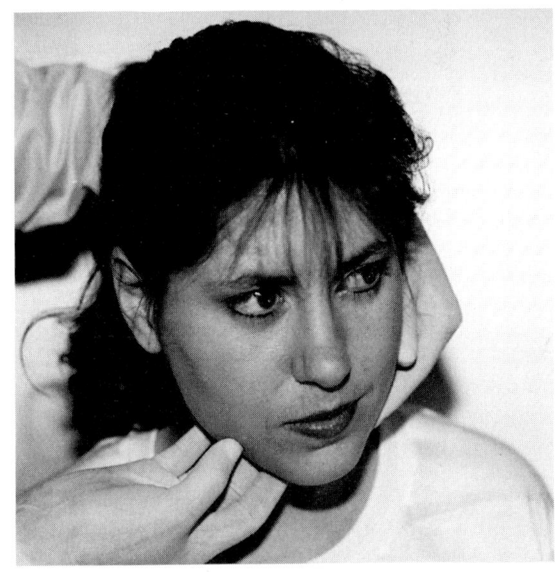

▼ **Figure 10–7**
Palpating the cervical lymph nodes inferior to the chin by having the patient's chin lowered.

▼ ANTERIOR JUGULAR LYMPH NODES

The **anterior jugular lymph nodes** (**jug**-you-lar) or superficial anterior cervical lymph nodes are located on each side of the neck along the length of the anterior jugular vein, anterior to the sternocleidomastoid muscle, to drain the infrahyoid region of the neck. The anterior jugular nodes then empty into the inferior deep cervical nodes.

To palpate effectively for the external and anterior jugular nodes in the mid portion of the neck, having the patient turn the head to the side makes the important landmark of the sternocleidomastoid muscle more prominent (Figure 10–8). Palpation for these nodes should start below the ear and continue the whole length of the muscle's surface to the clavicles.

▼ Deep Cervical Lymph Nodes

The **deep cervical lymph nodes** (**ser**-vi-kal) (15-30 in number) are located along the length of the internal jugular vein on each side of the neck, deep to the sternocleidomastoid muscle (Figure 10–9). The deep cervical nodes extend from the base of the skull to the root of the neck, adjacent to the pharynx, esophagus, and trachea.

Again, having the patient turn the head to the side makes the important landmark of the sternocleidomastoid muscle more prominent and increases the accessibility for effective palpation of these nodes (see Figure 10–8). Palpation for the deep cervical nodes is on the underside of the anterior and posterior aspects of the muscle in contrast with the superficial cervical nodes that are on the muscle's surface. Palpation should start below the ear and continue down the length of the muscle to the clavicles.

For those lymph nodes in the most inferior portion of the neck in the area of the clavicles, having the patient raise the shoulders up and forward allows for effective palpation during an extraoral examination (Figure 10–10).

The deep cervical lymph nodes are divided into two groups, the superior and inferior deep cervical nodes. This division is based on whether the nodes are superior or inferior to the point where the omohyoid muscle crosses the internal jugular vein.

▼ SUPERIOR DEEP CERVICAL LYMPH NODES

The superior deep cervical lymph nodes are located deep beneath the sternocleidomastoid muscle, superior to where the omohyoid muscle crosses the internal jugular vein. The superior deep cervical nodes are primary nodes for and drain the posterior nasal cavity, posterior portion of the hard palate, soft palate, base of the tongue, maxillary third molars and associated tissues, esophagus, trachea, and thyroid gland.

The superior deep cervical nodes may be secondary nodes for all other nodes of the head and neck, except the occipital nodes and inferior deep cervical nodes.

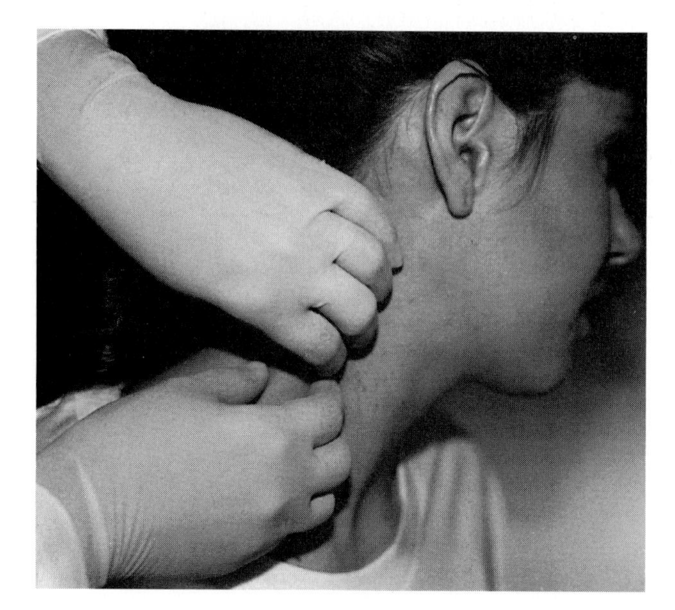

▼ **Figure 10–8**
Palpating the more inferior cervical lymph nodes by having the patient's head turned.

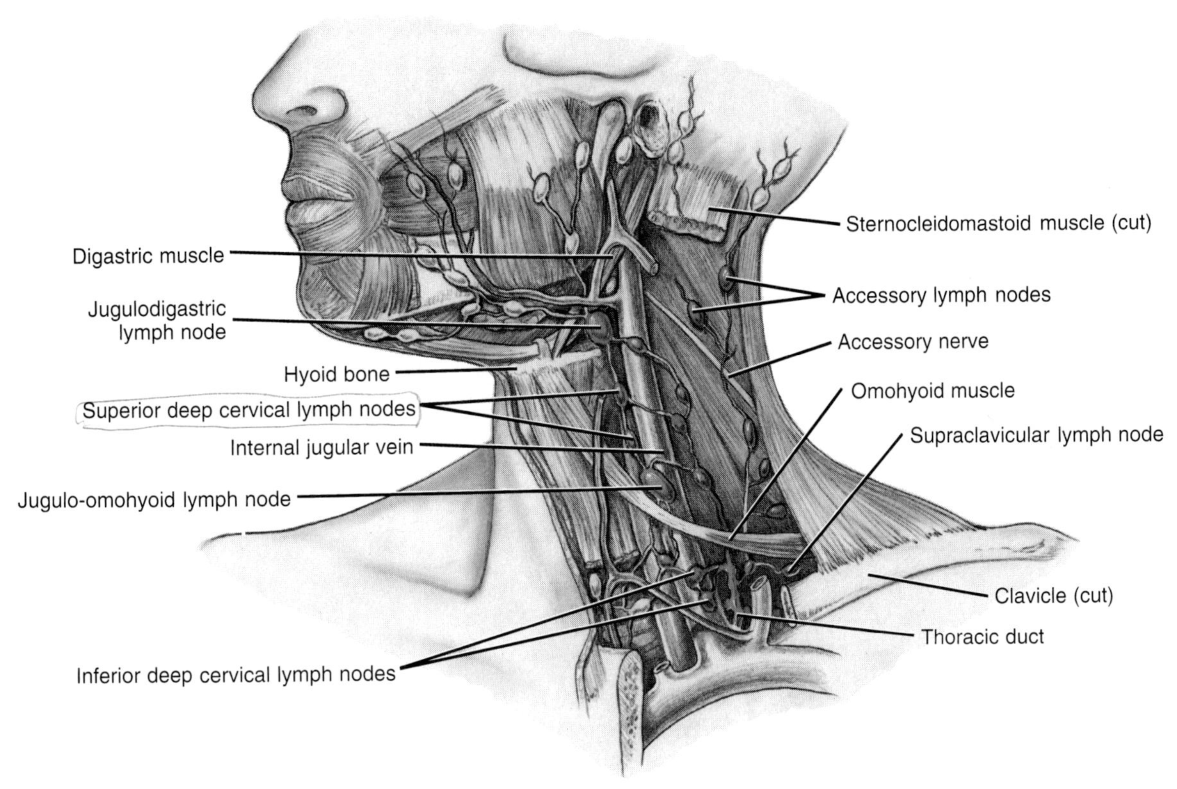

▼ **Figure 10–9**
Deep cervical lymph nodes and associated structures.

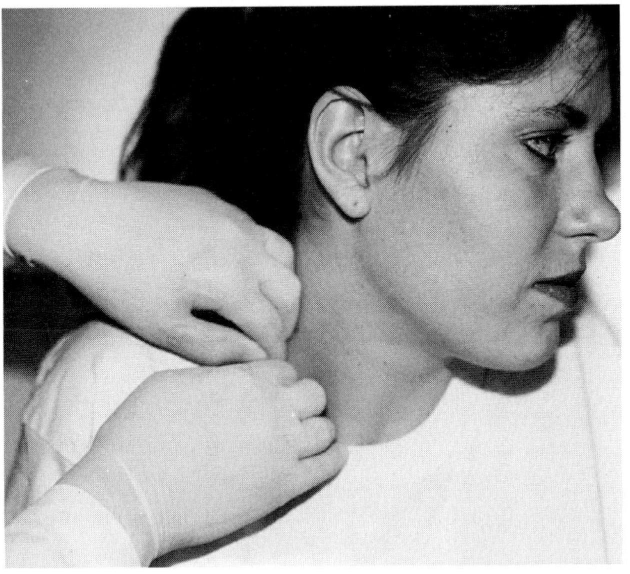

▼ **Figure 10–10**
Palpating the most inferior cervical lymph nodes by having the patient's shoulders raised up and forward.

The superior deep cervical nodes empty into the inferior deep cervical nodes or directly into the jugular trunk.

One node of the superior deep cervical nodes, the **jugulodigastric lymph node** (jug-you-lo-di-**gas**-tric) or tonsillar node, easily becomes palpable when the palatine tonsils or pharynx is inflamed. The jugulodigastric node is located below the posterior belly of the digastric muscle.

▼ INFERIOR DEEP CERVICAL LYMPH NODES

The inferior deep cervical lymph nodes are a continuation of the superior deep cervical group. The inferior deep cervical nodes are located deep to the sternocleidomastoid muscle, at the level to where the omohyoid muscle crosses the internal jugular vein or inferior to this point. They extend inferiorly into the supraclavicular fossa, superior to each clavicle. The inferior deep cervical nodes are primary nodes for and drain the posterior portion of the scalp and neck, the superficial pectoral region, and a portion of the arm.

The inferior deep cervical nodes may be secondary nodes for the occipital and superior deep cervical nodes. Their efferent vessels form the jugular trunk, which is one of the tributaries of the right lymphatic duct (on the right side) and the thoracic duct (on the left). The inferior deep cervical nodes communicate with the axillary lymph nodes that drain the breast region. These axillary nodes in the area of the armpit may be involved when the patient has breast cancer (adenocarcinoma), which is discussed later in this chapter.

A sometimes prominent node of the inferior deep cervical nodes, the **jugulo-omohyoid lymph node** (jug-you-lo-o-mo-**hi**-oid), is located at the crossing of the omohyoid muscle and internal jugular vein. The jugulo-omohyoid node drains the tongue and submental region.

▼ Accessory and Supraclavicular Lymph Nodes

In addition to the deep cervical lymph nodes, there are also the accessory and supraclavicular node groups in the most inferior portion of the neck. The **accessory lymph nodes** (ak-**ses**-o-ree) (2-6 in number) are located along the accessory nerve and drain the scalp and neck and then drain into the supraclavicular nodes. The other **supraclavicular lymph nodes** (soo-prah-klah-**vik**-you-ler) or transverse cervical nodes (1-10 in number) are located along the clavicle and drain the lateral cervical triangles. The supraclavicular nodes may empty into one of the jugular trunks or directly into the right lymphatic duct or thoracic duct.

Tonsils

The **tonsils** (**ton**-sils) are masses of lymphoid tissue located in the oral cavity and pharynx. Unlike lymph nodes, tonsils are not located along lymphatic vessels. All the tonsillar tissue drains into the superior deep cervical lymph nodes, particularly the jugulodigastric lymph node (see Tables 10–1 and 10–2).

▼ PALATINE AND LINGUAL TONSILS

The **palatine tonsils** (pal-**ah**-tine) are two rounded masses of variable size located in the oral cavity between the anterior and posterior tonsillar pillars (Figure 10–11). The **lingual tonsil** is an indistinct layer of lymphoid tissue located intraorally on the base of the dorsal surface of the tongue (Figure 10–12).

▼ PHARYNGEAL AND TUBAL TONSILS

The **pharyngeal tonsil** (fah-**rin**-je-il) is located on the posterior wall of the nasopharynx (Figure 10–13). This tonsil is also called **adenoids** (**ad**-in-oidz) and is normally enlarged in children. The **tubal tonsil** (**tube**-al) is also located in the nasopharynx, posterior to the openings of the eustachian or auditory tube (Figure 10–13).

Lymphadenopathy

When a patient has a disease process active in a region, such as cancer or infection, the region's lymph nodes respond. The resultant increase in size and change in consistency of the lymphoid tissue is termed ***lymphadenopathy*** (lim-fad-in-**op**-ah-thee). Lymphadenopathy results from an increase in both the size of each individual lymphocyte and the overall cell count

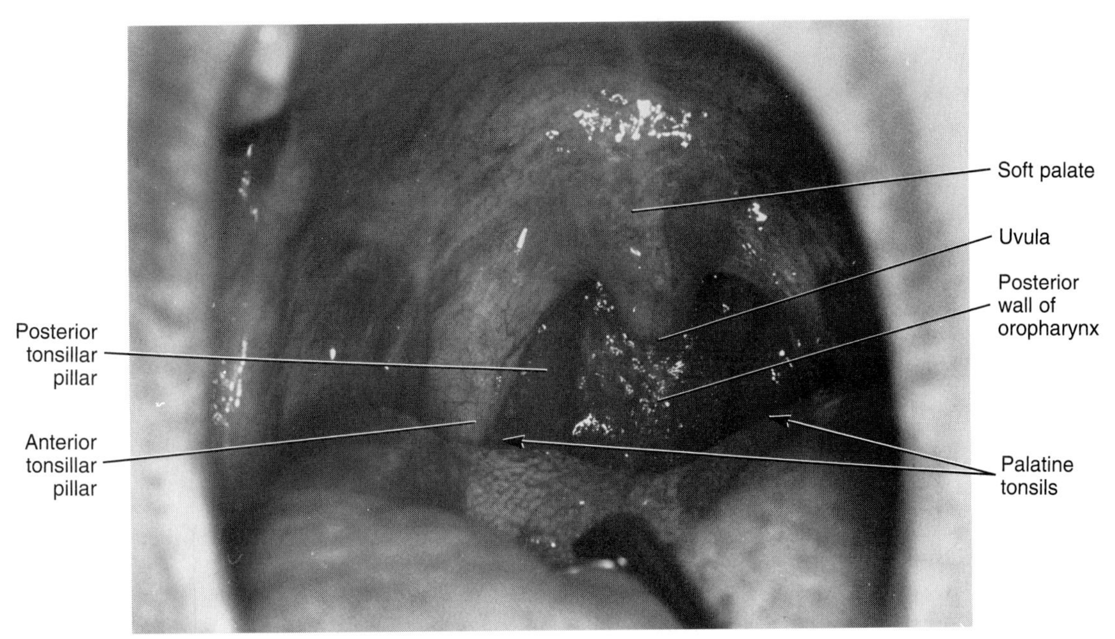

▼ **Figure 10–11**
Palatine tonsils in oral cavity.

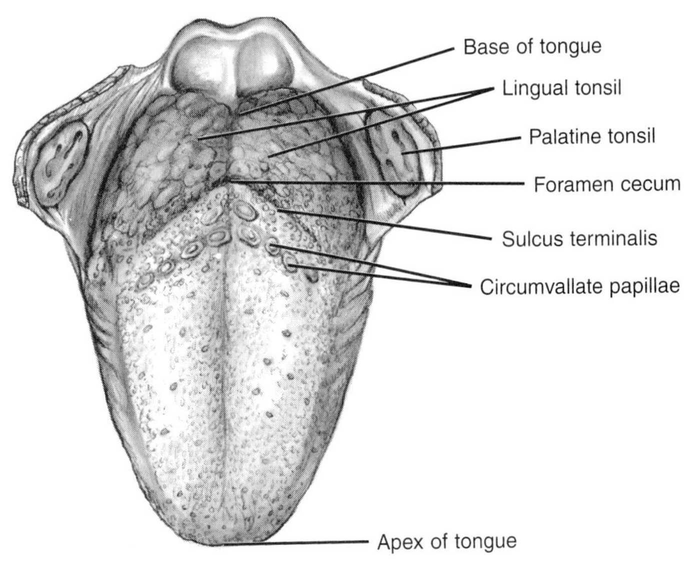

▼ **Figure 10–12**
Lingual tonsil on the base of the tongue's dorsal surface.

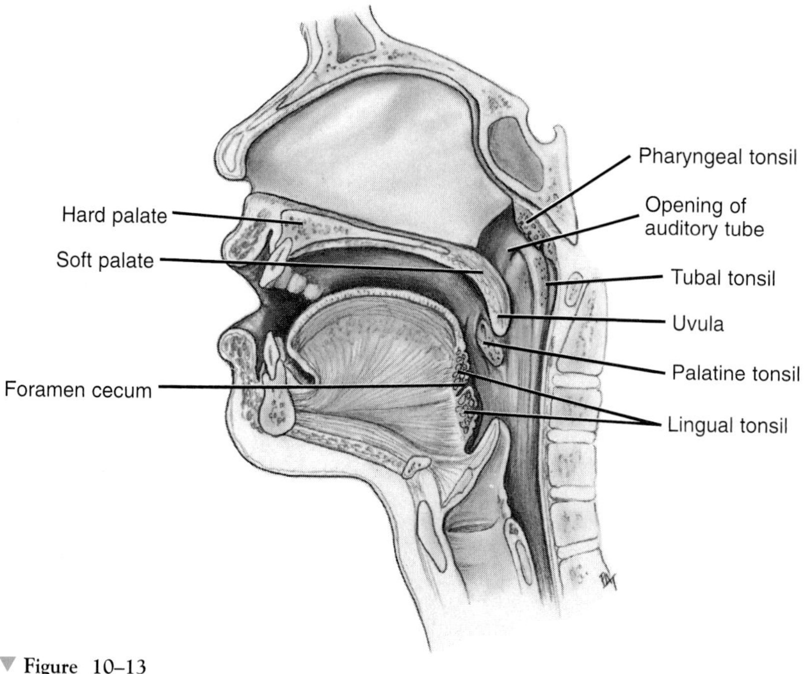

▼ **Figure 10–13**
Pharyngeal tonsil and tubal tonsil.

in the lymphoid tissue. With more and larger lymphocytes, the lymphoid tissue is better able to fight the disease process.

The lymph nodes can also be involved in the spread of infection, such as dental infection from the teeth, which is covered in Chapter 12. This spread of infection occurs along the connecting lymphatic vessels of the involved nodes.

This lymphadenopathy may allow the node to be viewed during an extraoral examination. More important, changes in consistency allow the lymph node to be palpated during the extraoral examination along the even firmer backdrop of underlying bones and muscles or the examiner's hands.

This change in lymph node consistency can range from firm to bony hard. Lymph nodes can remain free or mobile from the surrounding tissue during the disease. However, they can also become attached or fixed to the surrounding tissues, such as skin, bone, or muscle, as the disease process progresses to involve the regional tissues. When the lymph nodes are involved with lymphadenopathy, the lymph node can also feel tender to the patient when palpated. This tenderness is due to the pressure on the area nerves from the node enlargement.

Again, any palpable lymph nodes found in a patient need to be recorded, and any appropriate physician referrals made by the dental professional. The changes in a lymph node when it is involved with infection are discussed in Chapter 12.

Lymphadenopathy can also occur to the tonsils, causing tissue enlargement. In most cases, this enlargement of the tonsils, with the exception of the tonsils in the pharynx, can be viewed on an intraoral examination of the patient (compare Figures 10–14 and 10–11). The intraoral tonsils may also be tender when palpated. Lymphadenopathy of the tonsils in both the oral cavity and pharynx may cause airway obstruction with its complications and lead to infection of the tonsillar tissue. A dental professional may need to refer the patient to a physician if lymphadenopathy and infection of intraoral tonsillar tissue are noted.

Metastasis and Cancer

Even though the lymph nodes usually assist in fighting the disease process, the lymph nodes can become involved in the spread of a cancer from the region they filter. The spread of a cancer from the original or primary site of the tumor to another or secondary site

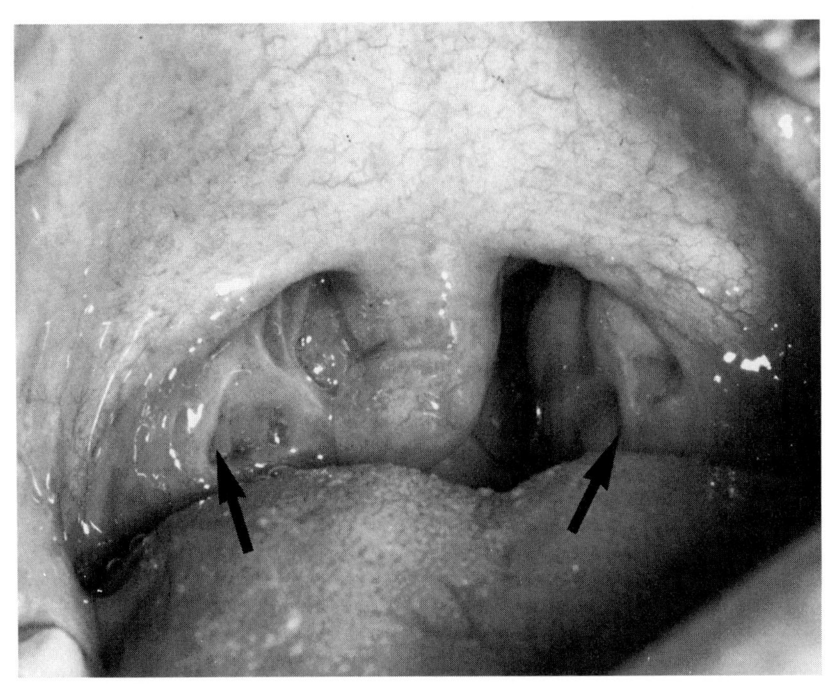

▼ **Figure 10–14**
Enlarged palatine tonsils undergoing lymphadenopathy (note arrows).

is called ***metastasis*** (meh-**tas**-tah-sis). Primary nodes are the initial secondary site in which the cancer will metastasize from the tumor.

Many times if the cancer is caught early enough at the primary tumor site or even at the initial secondary site of the primary lymph nodes, surgery to remove the tumor as well as the primary nodes may successfully stop metastasis. If the cancer is not caught early or stopped by the primary nodes, the cancer will spread to secondary nodes, and metastasis will continue to progress. Cancerous cells can slowly travel unchecked in the lymph from node to node if they are not stopped by any of the nodes along the lymphatic vessels.

If the cancer metastasizes past all the nodes, the cancer cells can enter the blood system by way of the lymphatic ducts. The spread of cancer or metastasis by way of the blood vessels is quicker than by way of the lymph nodes, so the cancer can quickly metastasize to the rest of the body, causing possibly fatal systemic involvement with the cancer. Thus an increase in the number of nodes involved with the cancer before involving the lymphatic duct and associated blood vessels may mean a better outcome for the patient.

When they are involved with cancer, the lymph nodes can become bony hard and possibly fixed to surrounding tissues as the cancer grows and spreads. The cancerous nodes are usually not tender. In comparison, those nodes involved with acute infection are firm, mobile, and tender (see Chapter 12 for more information). Metastasis in lymph nodes occurs mainly with those cancers that are carcinomas from epithelial tissues.

Identification Exercise

Identify the structures on the following diagrams by filling in the blanks with the correct anatomical term. You can check your answers by looking back at the figure indicated in parentheses for each identification diagram.

1. (Figure 10–1)

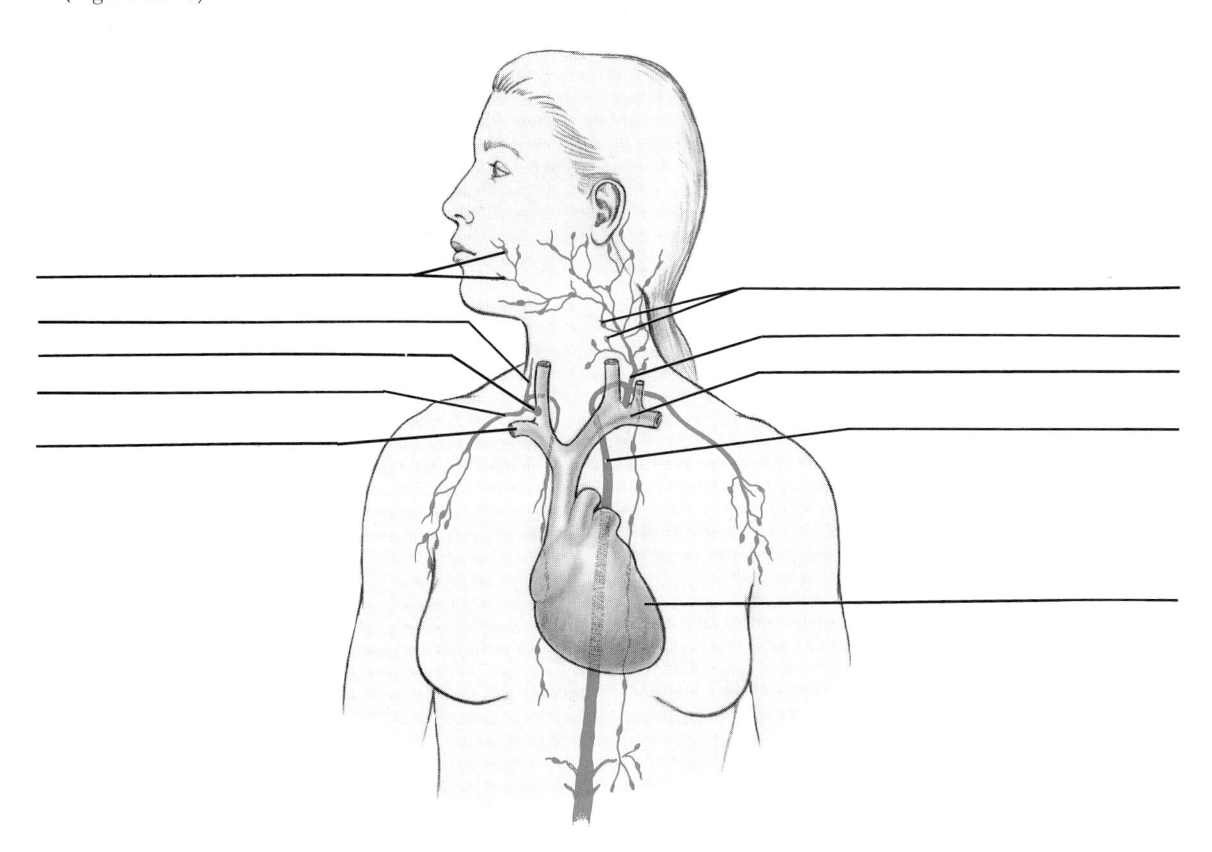

Identification Exercise *Continued*

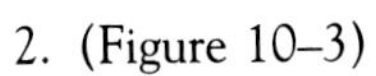

2. (Figure 10–3)

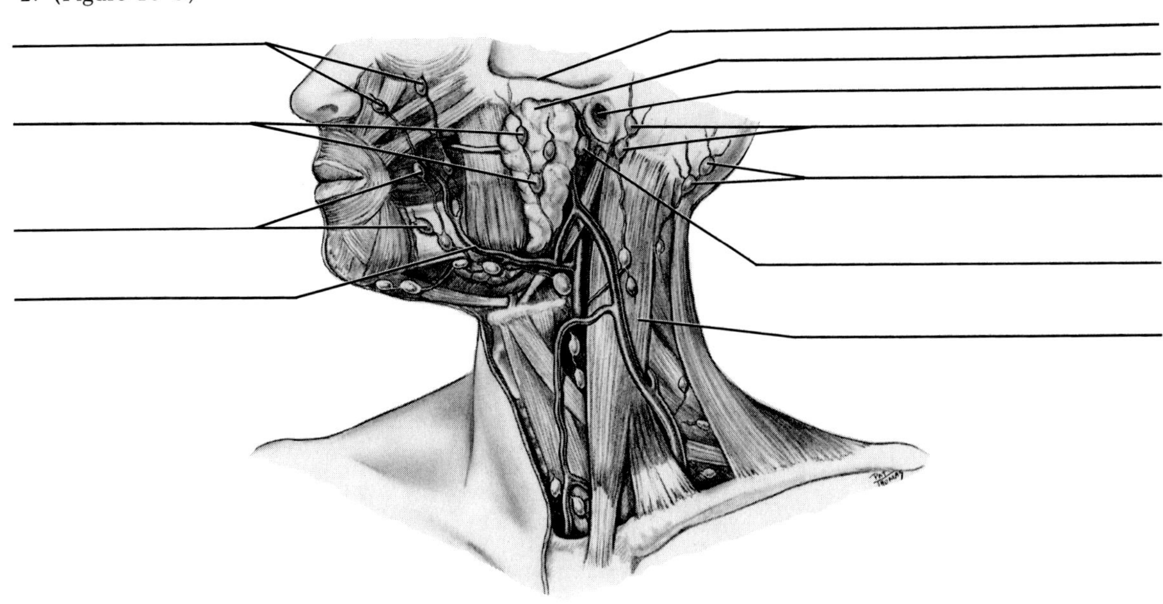

3. (Figure 10–5)

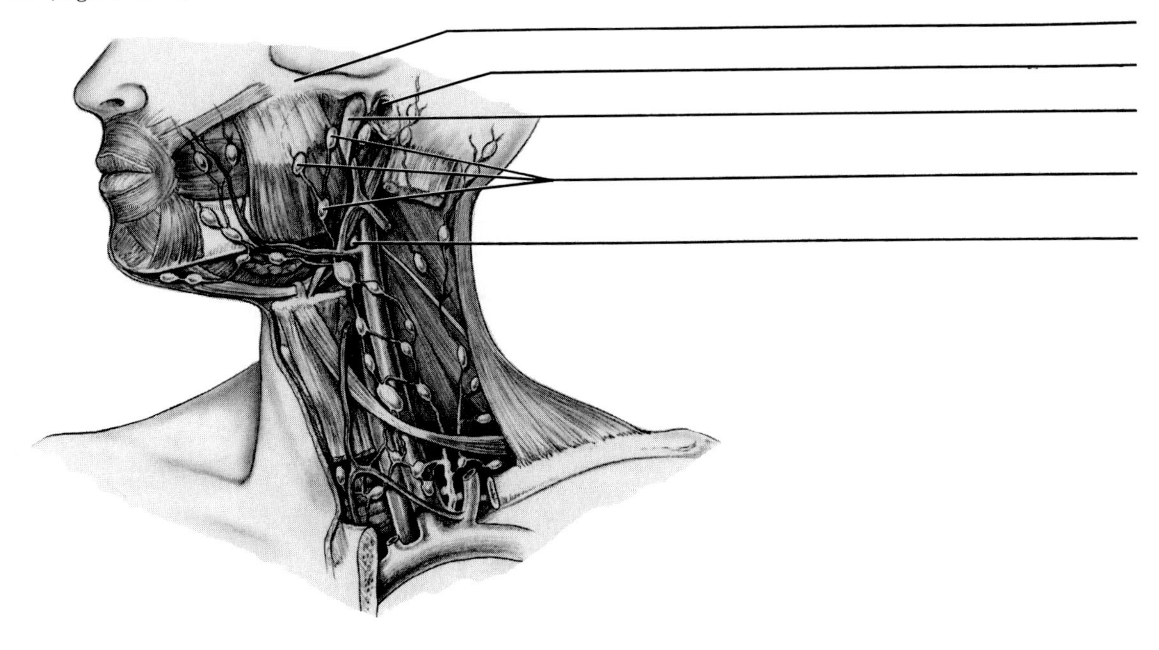

Identification Exercise *Continued*

4. (Figure 10–6)

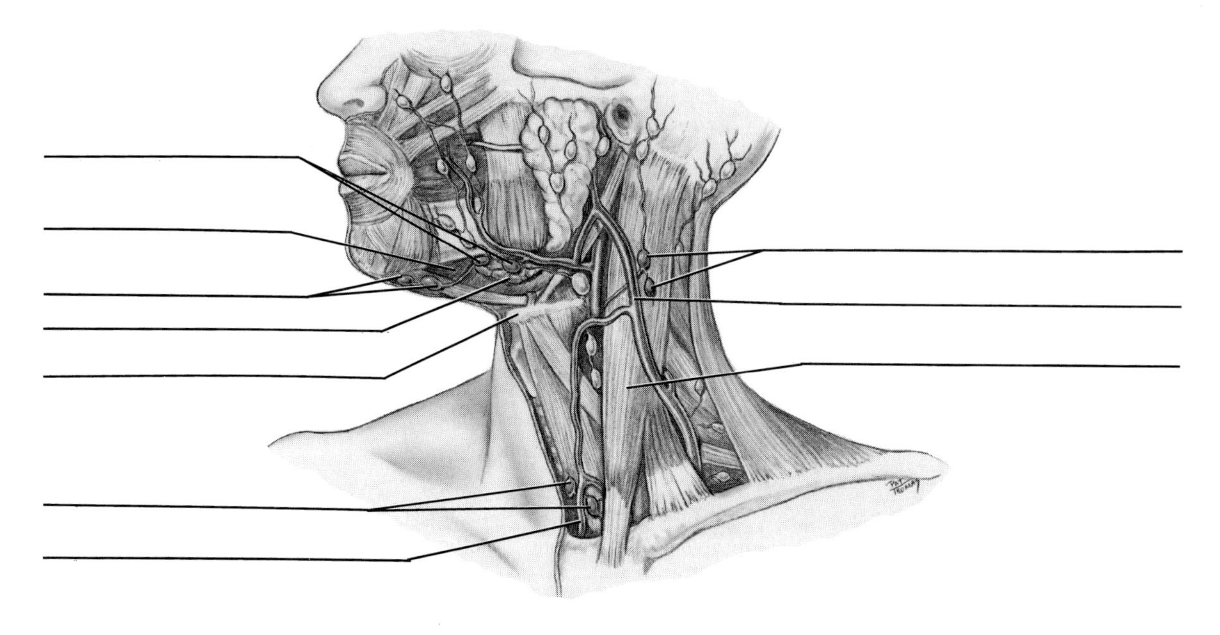

5. (Figure 10–9)

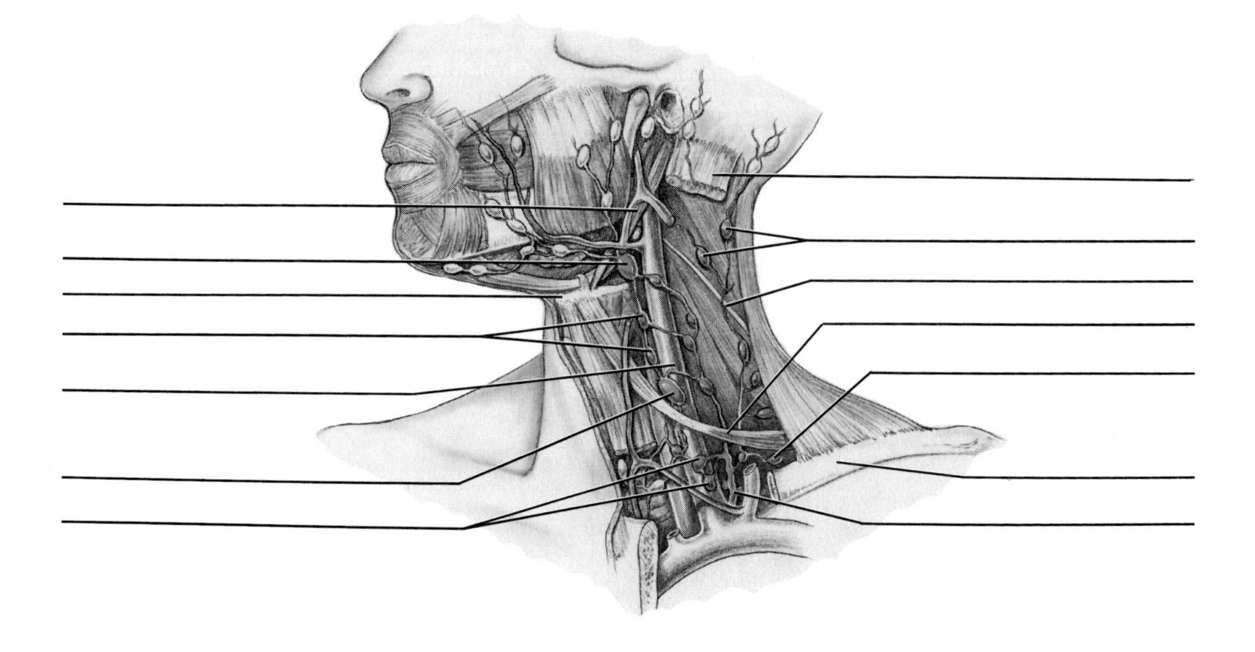

Identification Exercise *Continued*

6. (Figure 10–12)

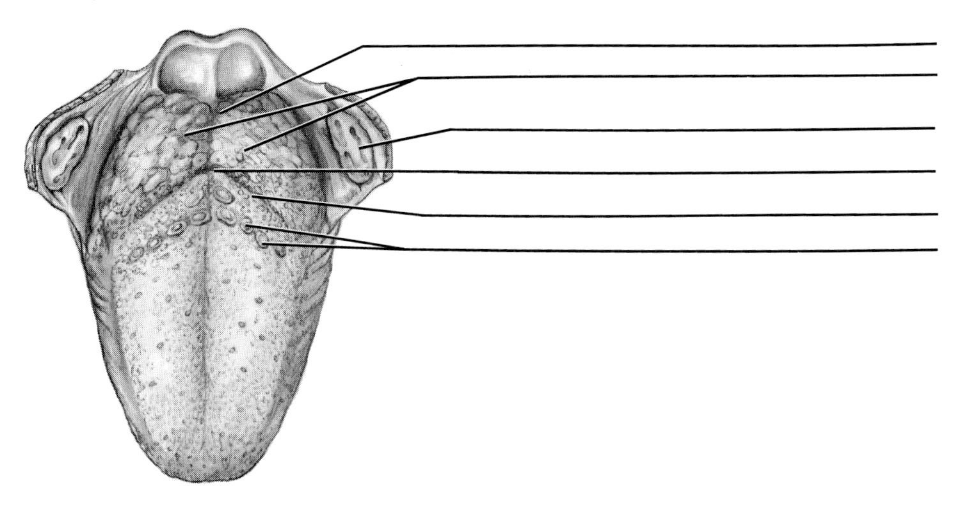

REVIEW QUESTIONS

1. Which of the following lymph node groups have both superficial and deep nodes within the group?
   A. Facial nodes
   B. Buccal nodes
   C. Parotid nodes
   D. Occipital nodes
   E. Submandibular nodes

2. Which of the following structures leave each individual lymph node at the hilus?
   A. Lymphatic ducts
   B. Tonsillar tissue
   C. Efferent lymphatic vessel
   D. Afferent lymphatic vessels

3. Which of the following lymph nodes are considered part of the facial lymph nodes?
   A. Sublingual, submandibular, zygomatic, and buccal nodes
   B. Infraorbital, nasal, buccal, and submental nodes
   C. Mandibular, lingual, malar, and zygomatic nodes
   D. Malar, buccal, nasolabial, and mandibular nodes
   E. Zygomatic, nasolabial, masseteric, and submental nodes

4. Which of the following components of the lymphatic system have one-way valves?
   A. Arteries
   B. Veins
   C. Vessels
   D. Nodes
   E. Ducts

5. Which of the following nodes drains lymph from a local region before the lymph flows to a more distant region?
   A. Primary
   B. Secondary
   C. Central
   D. Tertiary

6. The buccal lymph nodes are located superficial to the:
   A. sublingual gland.
   B. buccinator muscle.
   C. sternocleidomastoid muscle.
   D. parotid gland.
   E. submandibular gland.

7. Which of the following lymph node groups extend from the base of the skull to the root of the neck?
   A. Facial nodes
   B. Deep cervical nodes
   C. Occipital nodes
   D. Jugulodigastric nodes
   E. Anterior jugular nodes

REVIEW QUESTIONS *Continued*

8. The external jugular lymph nodes are located:
   A. anterior to the hyoid bone.
   B. along the external jugular vein.
   C. deep to the sternocleidomastoid muscle.
   D. close to the symphysis of the mandible.

9. The thoracic duct empties into the:
   A. aortic arch of the body.
   B. superior vena cava of the body.
   C. junction of the right and left brachiocephalic veins.
   D. junction of the left internal jugular and subclavian veins.

10. Which of the following nodes are the primary lymph nodes that drain the skin and mucous membranes of the lower face?
    A. Occipital nodes
    B. Malar nodes
    C. Submandibular nodes
    D. Superficial parotid nodes
    E. Deep parotid nodes

11. Which of the following nodes are secondary lymph nodes for the occipital nodes?
    A. Buccal nodes
    B. Submental nodes
    C. Submandibular nodes
    D. Superior deep cervical nodes
    E. Inferior deep cervical nodes

12. Which of the following pairs of lymph nodes are both considered part of the superficial cervical lymph node group?
    A. External and anterior jugular nodes
    B. Superficial and deep jugular nodes
    Internal and external jugular nodes

13. Medial and lateral jugular nodes
    A. Which of the following statements concerning the submental lymph nodes is correct?
    B. Located deep to the mylohyoid muscle
    C. Located between the mandible's symphysis and hyoid bone
    D. Drains the labial commissure and base of tongue
    E. Secondary node for superior deep cervical nodes

14. Which muscle needs to be made more prominent on a patient to achieve effective palpation of the region where the superior deep cervical lymph nodes are located?
    A. Masseter muscle
    B. Trapezius muscle
    C. Sternocleidomastoid muscle
    D. Epicranial muscle

15. The lingual tonsil is located:
    A. posterior to the auditory tube's opening.
    B. on the superior posterior wall of the nasopharynx.
    C. at the base of the tongue.
    D. between the anterior and posterior tonsillar pillars.

▼ *This chapter discusses the following topics:*

▼ *Key words*

Fascia, fasciae (**fash**-e-ah, **fash**-e-ay) Layers of fibrous connective tissue that underlie the skin and also surround the muscles, bones, vessels, nerves, organs, and other structures of the body.

Fascial spaces (**fash**-e-al) Potential spaces between the layers of fascia in the body.

Chapter 11

Fascia and Spaces

▼ ***After studying this chapter, the reader should be able to:***

1. Define and pronounce all the key words and anatomical terms in this chapter.
2. Locate and identify the fascia of the head and neck on a diagram, a skull, and a patient.
3. Locate and identify the major spaces of the head and neck on a diagram, a skull, and a patient.
4. Discuss the communication between the major spaces of the head and neck.
5. Correctly complete the review questions for this chapter.
6. Integrate the knowledge of head and neck fascia and spaces into the clinical practice of patient examination, administration of local anesthetics, and management of the spread of dental infection.

Fascia

The ***fascia*** (plural ***fasciae***) (**fash**-e-ah, **fash**-e-ay) consists of layers of fibrous connective tissue. The fascia lies underneath the skin and also surrounds the muscles, bones, vessels, nerves, organs, and other structures of the body. The fasciae of the body can be divided into the superficial fascia and deep fascia.

SUPERFICIAL FASCIA

In all areas of the body, the superficial fascia is found just beneath and attached to the skin. The superficial fascia generally separates the skin from the deeper structures of the body, allowing the skin to move independently of these deeper structures. The superficial fascia varies in thickness in different parts of the body. The superficial fascia is composed of fat, as well as irregularly arranged connective tissue. The vessels and nerves of the skin travel in the superficial fascia.

Superficial Fascia of the Head and Neck

The superficial fascia of the body does not usually enclose muscles, except in the superficial fascia of the head and neck (Figure 11–1A and B). In the face, the superficial fascia encloses the muscles of facial expression. The superficial cervical fascia of the neck contains the platysma muscle, which covers most of the anterior cervical triangle.

DEEP FASCIA

The deep fascia covers the deeper structures of the body, such as the bones, muscles, vessels, and nerves. The deep fascia consists of a dense and inelastic fibrous tissue forming sheaths around these deeper structures.

Deep Fascia of the Face and Jaws

The deep fascia of the face and jaws is divided into fascial layers that are continuous with each other and

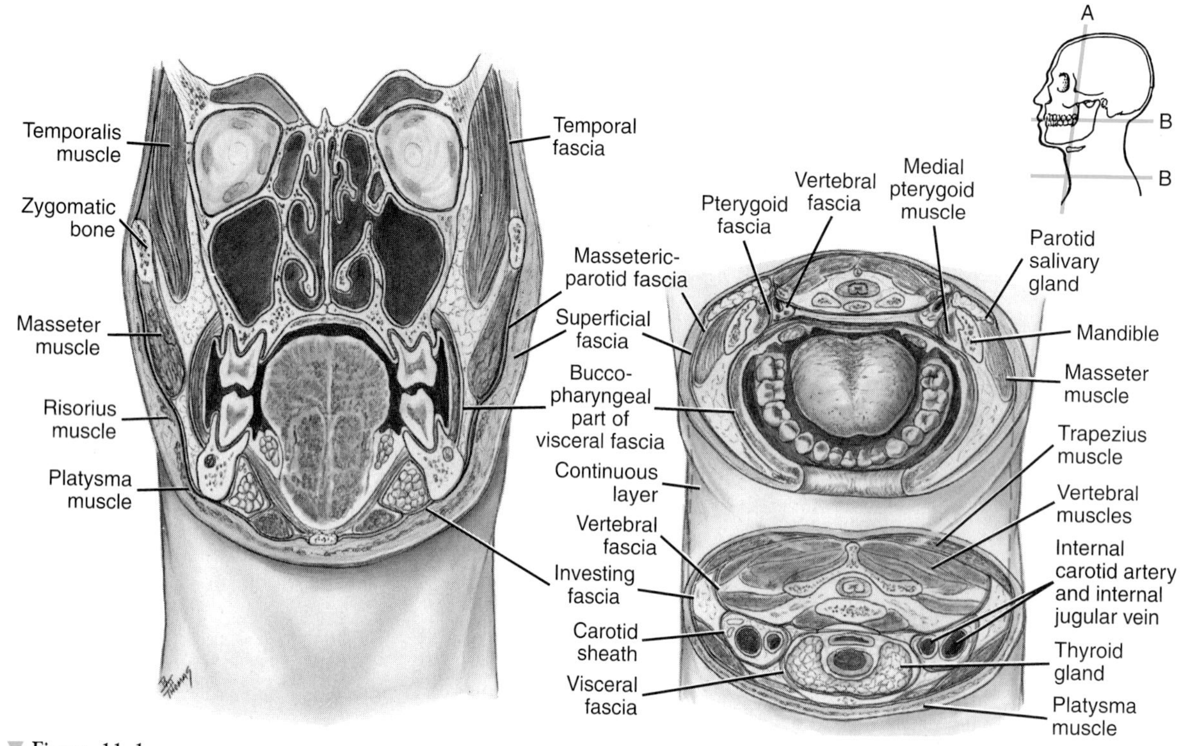

Figure 11–1
Frontal section (A) of the head highlighting the fasciae of the face. Transverse sections (B) at the oral cavity and neck highlighting the continuous nature of the investing fascia.

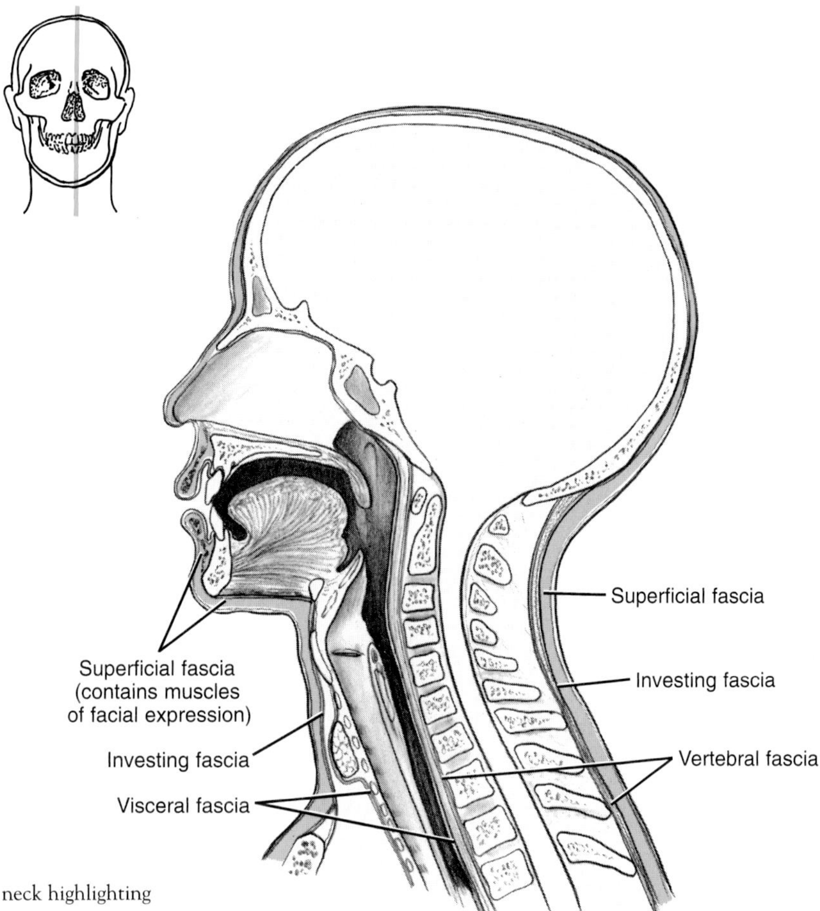

▼ **Figure 11–2**
Midsagittal section of the head and neck highlighting the deep cervical fasciae.

with the deep cervical fascia (see Figure 11–1). The deep fasciae of the face and jaws include the temporal, masseteric-parotid, and pterygoid fasciae.

The **temporal fascia** (**tem**-poh-ral) covers the temporalis muscle down to the zygomatic arch. The **masseteric-parotid fascia** (mass-et-**tehr**-ik-pah-**rot**-id) is located inferior to the zygomatic arch and over the masseter muscle and also surrounds the parotid salivary gland. The **pterygoid fascia** (**teh**-ri-goid) is located on the medial surface of the medial pterygoid muscle. These fasciae are all continuous with the investing layer of the deep cervical fascia.

▼ Deep Cervical Fascia

The deep cervical fascia (**ser**-vi-kal) is composed of layers that include the investing fascia, carotid sheath, visceral fascia, and vertebral fascia (Figures 11–2 and 11–3). Again, it is important to note that these layers of deep cervical fascia are continuous with each other and with the deep fascia of the face and jaws.

▼ INVESTING FASCIA

The **investing fascia** (in-**vest**-ing) is the most external layer of deep cervical fascia. The investing fascia surrounds the neck (hence the name), continuing onto the masseteric-parotid fascia. The investing fascia splits around two glands (submandibular and parotid) and two muscles (sternocleidomastoid and trapezius), enclosing them completely. Branching laminae from the investing fascia also provide the deep fasciae that surround the infrahyoid muscles, from the hyoid bone down to the sternum.

▼ CAROTID SHEATH AND VISCERAL FASCIA

The **carotid sheath** (kah-**rot**-id sheeth) is a tube of deep cervical fascia deep to the investing fascia and sternocleidomastoid muscle, running down each side of the neck from the base of the skull to the thorax. The

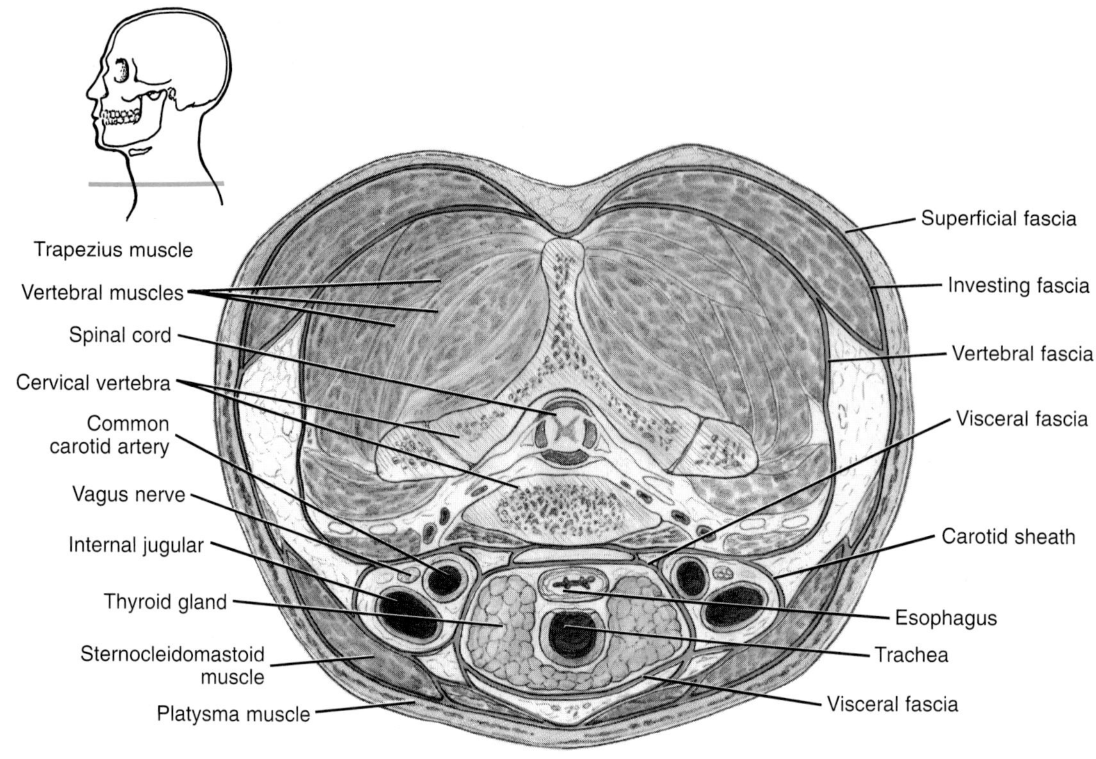

▼ **Figure 11–3**
Transverse section of the neck diagramming the deep cervical fasciae.

carotid sheath houses the internal carotid and common carotid arteries and the internal jugular vein, as well as the tenth cranial or vagus nerve. All these structures travel between the braincase and thorax.

Deep to the carotid sheath is the **visceral fascia** (**vis**-er-al) or pretracheal fascia, which is a single, midline tube of deep cervical fascia running down the neck. The visceral fascia surrounds the airway and food way, including the trachea, esophagus, and thyroid gland.

Nearer to the skull, the visceral fascia located posterior and lateral to the pharynx is known as the **buccopharyngeal fascia** (buk-o-fah-**rin**-je-al). This deep cervical fascial layer encloses the entire upper part of the alimentary canal and is continuous with the fascia on the surface of the buccinator muscle, where that muscle and the superior pharyngeal constrictor muscle come together at the pterygomandibular raphe (see Chapter 4 for more information).

▼ VERTEBRAL FASCIA

The deepest layer of the deep cervical fascia, the **vertebral fascia** (**ver**-teh-brahl) or prevertebral fascia, covers the vertebrae, spinal column, and associated muscles. Some clinicians distinguish a branch of the vertebral fascia, called the alar fascia, which is said to run from the base of the skull to connect with the visceral fascia low in the neck.

Spaces

Potential spaces are created between the layers of fascia of the body because of the sheetlike nature of the fasciae. These potential spaces are termed ***fascial spaces*** (**fash**-e-al) or planes. It is important to remember that these fascial spaces are not actually spaces in healthy patients since they contain loose connective tissue. Other spaces are present in the head and neck but not created necessarily by the fascia.

▼ SPACES OF THE HEAD AND NECK

A dental professional must have knowledge of the anatomical aspects of the spaces of the head and neck when examining a patient. These spaces are important because they can be involved in infections from dental tissues. These spaces also communicate with each other directly, as well as through their blood and lymph vessels. This communication may allow the spread of dental infection from an initial superficial area to more vital deeper structures. The spread of infection by way of these spaces may result in serious consequences. The role of spaces in the spread of dental infection is discussed further in Chapter 12.

Spaces of the Face and Jaws

The major spaces of the face and jaws include the vestibular spaces of the maxilla and mandible; the canine, parotid, buccal, and masticatory spaces; the space of the body of the mandible; and the submental, submandibular, and sublingual spaces (Table 11–1). It is important to remember that these spaces of the face and jaw can communicate with each other and also with the cervical fascial spaces. Unlike the neck, the spaces of the face and jaws are often defined by the arrangement of muscles and bones, in addition to the fasciae. Thus many of the spaces are not fascial spaces.

▼ VESTIBULAR SPACE OF THE MAXILLA

The space of the upper jaw, the **vestibular space of the maxilla** (mak-**sil**-ah), is located medial to the buccinator muscle and inferior to the attachment of this muscle along the alveolar process of the maxilla (Figure 11–4). Its lateral wall is the oral mucosa. This space communicates with the maxillary molar teeth and periodontium and thus can become involved with infections of these tissues.

▼ VESTIBULAR SPACE OF THE MANDIBLE

The **vestibular space of the mandible** (**man**-di-bl) is located between the buccinator muscle and overlying oral mucosa (see Figure 11–4). This space is bordered by the attachment of the buccinator muscle onto the mandible. This important space of the lower jaw communicates with the mandibular teeth and periodontium, as well as the space of the body of the mandible.

▼ CANINE SPACE

The **canine space** (**kay**-nine) is located superior to the upper lip and lateral to the apex of the maxillary canine (Figure 11–5). This space is deep to the overlying skin and muscles of facial expression that elevate the upper lip (levator labii superioris and zygomaticus minor). The floor of the space is the canine fossa, which is covered by periosteum. This space is bordered anteriorly by the orbicularis oris muscle and posteriorly by the levator anguli oris muscle. The canine space communicates with the buccal space.

▼ BUCCAL SPACE

The **buccal space** is the fascial space formed between the buccinator muscle (actually the buccopharyngeal fascia) and masseter muscle (see Figure 11–5). Therefore, the buccal space is inferior to the zygomatic arch, superior to the mandible, lateral to the buccinator muscle, and medial and anterior to the masseter muscle.

This bilateral space is partially covered by the platysma muscle, as well as by an extension of fascia from the parotid salivary gland capsule. The space contains the buccal fat pad. The buccal space communicates with the canine space, pterygomandibular space, and space of the body of the mandible.

▼ PAROTID SPACE

The **parotid space** (pah-**rot**-id) is a fascial space created inside the investing fascial layer of the deep cervical fascia as it envelops the parotid salivary gland (Figure 11–6). The space contains not only the entire parotid gland but also much of the facial or seventh cranial nerve and a portion of the external carotid artery and retromandibular vein. The fascial boundaries of this space help to keep infections of the parotid gland from spreading to other sites.

▼ MASTICATOR SPACE

The **masticator space** (mass-ti-**kay**-tor) is a general term used to include the entire area of the mandible and muscles of mastication. Thus the masticator space includes the temporal, infratemporal, and submasseteric spaces, as well as the masseter muscle and ramus and body of the mandible. All portions of the masticator space communicate with each other, as well as with the submandibular space and a cervical fascial space, the parapharyngeal space (discussed later in this chapter).

A portion of the masticator space is the **temporal space** (**tem**-poh-ral), which is formed by the temporal

Table 11–1

MAJOR SPACES OF THE FACE AND JAWS WITH THEIR LOCATIONS, CONTENTS, AND COMMUNICATION PATTERNS

Space	Location	Contents	Communication Patterns
Maxillary vestibular space	Between buccinator muscle and oral mucosa		Maxillary teeth and periodontium
Mandibular vestibular space	Between buccinator muscle and oral mucosa		Mandibular teeth and periodontium, and body of mandible
Canine	Within superficial fascia over canine fossa		Buccal
Buccal	Lateral to buccinator muscle	Buccal fat pad	Canine, pterygomandibular, and body of mandible
Parotid	Within parotid gland	Parotid gland, facial nerve, external carotid artery, and retromandibular vein	
Masticator	Area of mandible and muscles of mastication	Temporal, submasseteric, and infratemporal spaces	All portions communicate with each other and submandibular and parapharyngeal
Temporal	Portion of masticator space between temporal fascia and temporalis muscle	Fat	Infratemporal and submasseteric
Infratemporal	Portion of masticator space between lateral pterygoid plate, maxillary tuberosity, and ramus	Maxillary artery and branches, mandibular nerve and branches, and pterygoid plexus	Temporal, submasseteric, submandibular, and parapharyngeal
Pterygomandibular	Portion of infratemporal space between medial pterygoid muscle and ramus	Inferior alveolar nerve and vessels	Submandibular and parapharyngeal
Submasseteric	Portion of masticator space between masseter muscle and external surface of ramus		Temporal and infratemporal
Body of mandible	Periosteum covering mandible	Mandible and inferior alveolar nerve, artery, and vein	Vestibular space of mandible, buccal, submental, submandibular, and sublingual
Submental	Midline between symphysis and hyoid bone	Submental nodes and anterior jugular vein	Body of mandible, submandibular, and sublingual
Submandibular	Medial to mandible and below mylohyoid muscle	Submandibular nodes and gland and facial artery	Infratemporal, body of mandible, submental, sublingual and parapharyngeal
Sublingual	Medial to mandible and above mylohyoid muscle	Sublingual gland and ducts, submandibular duct, lingual nerve and artery, and hypoglossal nerve	Body of mandible, submental, and submandibular

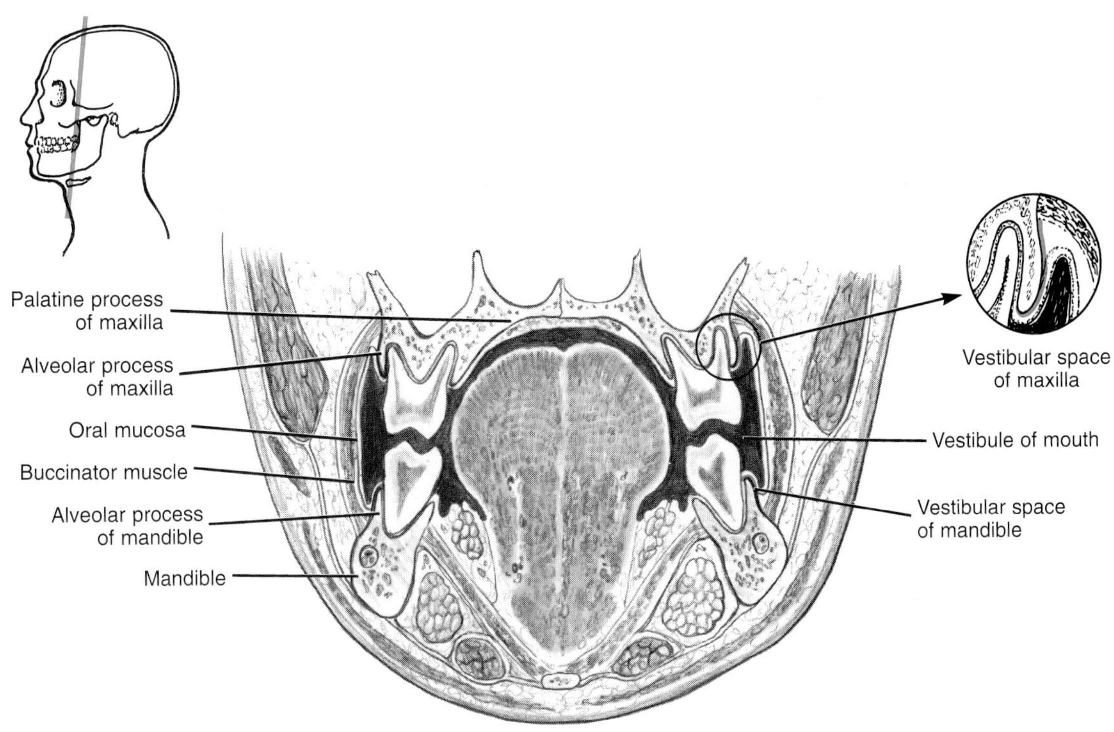

▼ **Figure 11–4**
Frontal section of the head highlighting the maxillary vestibular space and mandibular vestibular space.

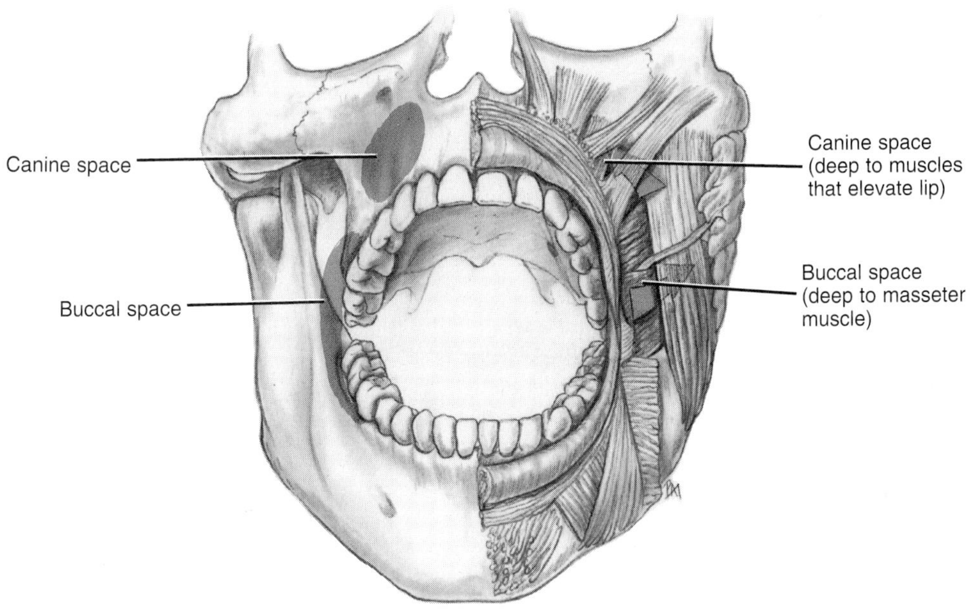

▼ **Figure 11–5**
Frontal view of the head highlighting the canine space and buccal space.

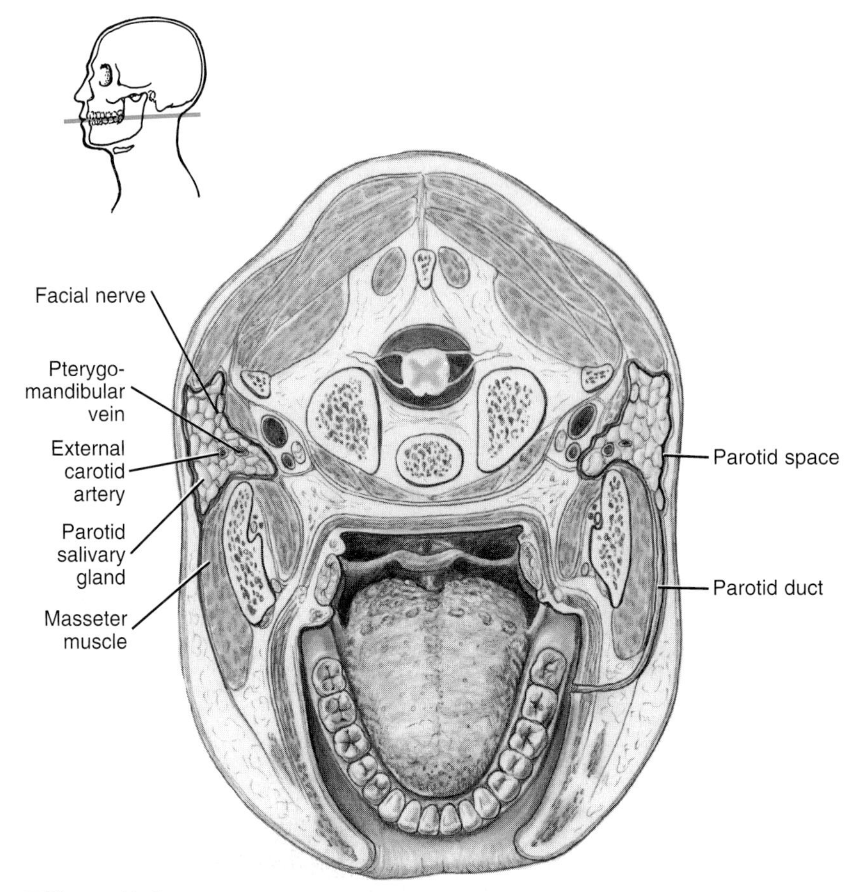

▼ **Figure 11–6**
Transverse section of the head highlighting the parotid space.

fascia covering the temporalis muscle (Figure 11–7). This space is between the fascia and muscle and therefore extends from the superior temporal line down to the zygomatic arch and infratemporal crest. The temporal space contains fat tissue and communicates with the infratemporal and submasseteric spaces.

The **infratemporal space** (in-frah-**tem**-poh-ral) occupies the infratemporal fossa, an area adjacent to the lateral pterygoid plate and maxillary tuberosity (Figure 11–7 and 11–8A). The space is bordered laterally by the medial surface of the mandible and the temporalis muscle. Its roof is formed by the infratemporal surface of the greater wing of the sphenoid bone. Medially, the space is bordered by the lateral pterygoid plate anteriorly and by the pharynx with its visceral layer of deep fascia posteriorly. There is no boundary inferiorly and posteriorly, where the infratemporal space is continuous with the cervical fascial space, the parapharyngeal space (discussed later).

The infratemporal space contains a portion of the maxillary artery as it branches, the mandibular nerve and its branches, and the pterygoid plexus of veins. It also houses the medial and lateral pterygoid muscles. This space communicates with the temporal and submasseteric spaces, as well as the submandibular and parapharyngeal spaces.

The **pterygomandibular** (teh-ri-go-man-**dib**-you-lar) **space** is a portion of the infratemporal space, formed by the lateral pterygoid muscle (roof), medial pterygoid muscle (medial wall), and mandibular ramus (lateral wall) (Figure 11–8A and B).

The pterygomandibular space is important because it contains the inferior alveolar nerve and vessels and is the site for the inferior alveolar nerve anesthetic

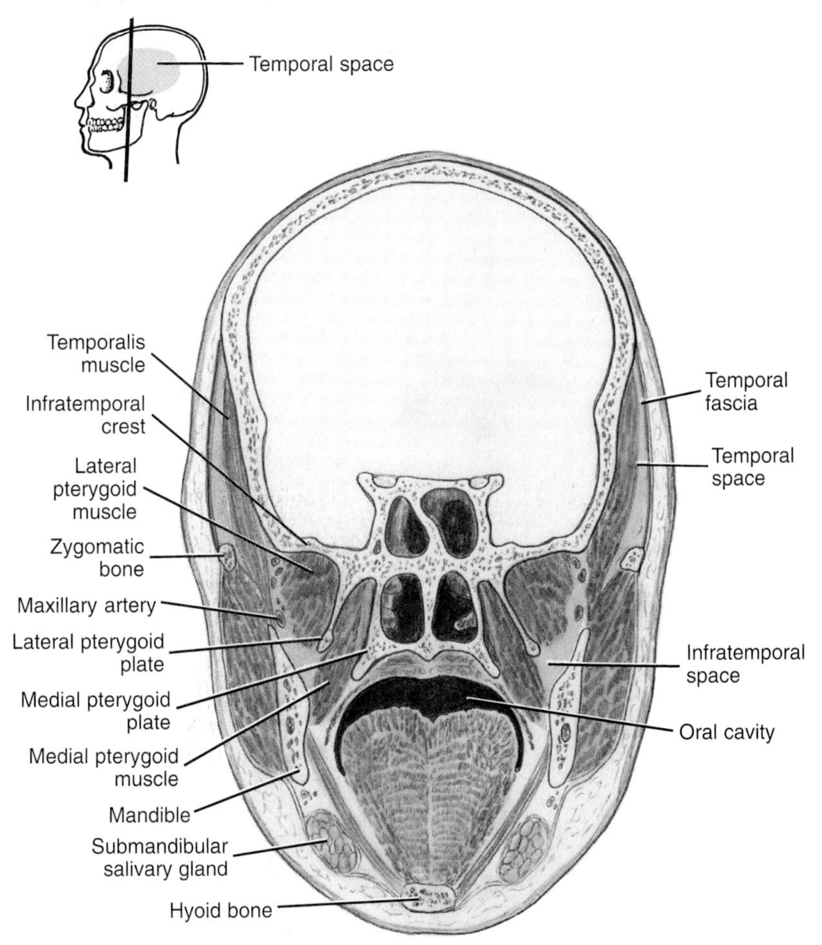

Figure 11–7
Frontal section of the head highlighting the temporal space and infratemporal space.

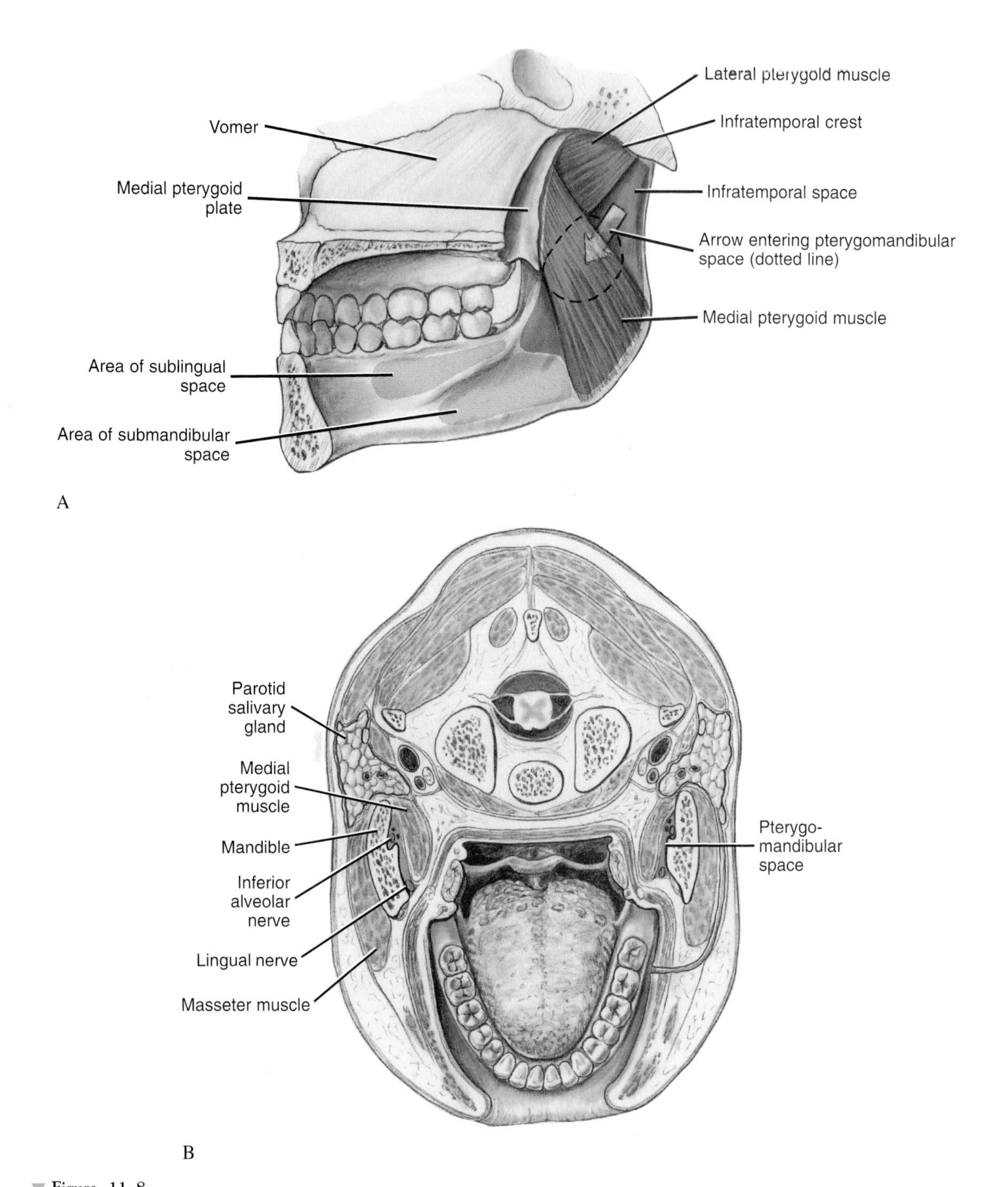

Figure 11–8
A: Median section of the skull highlighting the infratemporal space and pterygomandibular space. **B:** Transverse section of the head highlighting the pterygomandibular space.

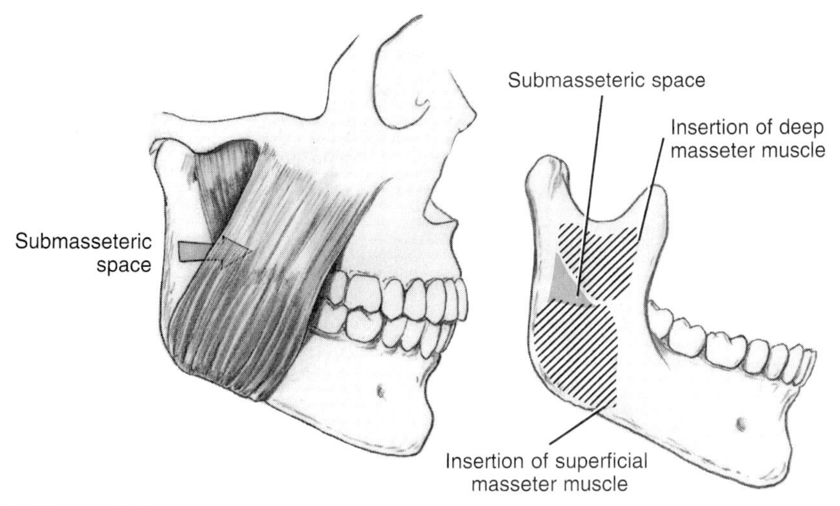

Figure 11–9
Lateral views of the face and mandible highlighting the submasseteric space.

block (see Chapter 9). This space communicates with the submandibular space and the parapharyngeal space of the neck (discussed later).

Another portion of the masticator space, the **submasseteric space** (sub-mas-et-**tehr**-ik) is located between the masseter muscle and external surface of the vertical ramus (Figure 11–9). This space communicates with the temporal and infratemporal spaces.

▼ SPACE OF THE BODY OF THE MANDIBLE

The **space of the body of the mandible** (**man**-di-bl) is formed by the periosteum, covering the body of the mandible from its symphysis to the anterior borders of the masseter and medial pterygoid muscles (Figure 11–10).

This potential space of the body of the mandible contains the mandible, a portion of the inferior alveolar

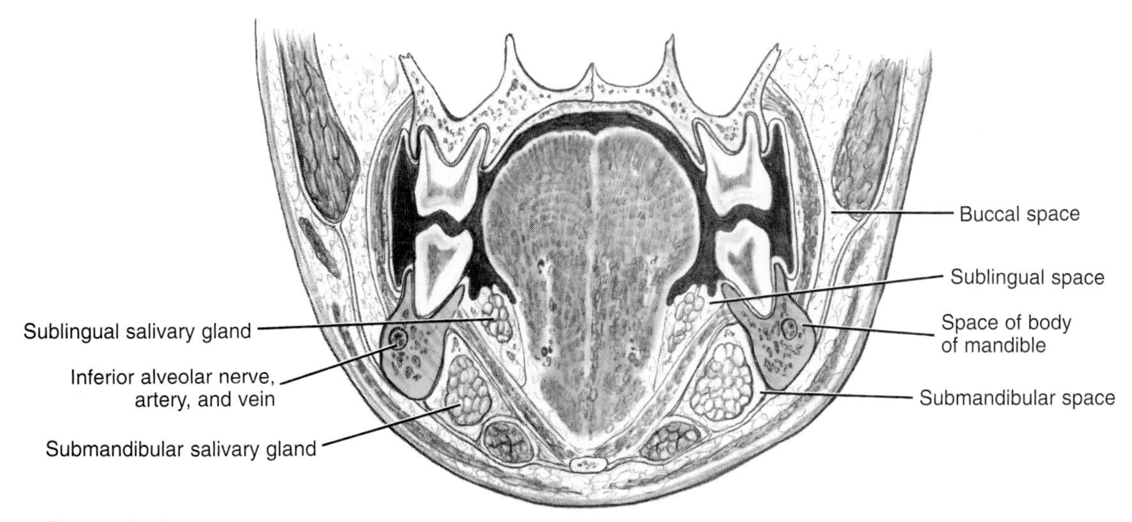

Figure 11–10
Frontal section of the head highlighting the space of the body of the mandible.

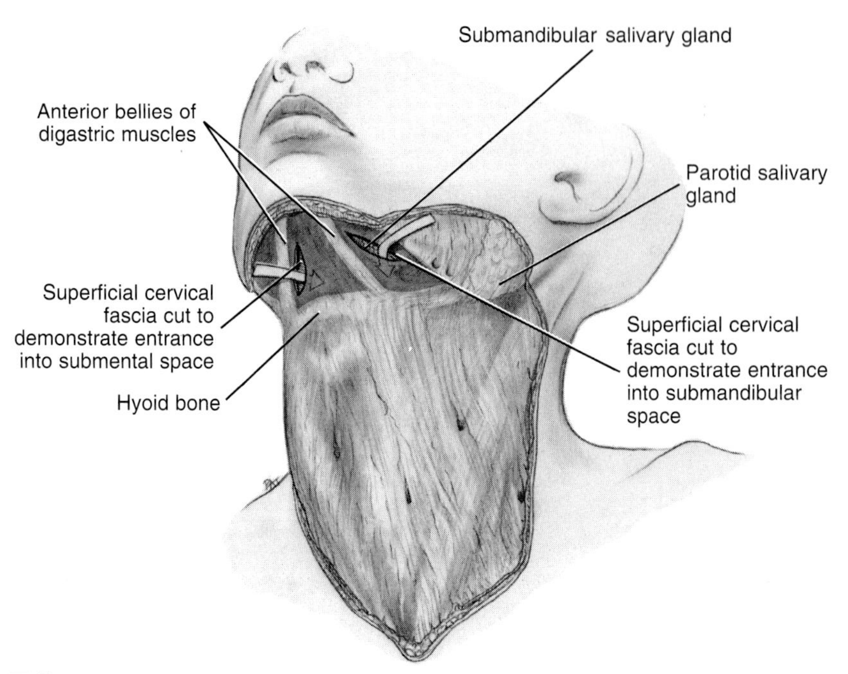

Figure 11–11
Anterolateral view of the neck. The skin has been removed, leaving the superficial cervical fascia in place (platysma has been omitted). The location of the submental space at the midline and the two lateral submandibular spaces are indicated.

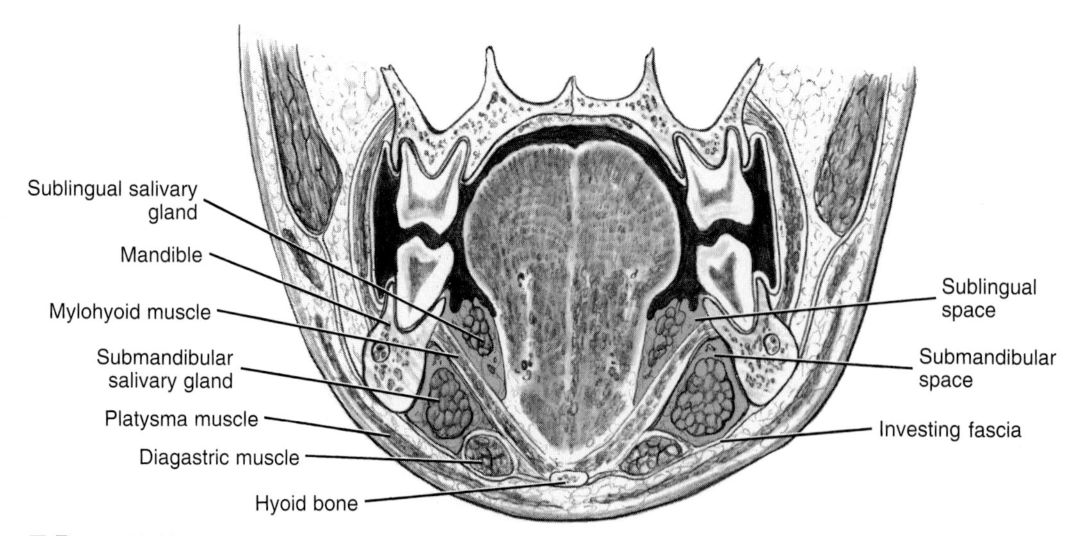

Figure 11–12
Frontal section of the head and neck highlighting the submandibular space and sublingual space.

nerve, artery, and vein, and the dental and alveolar branches of these vessels, as well as the mental and incisive branches. The space of the mandible communicates with the vestibular space of the mandible, as well as the buccal, submental, submandibular, and sublingual spaces.

▼ SUBMENTAL SPACE

The **submental space** (sub-**men**-tal) is located in the midline between the mandibular symphysis and hyoid bone (Figure 11–11). The floor of this space is the superficial cervical fascia over the suprahyoid muscles. The roof is the mylohyoid muscle, covered by the investing fascia. Forming the lateral boundaries of this space are the diverging anterior bellies of the digastric muscles.

The submental space contains the submental lymph nodes and the origin of the anterior jugular vein. The submental space communicates with the space of the body of the mandible and the submandibular and sublingual spaces.

▼ SUBMANDIBULAR SPACE

The **submandibular space** (sub-man-**dib**-you-lar) is located lateral and posterior to the submental space on each side of the jaws (Figure 11–11 and 11–12). The cross-sectional shape of this bilateral potential space is triangular, with the mylohyoid line of the mandible being its superior boundary. The mylohyoid muscle forms the medial as well as the superior boundary of the space, and the hyoid bone creates its medial apex.

The submandibular space contains the submandibular lymph nodes, most of the submandibular salivary gland, and portions of the facial artery. This space communicates with the infratemporal, submental, and sublingual spaces and the parapharyngeal space of the neck (discussed later in this chapter).

▼ SUBLINGUAL SPACE

The **sublingual space** (sub-**ling**-gwal) is located below the oral mucosa, thus making this tissue its roof (see Figure 11–12). The floor of this space is the mylohyoid muscle. Thus this muscle creates the division between the submandibular and sublingual spaces. The tongue and its intrinsic muscles form the medial boundary of the sublingual space, and the mandible forms its lateral wall.

The sublingual space contains the sublingual salivary gland and ducts, the duct of the submandibular salivary gland, a portion of the lingual nerve and artery, and the twelfth cranial or hypoglossal nerve. The sublingual space communicates with the submental and submandibular spaces and the space of the body of the mandible.

▼ Cervical Fascial Spaces

The major cervical fascial spaces (**ser**-vi-kal) include the parapharyngeal, retropharyngeal, and previsceral spaces (Figures 11–13 and 11–14A and B, Table 11–2). These fascial spaces of the neck can communicate with the spaces of the face and jaws, as well as with each other. Most important, these spaces connect the spaces of the head and neck with those of the thorax, allowing dental infections to spread to vital organs, such as the heart and lungs.

▼ PARAPHARYNGEAL SPACE

The **parapharyngeal space** (pare-ah-fah-**rin**-je-al) is a fascial space lateral to the pharynx and medial to the medial pterygoid muscle, paralleling the carotid sheath. The parapharyngeal space in its posterior portion is adjacent to the carotid sheath, which contains the internal and common carotid arteries and the internal jugular vein, as well as the tenth cranial or vagus nerve. It is also adjacent to the ninth, eleventh, and twelfth cranial nerves as they exit the cranial cavity.

Anteriorly, the parapharyngeal space extends to the pterygomandibular raphe, where it is continuous with the infratemporal and buccal spaces. The parapharyngeal space anteriorly contains a few lymph nodes. Posteriorly, the space extends around the pharynx, where it is continuous with another cervical fascial space, the retropharyngeal space. Dental infections become dangerous when they reach the parapharyngeal space also because of its connection to the retropharyngeal space (see Chapter 12 for more information).

▼ RETROPHARYNGEAL SPACE

The **retropharyngeal space** (re-troh-fah-**rin**-je-al) is a fascial space located immediately posterior to the pharynx, between the vertebral and visceral fasciae. The retropharyngeal space extends from the base of

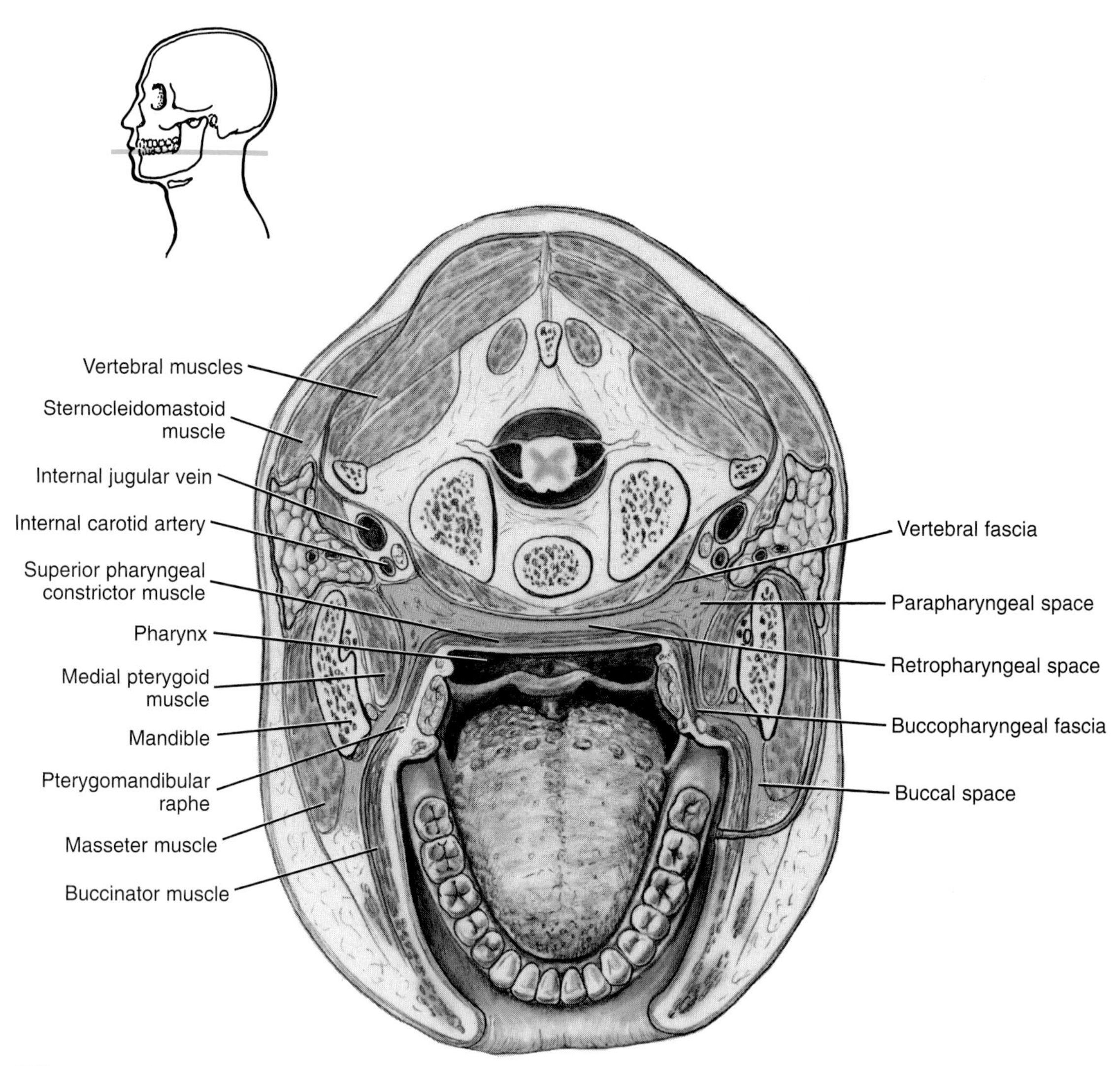

Figure 11–13
Transverse section of the oral cavity highlighting the retropharyngeal space and parapharyngeal space.

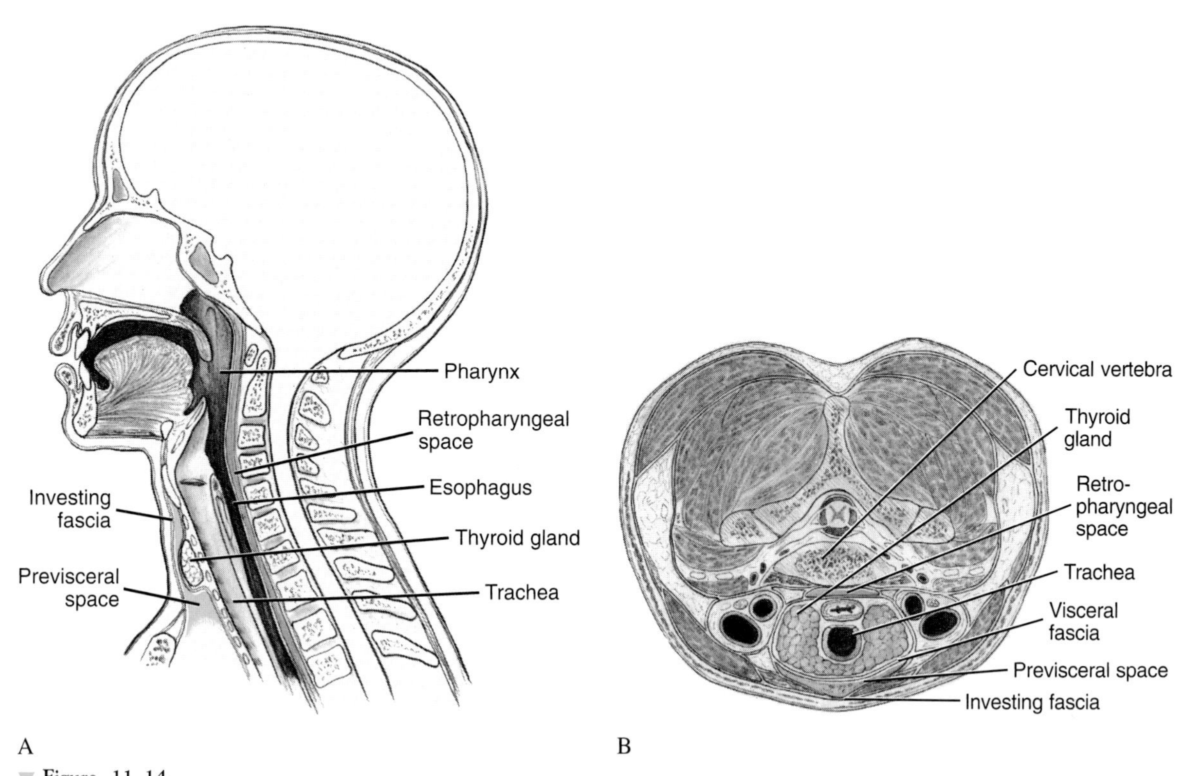

Figure 11–14
A: Midsagittal section of the head and neck highlighting the retropharyngeal space and previsceral space. **B:** Transverse section of the neck highlighting the retropharyngeal space and previsceral space.

the skull, where it is posterior to the superior pharyngeal constrictor muscle, down to the thorax.

Because of the rapidity with which dental infections can travel down the retropharyngeal space, it is also known as the danger space (see Chapter 12 for more information). This space communicates with the parapharyngeal spaces.

▼ PREVISCERAL SPACE

The **previsceral space** (pre-**vis**-er-al) is located between the visceral and investing fasciae, anterior to the trachea. The previsceral space communicates with the parapharyngeal spaces.

Table 11–2

MAJOR CERVICAL FASCIAL SPACES WITH THEIR LOCATIONS, CONTENTS, AND COMMUNICATION PATTERNS

Fascial Space	Location	Contents	Communication Patterns
Parapharyngeal	Lateral to the visceral fascia around pharynx	Nodes	Masticator, submandibular, retropharyngeal, previsceral, and adjacent to carotid sheath
Retropharyngeal	Between vertebral and visceral fasciae		Parapharyngeal
Previsceral	Between visceral and investing fasciae	Nodes and cervical vessels	Parapharyngeal

REVIEW QUESTIONS

1. Which of the following describes deep fascia?
   A. Dense and inclastic tissue forming sheaths around deep structures
   B. Fatty and elastic fibrous tissue found just beneath the skin
   C. Potential spaces containing loose connective tissue
   D. Fatty and elastic fibrous tissue forming spaces under the skin
   E. Dense and inelastic deep structures of the vascular system

2. Which of the following is covered by the carotid sheath?
   A. Internal and common carotid arteries and tenth cranial nerve
   B. External and common carotid arteries and fifth cranial nerve
   C. External jugular vein and tenth cranial nerve
   D. Internal jugular vein and fifth cranial nerve

3. In which of the following spaces is the pterygoid plexus of veins located?
   A. Parotid space
   B. Temporal space
   C. Infratemporal space
   D. Buccal space
   E. Parapharyngeal space

4. The masticator space includes the submasseteric space and the:
   A. submental space.
   B. submandibular space.
   C. sublingual space.
   D. pterygomandibular space.
   E. retropharyngeal space.

5. Which of the following tissues surrounds the space of the body of the mandible?
   A. Fatty tissue
   B. Loose connective tissue
   C. Periosteum
   D. Elastic tissue

6. The submandibular space communicates most directly with the:
   A. temporal space.
   B. parotid space.
   C. buccal space.
   D. sublingual space.
   E. canine space.

7. The parapharyngeal space is located between the superior pharyngeal constrictor muscle and the:
   A. lateral pterygoid muscle.
   B. medial pterygoid muscle.
   C. mylohyoid muscle.
   D. masseter muscle.
   E. buccinator muscle.

8. Which space is located in the midline between the mandibular symphysis and the hyoid bone?
   A. Retropharyngeal space
   B. Sublingual space
   C. Submandibular space
   D. Submental space
   E. Submasseteric space

9. Which of the following nerves is located in the pterygomandibular space?
   A. Infraorbital nerve
   B. Posterior superior alveolar nerve
   C. Inferior alveolar nerve
   D. Anterior superior alveolar nerve

10. Which of the following areas directly communicates with the retropharyngeal space?
    A. Masticator space
    B. Submandibular space
    C. Previsceral space
    D. Parapharyngeal space

Chapter 12

Spread of Dental Infection

▼ *This chapter discusses the following topics:*

▼ *Key words*

Abducens nerve paralysis (ab-**doo**-senz pay-**ral**-i-sis) Loss of function of the sixth cranial nerve.

Abscess (**ab**-ses) Type of infection from the entrapment of pathogens with suppuration in a contained space.

Bacteremia (bak-ter-**ee**-me-ah) Bacteria traveling within the blood system.

Cavernous sinus thrombosis (**kav**-er-nus **sy**-nus **throm**-bo-sus) Infection of the cavernous venous sinus.

Cellulitis (sel-you-**lie**-tis) Diffuse inflammation of soft tissue.

Embolus, emboli (**em**-bol-us, **em**-bol-eye) Foreign material, such as a thrombus, that travels in the blood and can block the vessel.

Fistula, fistulae (**fis**-chool-ah, **fis**-chool-ay) Passageway in the skin, mucosa, or even bone allowing drainage of an abscess at the surface.

Ludwig's angina (**lood**-vigz an-**ji**-nah) Serious infection of the submandibular space, with a risk of spread to the neck and chest.

Lymphadenopathy (lim-fad-in-**op**-ah-thee) Process in which there is an increase in the size and a change in the consistency of lymphoid tissue.

Maxillary sinusitis Infection of the maxillary sinus.

Meningitis (men-in-**jite**-is) Inflammation of the meninges of the brain or spinal cord.

Normal flora (**flor**-ah) Resident microorganisms that usually do not cause infections.

Opportunistic infections (op-or-tu-**nis**-tik) Normal flora creating an infectious process because the body's defenses are compromised.

Osteomyelitis (os-tee-o-my-il-**ite**-is) Inflammation of bone marrow.

Paresthesia (par-es-**the**-ze-ah) Abnormal sensation from an area, such as burning or prickling.

Pathogens (**path**-ah-jens) Flora that are not normal body residents and can cause an infection.

Perforation (per-fo-**ray**-shun) Abnormal hole in a hollow organ, such as in the wall of a sinus.

Primary node Lymph node that drains lymph from a particular region.

Pustule (**pus**-tule) Small, elevated, circumscribed suppuration-containing lesion of either the skin or oral mucosa.

Secondary node Lymph node that drains lymph from a primary node.

Stoma (**stow**-mah) Opening, such as with a fistula.

Thrombus, thrombi (**throm**-bus, **throm**-by) Clot that forms on the inner blood vessel wall.

Chapter 12

Spread of Dental Infection

▼ ***After studying this chapter, the reader should be able to:***

1. Define and pronounce all the key words and anatomical terms in this chapter.
2. Discuss the spread of infection to the sinuses and by the blood system, lymphatics, and spaces to other areas in the head and neck region.
3. Trace the routes of the spread of dental infection in the head and neck region on a diagram, a skull, and a patient.
4. Discuss the lesions and complications that can occur with the spread of dental infection in the head and neck region.
5. Discuss the prevention of the spread of dental infection during patient care.
6. Correctly complete the review questions for this chapter.
7. Integrate the knowledge of the spread of dental infection into the prevention and management of the spread of dental infection.

Infectious Process

A dental professional should understand the **infectious process** that allows a microorganism to create disease. The healthy body usually lives in balance with a number of resident microorganisms or ***normal flora*** (**flor**-ah). Certain microorganisms called ***pathogens*** (**path**-ah-jens) are not normal body residents and can cause an infectious process in the body. Pathogens have certain factors that help further the infectious process, such as capsules, spores, and toxins.

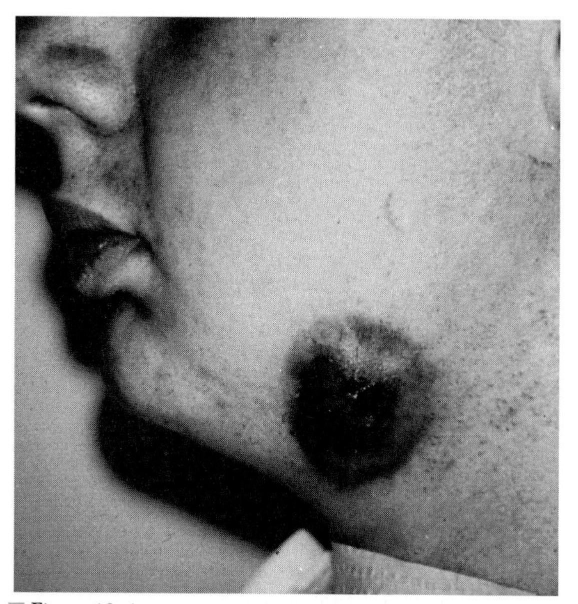

Figure 12–1
Abscess on the buccal skin surface from the formation of a fistula and stoma from periapical involvement of the mandibular second molar. (Courtesy of Dr. Mark Gabrielson.)

Dental Infections

Dental infections or odontogenic infections involve the teeth or associated tissues. The oral pathogens involved are mainly anaerobic and usually of more than one species. These infections can be of dental origin or from a nonodontogenic source.

Those of dental origin are usually from progressive dental caries or extensive periodontal disease. Pathogens can also be introduced deeper into the oral tissues by the trauma caused by dental procedures. Possible examples include the contamination of dental surgical sites (e.g., tooth extraction) and needle tracks during administration of a local anesthetic.

Some dental infections are secondary infections from a nonodontogenic source. These sources of infection can come from the tissues surrounding the oral cavity, such as the skin, tonsils, ear, or sinuses. These other sources of infections must be diagnosed and treated early to prevent further spread and complications. Prompt referral to the patient's physician is indicated in these circumstances.

DENTAL INFECTION LESIONS

Dental infections can result in an abscess, cellulitis, or osteomyelitis in the head and neck region. The type of lesion depends on the location of the infection and thus the type of tissue involved.

Abscess

An oral ***abscess*** (**ab**-ses) occurs when there is entrapment of pathogens with suppuration (pus) from a dental infection in a closed space, such as that created by the oral mucosa (Figures 12–1, 12–2, and 12–3). Thus abscess formation can occur with progressive caries. During decay, pathogens invade the dental pulp, followed by the infection spreading deep into the periapical areas, causing a periapical abscess.

Deep pockets around the teeth with severe periodontal disease or around an erupting third molar can also allow entrapment of pathogens, causing periodontal and pericoronal abscesses (pericoronitis), respec-

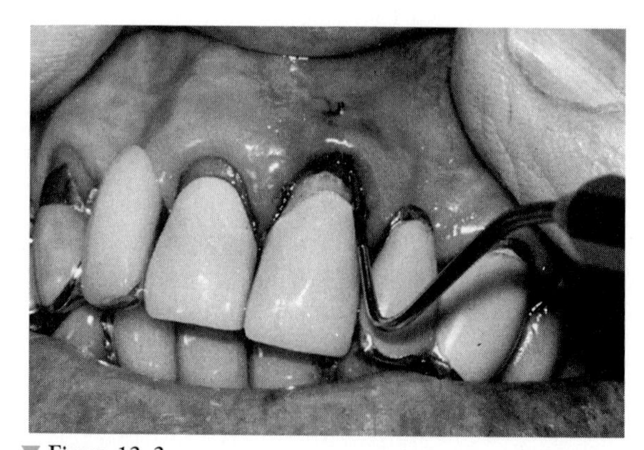

Figure 12–2
Periodontal abscess of the maxillary central incisor with fistula and stoma formation (probe inserted) in the maxillary vestibule.

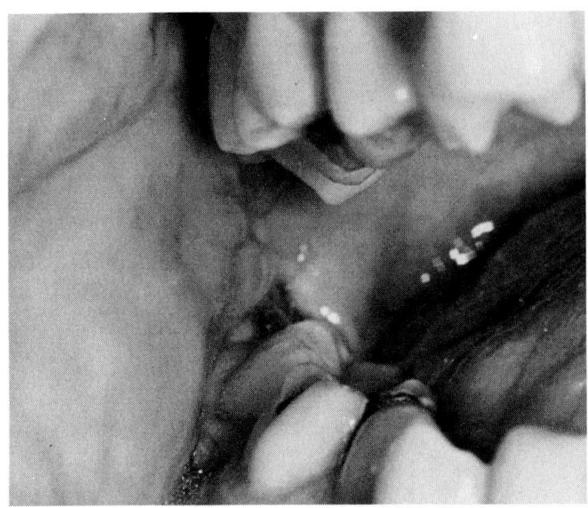

▼ Figure 12–3
Pericoronitis of the mandibular third molar with abscess formation.

tively. Abscess formation may not be visible radiographically during the early stages of the lesion.

Abscess formation can also lead to the formation of a passageway or ***fistula*** (plural, ***fistulae***)(**fis**-chool-ah, **fis**-chool-ay) in the skin, oral mucosa, or even bone in order to drain the infection and suppuration at a surface (see Figures 12–1 and 12–2). The infectious process causes the overlying tissue to undergo necrosis, forming this tract or canal in the tissue. The opening of the fistula is called a ***stoma*** (**stow**-mah). The abscess can be acute or chronic.

If the dental infection is surrounded by the alveolar bone, it will break down the bone in its thinnest portion (either the facial or lingual cortical plate), following the path of least resistance. The soft tissue over a fistula in the alveolar bone may also have an extraoral or intraoral localized area of swelling or pustule. A ***pustule*** (**pus**-tule) is a small, elevated, circumscribed suppuration-containing lesion (pus) of either the skin or oral mucosa. The position of the abscess's pustule (or parvulis on the gingiva) is largely determined by the relationship between the fistula and overlying muscle attachments, again following the path of least resistance (Table 12–1). The attachments of the muscles to the bones serve as barriers to the spread of infection, unlike the other soft tissues of the face.

▼ Table 12–1

MOST COMMON TEETH AND ASSOCIATED PERIODONTIUM INVOLVED IN CLINICAL PRESENTATIONS OF ABSCESSES AND FISTULAE

Clinical Presentation of Lesion	Most Common Teeth and Associated Periodontium Involved
Maxillary vestibule	Maxillary central or lateral incisor, all surfaces and root Maxillary canine, all surfaces and root (short root below levator anguli oris) Maxillary premolars, buccal surfaces and roots Maxillary molars, buccal surfaces or buccal roots (short roots below buccinator)
Penetration of nasal floor	Maxillary central incisor, root
Nasolabial skin region	Maxillary canine, all surfaces and root (long root above levator anguli oris)
Palate	Maxillary lateral incisor, lingual surface and root Maxillary premolars, lingual surfaces and roots Maxillary molars, lingual surfaces or palatal roots
Perforation into maxillary sinus	Maxillary molars, buccal surface and buccal roots (long roots)
Buccal skin surface	Maxillary molars, buccal surfaces and buccal roots (long roots above buccinator) Mandibular first and second molars, buccal surfaces and buccal roots (long roots below buccinator)
Mandibular vestibule	Mandibular incisors, all surfaces and roots (short roots above mentalis) Mandibular canine and premolars, all surfaces and roots (all roots above depressors) Mandibular first and second molars, buccal surfaces and roots (short roots above buccinator)
Submental skin region	Mandibular incisors, root (long roots below mentalis)
Sublingual region	Mandibular first molar, lingual surface and roots (all roots above mylohyoid) Mandibular second molar, lingual surface and roots (short roots above mylohyoid)
Submandibular skin region	Mandibular second molar, lingual surface and roots (long roots below mylohyoid) Mandibular third molars, all surfaces and roots (all roots below mylohyoid)

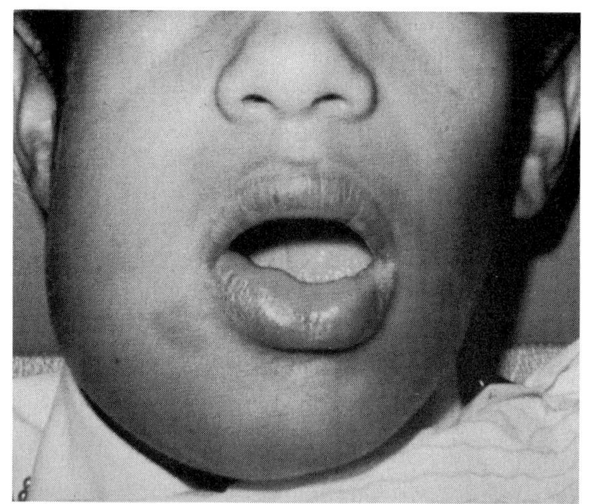

Figure 12–4
Cellulitis involving the buccal space from an abscess of the mandibular first molar with swelling noted. (Courtesy of Dr. Mark Gabrielson.)

Cellulitis

Cellulitis (sel-you-**lie**-tis) is the diffuse inflammation of soft tissue spaces. Cellulitis of the face and neck can also occur with dental infections (Figure 12–4). The clinical signs and symptoms are pain, tenderness, redness, and diffuse edema of the soft tissue spaces, causing a massive and firm swelling (Table 12–2). There also may be difficulty swallowing (dysphagia) or restricted eye opening (ptosis) if the cellulitis occurs within the pharynx or orbital regions, respectively.

Osteomyelitis

Another type of lesion related to dental infections is ***osteomyelitis*** (os-tee-o-my-il-**ite**-is), an inflammation of the bone marrow (Figure 12–5). This inflammation can involve any portion of the jaw bone: the alveolar bone proper, supporting cortical bone, or trabecular bone. This inflammation develops from the invasion of the bone tissue by pathogens from a periapical abscess, an extension of cellulitis, or contamination of surgical sites. Osteomyelitis most frequently occurs in the mandible and only rarely occurs in the maxilla because of the mandible's thicker cortical plates and reduced vascularization.

Continuation of osteomyelitis leads to bone resorption and sequestra formation (portions of necrotic bone separated from normal bone tissue), with radiographs showing the damage to the bone. ***Paresthesia*** (par-es-**the**-ze-ah), an abnormal sensation from the area, such as burning or prickling, may develop in the mandible if the infection involves the mandibular canal carrying the inferior alveolar nerve. Localized paresthesia of the lower lip may occur if the infection is distal to the mental foramen where the mental nerve exits. Osteomyelitis of the jaws is uncommon since dental care and antibiotics are readily available.

MEDICALLY COMPROMISED PATIENTS

Normal flora usually do not create an infectious process unless the body's natural defenses are com-

Table 12–2
POSSIBLE SPACE, TEETH, AND PERIODONTIUM INVOLVED WITH A CLINICAL PRESENTATION OF CELLULITIS FROM THE SPREAD OF DENTAL INFECTION

Clinical Presentation of Lesion	Space Involved	Most Common Teeth and Associated Periodontium Involved in Infection
Infraorbital, zygomatic, and buccal regions	Buccal space	Maxillary premolars and maxillary and mandibular molars
Posterior border of mandible	Parotid space	Not generally of odontogenic origin
Submental region	Submental space	Mandibular anterior teeth
Unilateral submandibular region	Submandibular space	Mandibular posterior teeth
Bilateral submandibular region	Submental, sublingual, and submandibular spaces with Ludwig's angina	Spread of mandibular dental infection
Lateral cervical region	Parapharyngeal space	Spread of mandibular dental infection

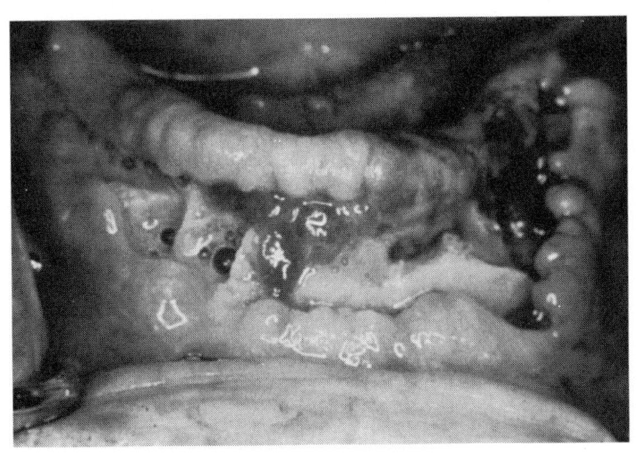

Figure 12–5
Osteomyelitis of the mandible with swelling noted.

promised, and then they can create ***opportunistic infections*** (op-or-tu-**nis**-tik). Medically compromised individuals include those with acquired immunodeficiency syndrome (AIDS) and those with Type I or juvenile diabetes, which is insulin-dependent. Patients undergoing radiation therapy for an oral malignancy are also at risk for serious infections.

Some patients also have a higher risk of complications from dental infections due to their medical history. Patients in this category include those at risk for infective endocarditis and those with a prosthetic heart valve.

Spread of Dental Infections

Many infections that initially start in the teeth and associated oral tissues can have serious consequences if the dental infections spread to vital tissues or organs. Usually a localized abscess establishes a fistula in the skin, oral mucosa, or bone, so natural drainage of the infection occurs and the risk of spread of the infection diminishes. Yet fistula formation or even drainage does not always occur, and the dental infection can spread.

A spread of dental infection can occur to the paranasal sinuses and other vital tissues or organs through the blood system, lymphatics, or spaces in the head and neck. It is important to review the communication patterns that are possible in these various tissues to understand the possible routes of infection that may occur (see Chapters 3, 6, 10, and 11 for more information).

SPREAD TO THE PARANASAL SINUSES

The **paranasal sinuses** of the skull can become infected from the direct spread of infection from the teeth and associated oral tissues. This spread of infection to the paranasal sinuses from another region is called ***secondary sinusitis*** (sy-nu-**si**-tis). A ***perforation*** (per-fo-**ray**-shun), an abnormal hole in the wall of the sinus, also can occur due to infection.

Maxillary Sinusitis

Secondary sinusitis of dental origin occurs mainly with the **maxillary sinuses** since the maxillary posterior teeth and associated tissues are in close proximity to the sinuses (see Chapter 3 for more information). Thus ***maxillary sinusitis*** can occur from a spread of infection from a periapical abscess of a maxillary posterior tooth that perforates the sinus floor to involve the sinus mucosa. A contaminated tooth or root fragments also can be surgically displaced into the maxillary sinus during an extraction.

Most infections of the maxillary sinuses are not of dental origin but are due to an upper respiratory infection, when infection in the nasal region spreads to the sinuses. An infection in one sinus can also travel through the nasal cavity to other sinuses, leading to serious complications for the patient. There also can be a spread of infection from the sinuses to the cranial cavity and brain. It is therefore important that sinusitis be treated aggressively to eliminate the initial infection.

The symptoms of sinusitis are headache, usually near the involved sinus, and foul-smelling nasal or pharyngeal discharge, possibly with some systemic signs of infection, such as fever and weakness. The skin over the involved sinus can be tender, hot, and red due to the inflammatory process in the area. Difficulty in breathing (dyspnea) occurs, as well as pain, when the nasal passages and sinus ostia become blocked by the effects of tissue inflammation.

Early radiographic evidence of the sinusitis is thickening of the walls of the sinus. Later radiographic evaluation of an infected sinus will often show increased opacity, and possibly perforation. A panoramic view of the skull is helpful because bilateral comparisons can be made.

Acute sinusitis usually responds to antibiotic therapy. Drainage from the sinuses can be helped through the use of decongestants. Surgery may be needed in the cases of chronic maxillary sinusitis to enlarge the ostia in the lateral walls in the nasal cavity to create adequate drainage and diminish the effects of the infection. The surgical approach to the maxillary sinus is through the thin bone of the canine fossa.

SPREAD BY BLOOD SYSTEM

The blood system of the head and neck can allow the spread of infection from the teeth and associated oral tissues. This occurs because the pathogens can travel in the veins, draining the infected oral site to other tissues or organs (see Chapter 6 for more information). The spread of dental infection by way of the blood system can occur from a bacteremia or an infected thrombus.

Bacteremia

Bacteria that travel in the blood are referred to as ***bacteremia*** (bak-ter-**ee**-me-ah). Transient bacteremia occurs during dental treatment, surgery, or trauma. Individuals with a high risk for infective endocarditis and those with a prosthetic heart valve or joint may have these bacteria lodge in the compromised tissues and set up serious infection deep in the heart, which can result in massive and fatal heart damage. These patients may need antibiotic premedication to prevent bacteremia from occurring during dental treatment.

Cavernous Sinus Thrombosis

An infected intravascular clot or ***thrombus*** (plural ***thrombi***) (**throm**-bus, **throm**-by) can dislodge from the inner blood vessel wall and travel as an ***embolus*** (plural **emboli**) (**em**-bol-us, **em**-bol-eye). The emboli or foreign material can travel in the veins, draining the oral cavity to areas such as the dural **venous sinuses** within the cranial cavity (see Chapter 6 for more information). These dural sinuses are channels by which blood is conveyed from the cerebral veins into the veins of the neck, particularly the internal jugular vein. However, because these veins lack valves, blood can flow both into and out of the cranial cavity.

The venous sinus that is most likely to be involved in the possibly fatal spread of dental infection is the cavernous sinus. An infection of this venous sinus is called ***cavernous sinus thrombosis*** (**kav**-er-nus **sy**-nus **throm**-bo-sus). The cavernous sinus is located on the side of the body of the sphenoid bone. Each cavernous venous sinus communicates with the one on the opposite side and also with the **pterygoid plexus of veins** and **superior ophthalmic vein,** which anastomoses with the **facial vein.** These major veins drain teeth through the **posterior superior** and **inferior alveolar veins** and the lips through the **superior** and **inferior labial veins.**

None of these major veins that communicate with the cavernous sinus have valves to prevent retrograde blood flow back into the cavernous sinus. Thus dental infections that drain into these major veins may initiate an inflammatory response, resulting in increased blood stasis, thrombus formation, and increased extravascular fluid pressure. Increased pressure can reverse the direction of venous blood flow, causing transport of the infected thrombus into this venous sinus and thus causing cavernous sinus thrombosis.

Needle-track contamination can also result in a spread of infection to the pterygoid plexus of veins if a posterior superior alveolar anesthetic block is incorrectly administered. Nonodontogenic infections from what physicians consider the dangerous triangle of the face, the orbital region, nasal region, or paranasal sinuses, also may result in a spread of infection to the cavernous venous sinus.

The signs and symptoms of cavernous sinus thrombosis include those of serious systemic infection, such as fever, drowsiness, and rapid pulse. There also is loss of function of the sixth cranial nerve or **abducens** since the nerve runs through the cavernous venous sinus. This loss of function is called ***abducens nerve paralysis*** (ab-**doo**-senz pay-**ral**-i-sis). Because the muscle supplied by the abducens nerve moves the eyeball laterally, inability to perform this movement suggests nerve damage. The patient also will usually have double vision (diplopia) because of the restricted movement of the one eye. There will also be edema of the eyelids and conjunctivae, tearing (lacrimation), and extruded eyeballs (exophthalmus) depending on the course of the infection.

With cavernous sinus thrombosis, there may also be damage to the other cranial nerves, such as the **oculomotor nerve** (third) and **trochlear nerve** (fourth), as well as the ophthalmic and maxillary divisions of the **trigeminal nerve** (fifth) and changes in the tissues they innervate. All these nerves travel in the wall of the cavernous sinus. Finally, this infection can be fatal because it may lead to inflammation of the meninges of the brain or spinal cord, which results in ***meningitis*** (men-in-**jite**-is). Patients with meningitis require immediate hospitalization with intravenous antibiotics and anticoagulants.

SPREAD BY LYMPHATICS

The **lymphatics** of the head and neck can allow the spread of infection from the teeth and associated oral tissues. This occurs because the pathogens can travel in the lymph through the lymphatic vessels that connect the series of nodes from the oral cavity to other tissues or organs (see Chapter 10 for more information). Thus these pathogens can move from a ***primary node*** near the infected site to a ***secondary node*** at the distant site. A primary node drains lymph from a particular region, whereas secondary nodes drain lymph from a primary node.

The route of dental infection traveling through the lymph nodes varies according to the teeth involved. The **submental lymph nodes** drain the mandibular incisors and associated tissues. The submental nodes then empty into the submandibular nodes or directly into the deep cervical nodes. The **submandibular lymph nodes** are the primary nodes for all the teeth and associated tissues, except the mandibular incisors and maxillary third molars. The submandibular nodes then empty into the superior deep cervical nodes.

The superior deep cervical nodes are primary nodes for and drain the maxillary third molars and associated tissues. The superior deep cervical nodes empty either into the inferior deep cervical nodes or directly into the **jugular trunk** and then into the vascular system. Once in the vascular system, the infection can spread to all tissues and organs of the body (as discussed earlier).

Lymphadenopathy

A lymph node involved in infection undergoes ***lymphadenopathy*** (lim-fad-in-**op**-ah-thee), which results in an increase in size and a change in consistency of the lymph nodes. Lymphadenopathy is caused by an increase in both the size of each individual lymphocyte and the overall cell count in the lymphoid tissue. With more and larger lymphocytes, the lymphoid tissue is better able to fight the disease process.

Thus with lymphadenopathy, the node becomes tender and palpably firm, yet remains mobile. Evaluation of the nodes involved in lymphadenopathy can determine the degree of regional involvement of the infectious process. The amount of regional involvement is instrumental in the diagnosis and management of the infectious process.

SPREAD BY SPACES

The **fascial spaces** of the head and neck can allow the spread of infection from the teeth and associated oral tissues. This occurs because the pathogens can travel within the fascial planes from one space near the infected site to another distant space by the spread of the related inflammatory exudate (see Chapter 11 for more information). When involved in infections, the space can undergo swelling or cellulitis (see Table 12–2). This can cause a change in the normal proportions of the face (see Chapter 2 for more information).

If the maxillary teeth and associated tissues are infected, the infection can spread into the **vestibular space of the maxilla, buccal space,** or **canine space.** If the mandibular teeth and associated tissues are infected, the infection can spread into the **vestibular space of the mandible, buccal space, submental space, sublingual space, submandibular space,** or **space of the body of the mandible.** The infection can spread from these spaces into other spaces of the jaws and neck, possibly causing serious complications, such as Ludwig's angina.

Ludwig's Angina

One of the most serious lesions of the jaw region is ***Ludwig's angina*** (**lood**-vigz an-**ji**-nah) (Figure 12–6), a cellulitis of the **submandibular space.** This involves a spread of infection from any of the mandibular teeth or associated tissues to one space initially, either the submental, sublingual, or even submandibular space. The infection then spreads to the submandibular space bilaterally, with a risk of spread to the **parapharyngeal space** of the neck.

There is massive bilateral submandibular regional swelling, which extends down the anterior cervical triangle to the clavicles. Swallowing, speech, and breathing may be difficult, and high fever and drooling are evident. Respiratory obstruction may rapidly develop because the continued swelling displaces the

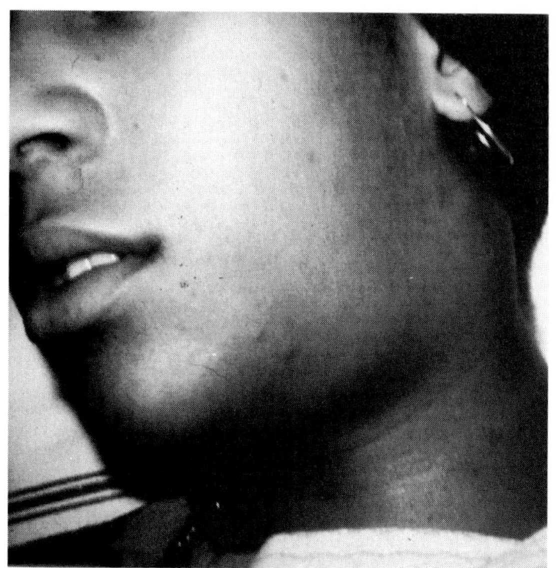

▼ **Figure 12–6**
Ludwig's angina showing involvement of the submandibular and submental spaces from an abscess of the mandibular third molar. (Courtesy of Dr. Mark Gabrielson.)

tongue upward and backward, thus blocking the pharyngeal airway. As the **retropharyngeal space** or danger space also becomes involved, edema of the larynx may cause complete respiratory obstruction, asphyxiation, and death.

Ludwig's angina is an acute medical emergency requiring immediate hospitalization and may necessitate an emergency cricothyrotomy to create a patent airway. With the advent of the earlier care of abscessed teeth and routine antibiotic treatment, Ludwig's angina is not a common dental emergency.

Prevention of the Spread of Dental Infections

Early diagnosis and treatment of dental infections must occur in all patients. There must be a strict adherence to aseptic protocol during treatment to prevent the spread of infection. Thus the removal of heavy plaque accumulations is recommended prior to all dental treatment, especially surgery. Care must also be taken not to contaminate surgical sites. Finally, it is important not to administer a local anesthetic through an area of dental infection, moving the pathogens deeper into the tissues.

A thorough medical history with periodic updates will allow the dental professional to perform safe treatment on medically compromised patients to avoid serious complications due to their dental diseases. These patients may need antibiotic premedication before dental treatment to prevent any serious sequelae.

REVIEW QUESTIONS

1. Which of the following complications is likely with untreated Ludwig's angina?
   A. Abducens nerve paralysis
   B. Meningitis
   C. Respiratory obstruction
   D. Sinus perforation
   E. Double vision

2. Which of the following cranial nerves may be involved with cavernous sinus thrombosis?
   A. Oculomotor and trochlear nerves
   B. Vagus and glossopharyngeal nerves
   C. Hypoglossal and accessory nerves
   D. Optic and olfactory nerves

3. Which of the following statements correctly describes an oral abscess?
   A. Inflammation of the bone
   B. Inflammation of the meninges
   C. Infection confined in oral mucosal space
   D. Diffuse inflammation of soft tissue

4. Which of the following statements concerning cavernous sinus thrombosis is correct?
   A. Associated major veins have valves
   B. Only dental infections can spread to the sinus
   C. Eye tissues are not affected
   D. Needle-track contamination may be involved

5. If an infection involves the lingual surface of the mandibular third molar, a swelling is most likely to be observed:
   A. in the mandibular vestibule.
   B. beneath the tongue.
   C. in the submandibular region.
   D. in the buccal and submental regions.

BIBLIOGRAPHY

Cawson RA. *Aids to Oral Pathology and Diagnosis.* Churchill and Livingstone, Edinburgh, London, Melbourne, and New York, 1981.

Clemente CD. *Anatomy,* 3rd edition. Urban and Schwarzenberg, Baltimore, 1987.

Dorland's Medical Dictionary, 28th edition. WB Saunders Company, Philadelphia, 1994.

DuBrul EL. *Sicher and DuBrul's Oral Anatomy,* 8th edition. Ishiyaku EuroAmerica, Inc, St. Louis, Missouri, 1988.

Evers H, Haegerstam G. *Introduction to Dental Local Anesthesia.* BC Decker, Ontario, Philadelphia, and Mediglobe, Switzerland, 1990.

Gray H. *Gray's Anatomy,* 37th edition. Churchill and Livingstone, Edinburgh, London, Melbourne, and New York, 1989.

Hohl TH, Whitacre RJ, Hooley JR, Williams BL. *Diagnosis and Treatment of Odontogenic Infections.* Stoma Press, Seattle, 1983.

Ibsen AC, Phelan JA. *Oral Pathology for the Dental Hygienist.* WB Saunders Company, Philadelphia, 1992.

Kasle MJ. *An Atlas of Dental Radiographic Anatomy,* 3rd edition. WB Saunders Company, Philadelphia, 1990.

Laskin D, Greenfield W, Elliot G, et al. *President's Conference on the Examination, Diagnosis, and Management of Temporomandibular Disorder.* American Dental Association, 1982.

Leonard PC. *Medical Terminology.* WB Saunders Company, Philadelphia, 1990.

Miles DA, Van Dis ML, Razmus TF. *Oral and Maxillofacial Radiology: Radiologic/Pathologic Correlations.* WB Saunders Company, Philadelphia, 1991.

Nomina Anatomica, 6th edition. Churchill and Livingstone, Edinburgh, London, Melbourne, and New York, 1989.

O'Toole M. *Miller-Keane Encyclopedia and Dictionary of Medicine, Nursing and Allied Health,* 5th edition. WB Saunders Company, Philadelphia, 1992.

Reed GM, Sheppard VF. *Basic Structures of the Head and Neck.* WB Saunders Company, Philadelphia, 1976.

Rohen JW, Yokochi C. *Color Atlas of Anatomy.* Igaku-Shoin, New York, 1983.

Schumacher G-H. *Anatomie für Stomatologen.* JA Barth, Leipzig, 1984.

Solomon EP. *Introduction to Human Anatomy and Physiology.* WB Saunders Company, Philadelphia, 1992.

Wilson-Pauwels LW, Akesson EJ, Stewart PA. *Cranial Nerves.* BC Decker, Toronto and Philadelphia, 1988.

ANSWERS TO THE REVIEW QUESTIONS

Chapter 1

1. B
2. B
3. D
4. B
5. B

Chapter 2

1. C
2. C
3. C
4. B
5. C

Chapter 3

1. C
2. D
3. A
4. D
5. A
6. D
7. D
8. B
9. C
10. B
11. B
12. A
13. B
14. C
15. C
16. B
17. E
18. D
19. C
20. A

Chapter 4

1. C
2. D
3. C
4. D
5. D
6. C
7. A
8. E
9. B
10. D
11. A
12. C
13. A
14. C
15. B
16. B
17. B
18. B
19. A
20. B

Chapter 5

1. C
2. B
3. C
4. C
5. C
6. E
7. A
8. B
9. D
10. C

Chapter 6

1. A
2. C
3. B
4. A
5. B
6. A
7. C
8. B
9. C
10. A

Chapter 7

1. A
2. B
3. B
4. B
5. A
6. B
7. D
8. A
9. D
10. C

Chapter 8

1. B
2. D
3. B
4. C
5. B
6. B
7. D
8. B
9. B
10. A
11. B
12. A
13. B
14. C
15. D

Chapter 9

1. B
2. E
3. A
4. C
5. C
6. C
7. A
8. B
9. C
10. D

Chapter 10

1. C
2. C
3. D
4. C
5. A
6. B
7. B
8. B
9. D
10. C
11. E
12. A
13. B
14. C
15. C

Chapter 11

1. A
2. A
3. C
4. D
5. C
6. D
7. B
8. D
9. C
10. D

Chapter 12

1. C
2. A
3. C
4. D
5. C

GLOSSARY OF KEY WORDS AND ANATOMICAL TERMS

A

Abducens nerve (ab-**doo**-senz) Sixth cranial nerve (VI) that serves an eye muscle.

Abducens nerve paralysis (ab-**doo**-senz pay-**ral**-i-sis) Loss of function of the sixth cranial nerve.

Abscess (**ab**-ses) Type of infection from the entrapment of pathogens with suppuration in a contained space.

Accessory lymph nodes (ak-**ses**-o-ree) Deep cervical nodes located along the accessory nerve.

Accessory nerve (ak-**ses**-o-ree) Eleventh cranial nerve (XI) that serves the trapezius and sternocleidomastoid muscles, as well as muscles of the soft palate and pharynx.

Action (**ak**-shun) Movement accomplished by the muscle when the muscle fibers contract.

Adenoids (**ad**-in-oidz) Another term for the pharyngeal tonsil.

Afferent nerve (**af**-er-int) Sensory nerve that carries information from the periphery of the body to the brain or spinal cord.

Afferent vessels (**af**-er-int) Type of lymphatic vessel in which lymph flows into the lymph node.

Ala, alae (**a**-lah, **a**-lay) Winglike cartilaginous structures that laterally bound the nares.

Alveolar mucosa (al-**ve**-o-lar) Mucosa that lines the vestibules of the oral region.

Alveolar process of the mandible (al-**ve**-o-lar) Portion of the mandible that contains the roots of the maxillary teeth.

Alveolar process of the maxilla (al-**ve**-o-lar) Ridge of maxillary bone that houses the roots of the maxillary teeth.

Anastomosis, anastomoses (ah-nas-tah-**moe**-sis, ah-nas-tah-**moe**-sees) Communication of a blood vessel with another by a connecting channel.

Anatomical nomenclature (an-ah-**tom**-ik-al **no**-men-kla-cher) System of names of anatomical structures.

Anatomical position (an-ah-**tom**-ik-al) Position where the body is erect, arms at the sides, palms and toes directed forward, and eyes looking forward.

Angle of the mandible Angle at the intersection of the posterior and inferior borders of the ramus.

Angular artery (**ang**-u-lar) Arterial branch that is a termination of the facial artery and supplies the tissues along the side of the nose.

Anterior (an-**tere**-ee-or) Front of an area of the body.

Anterior arch Arch of the atlas or first cervical vertebra.

Anterior auricular lymph nodes (aw-**rik**-you-lar) Superficial nodes located anterior to the ear.

Anterior cervical triangle (**ser**-vi-kal) Anterior region of the neck.

Anterior ethmoidal nerve (eth-**moy**-dal) Nerve from the nasal cavity and paranasal sinuses that converges with other orbital branches to form the nasociliary nerve.

Anterior jugular lymph nodes (**jug**-you-lar) Superficial cervical nodes located along the anterior jugular vein.

Anterior jugular vein (**jug**-you-lar) Vein that begins below the chin, descends near the midline, and drains into the external jugular vein.

Anterior superior alveolar artery (al-**ve**-o-lar) Arterial branch from the infraorbital artery that gives off dental and alveolar branches that supply the pulp tissue and periodontium of the anterior maxillary teeth.

Anterior superior alveolar (ASA) nerve (al-**ve**-o-lar) Nerve that serves the maxillary anterior teeth and tissues and is formed from dental and interdental branches and later joins the infraorbital nerve.

Anterior suprahyoid muscle group (soo-prah-**hi**-oid) Suprahyoid muscles located anterior to the hyoid bone that include the anterior belly of the digastric, mylohyoid, and geniohyoid muscles.

Anterior tonsillar pillar (**ton**-sil-ar **pil**-er) Vertical fold anterior to each palatine tonsil created by the palatoglossal muscle.

GLOSSARY OF KEY WORDS AND ANATOMICAL TERMS *Continued*

Antitragus (an-tie-**tra**-gus) Flap of tissue opposite the tragus of the ear.

Aorta (ay-**ort**-ah) Major artery that gives rise to the common carotid and subclavian arteries on the left side of the body and to the brachiocephalic artery on the right side.

Aperture (**ap**-er-cher) Opening or orifice in bone.

Apex (**ay**-peks) Pointed end of a conical structure.

Apex of the nose Tip of the nose.

Apex of the tongue Tip of the tongue.

Arch (arch) Prominent bridgelike bony structure.

Arteriole (ar-**ter**-ee-ole) Smaller artery that branches off an artery and connects with a capillary.

Artery (**art**-er-ee) Type of blood vessel that carries blood away from the heart.

Articular eminence (ar-**tik**-you-ler) Eminence on the temporal bone that articulates with the mandible at the temporomandibular joint.

Articular fossa (ar-**tik**-you-ler) Fossa on the temporal bone that articulates with the mandible at the temporomandibular joint.

Articulating surface of the condyle (ar-**tik**-you-late-ing) Portion of the head of the condyle that articulates with the temporal bone at the temporomandibular joint.

Articulation (ar-tik-you-**lay**-shin) Area where the bones are joined to each other.

Ascending palatine artery (ah-**send**-ing **pal**-ah-tine) Arterial branch from the facial artery that supplies the palatine muscles and tonsils.

Ascending pharyngeal artery (ah-**send**-ing fah-**rin**-je-al) Medial arterial branch from the external carotid artery that supplies the pharyngeal walls, soft palate, and brain tissue.

Atlas (**at**-lis) First cervical vertebra, which articulates with the occipital bone.

Attached gingiva (jin-**ji**-vah) Gingiva that tightly adheres to the bone over the roots of the teeth.

Auricle (**aw**-ri-kl) Oval flap of the external ear.

Auriculotemporal nerve (aw-**rik**-yule-lo-**tem**-poh-ral) Nerve that serves tissues of the ear and scalp and the parotid salivary gland and joins the posterior trunk of the mandibular division of the trigeminal nerve.

Autonomic nervous system (awt-o-**nom**-ik) Portion of the peripheral nervous system that operates without conscious control and is divided into the sympathetic and parasympathetic nervous systems.

Axis (**ak**-sis) Second cervical vertebra, which articulates with the first and third cervical vertebrae.

B

Bacteremia (bak-ter-**ee**-me-ah) Bacteria traveling within the blood system.

Base of the tongue The posterior one third or root of the tongue.

Bell's palsy (**pawl**-ze) Type of unilateral facial paralysis involving the facial nerve.

Body of the hyoid bone (**hi**-oid) Anterior midline portion of the hyoid bone.

Body of the mandible Horizontal portion of the mandible.

Body of the maxilla Portion of the maxilla that contains the maxillary sinus.

Body of the sphenoid bone (**sfe**-noid) Middle portion of the bone containing the sphenoid sinuses.

Body of the tongue Anterior two thirds of the tongue.

Bones (bones) Mineralized structures of the body that protect internal soft tissues and serve as the biomechanical basis for movement.

Brachiocephalic artery (bray-kee-oo-sah-**fal**-ik) Artery that branches directly off the aorta on the right side of the body and gives rise to the right common carotid and subclavian arteries.

Brachiocephalic vein (bray-kee-oo-sah-**fal**-ik) Vein that is formed from the merger of the internal jugular and subclavian veins with the right and

GLOSSARY OF KEY WORDS AND ANATOMICAL TERMS *Continued*

left brachiocephalic veins, forming the superior vena cava.

Bridge of the nose Bony structure inferior to the nasion in the nasal region.

Buccal (**buk**-al) Structures closest to the inner cheek.

Buccal artery Arterial branch from the maxillary artery that supplies the buccinator muscle and cheek tissues.

Buccal fascia Space formed between the buccinator and masseter muscles.

Buccal fat pad Dense pad of tissue covered by the buccal mucosa.

Buccal lymph nodes Superficial nodes of the face located at the mouth angle and superficial to the buccinator muscle.

Buccal mucosa Mucosa that lines the inner cheek.

Buccal nerve Nerve that serves the skin of the cheek and buccal tissue of the mandibular molar teeth and joins with the muscular nerve branches to form the anterior trunk of the mandibular division of the trigeminal nerve.

Buccal region (**buk**-al) Region of the head that is composed of the soft tissues of the cheek of the face.

Buccal space Fascial space between buccinator and masseter muscles.

Buccinator muscle (buck-**sin**-nay-tor) Muscle of facial expression that forms a portion of the cheek.

Buccopharyngeal fascia (buk-o-fah-**rin**-je-al) Deep cervical fascia that encloses the entire upper part of the alimentary canal.

C

Canal (kah-**nal**) Opening in bone that is long, narrow, and tubelike.

Canine eminence (**kay**-nine) Facial ridge of bone over the maxillary canine.

Canine fossa (**kay**-nine) Fossa at the roots of the maxillary canine teeth.

Canine space (**kay**-nine) Space located lateral to the apex of the maxillary canine.

Capillary (**kap**-i-lare-ee) Smaller blood vessel that branches off an arteriole to supply blood directly to tissue.

Carotid canal (kah-**rot**-id) Canal in the temporal bone that carries the internal carotid artery.

Carotid pulse (kah-**rot**-id) Reliable pulse palpated from the common carotid artery.

Carotid sheath (kah-**rot**-id sheeth) Deep cervical fascia forming a tube running down the side of the neck.

Carotid sinus (kah-**rot**-id **sy**-nus) Swelling in the artery just before the common carotid artery bifurcates into the internal and external carotid arteries.

Cavernous sinus thrombosis (**kav**-er-nus **sy**-nus **throm**-bo-sus) Infection of the cavernous venous sinus.

Cavernous venous sinus (**kav**-er-nus) Venous sinus located on the side of the sphenoid bone that communicates with the pterygoid plexus and superior ophthalmic vein.

Cellulitis (sel-you-**lie**-tis) Diffuse inflammation of soft tissue.

Central nervous system Portion of nervous system that consists of the spinal cord and brain.

Cervical muscles (**ser**-vi-kal) Muscles of the neck that include the sternocleidomastoid and trapezius muscles.

Cervical vertebrae (**ser**-vi-kal **ver**-teh-bray) Vertebrae in the vertebral column between the skull and thoracic vertebrae.

Chorda tympani nerve (**kor**-dah **tim**-pan-ee) Branch of the facial nerve that serves the submandibular and sublingual salivary glands and tongue.

Ciliary nerves (**sil**-eh-a-re) Nerves to or from the eyeball, with some ciliary nerves converging with branches from the nose to form the nasociliary nerve.

Circumvallate lingual papillae (serk-um-**val**-ate) Large lingual papillae anterior to the sulcus terminalis.

GLOSSARY OF KEY WORDS AND ANATOMICAL TERMS *Continued*

Common carotid artery (**kom**-in kah-**rot**-id) Artery that travels in the carotid sheath up the neck to branch into the internal and external carotid arteries.

Condyle (**kon**-dyl) Oval bony prominence typically found at articulations.

Condyle of the mandible Projection of bone from the ramus of the mandible that participates in the temporomandibular joint.

Conjunctiva (kon-junk-**ti**-vah) Membrane lining the inside of the eyelids and front of the eyeball.

Contralateral (kon-trah-**lat**-er-il) Structures on the opposite side of the body.

Cornu (**kor**-nu) Small hornlike prominence.

Coronal suture (kor-**oh**-nahl) Suture between the frontal and parietal bones.

Coronoid notch (**kor**-ah-noid) Notch in the anterior border of the ramus.

Coronoid process (**kor**-ah-noid) Anterior superior projection of the ramus of the mandible.

Corrugator supercilii muscle (cor-rew-**gay**-tor soo-per-**sili**-eye) Muscle of facial expression in the eye region that it used when frowning.

Cranial bones (**kray**-nee-al) Skull bones that form the cranium and include the occipital, frontal, parietal, temporal, sphenoid, and ethmoid bones.

Cranial nerves (**kray**-nee-al) Portion of the peripheral nervous system that is connected to the brain and carries information to and from the brain.

Cranium (**kray**-nee-um) Structure that is formed by the cranial skull bones and includes the occipital, frontal, parietal, temporal, sphenoid, and ethmoid bones.

Crest (krest) Roughened border or ridge on the bone surface.

Cribriform plate (**krib**-ri-form) Horizontal plate of the ethmoid bone that is perforated with foramina for the olfactory nerves.

Crista galli (**kris**-tah **gal**-lee) Vertical midline continuation of the perpendicular plate of the ethmoid bone into the cranial cavity.

D

Deep (deep) Structures located inward, away from the body surface.

Deep cervical lymph nodes (**ser**-vi-kal) Nodes located along the internal jugular vein that are divided into two groups, superior and inferior, based on the point where the omohyoid muscle crosses the vein.

Deep parotid lymph nodes (pah-**rot**-id) Nodes located deep to the parotid salivary gland.

Deep temporal arteries (**tem**-poh-ral) Arterial branches from the maxillary artery that supply the temporalis muscle.

Deep temporal nerves (**tem**-poh-ral) Muscular nerve branches that form the anterior trunk of the mandibular division of the trigeminal nerve and innervate the deep surface of the temporalis muscle.

Dens (denz) Odontoid process of the second cervical vertebra.

Depression of the mandible (de-**presh**-in) Lowering of the lower jaw.

Depressor anguli oris muscle (de-**pres**-er **an**-gu-lie **or**-is) Muscle of facial expression in the mouth region that depresses the angle of the mouth.

Depressor labii inferioris muscle (de-**pres**-er **lay**-be-eye in-**fere**-ee-o-ris) Muscle of facial expression in the mouth region that depresses the lower lip.

Digastric muscle (di-**gas**-trik) Suprahyoid muscle with an anterior and a posterior belly.

Disc of the temporomandibular joint (disk) Fibrous disc located between the temporal bone and condyle of the mandible.

Distal (**dis**-tl) Area that is farther away from the median plane of the body.

Dorsal (**dor**-sal) Back of an area of the body.

Dorsal surface of the tongue. Top surface of the tongue.

Duct (dukt) Passageway to carry the secretion from the exocrine gland to the location where it will be used.

GLOSSARY OF KEY WORDS AND ANATOMICAL TERMS *Continued*

E

Ebner's glands (**eeb**-ner) Minor salivary glands associated with the circumvallate lingual papilla.

Efferent nerve (**ef**-er-ent) Motor nerve that carries information away from the brain or spinal cord to the periphery of the body.

Efferent vessel (**ef**-er-ent) Type of lymphatic vessel in which lymph flows out of the lymph node in the area of the node's hilus.

Elevation of the mandible (el-eh-**vay**-shun) Raising of the lower jaw.

Embolus, emboli (**em**-bol-us, **em**-bol-eye) Foreign material, such as a thrombus, that travels in the blood and can block the vessel.

Eminence (**em**-i-nins) Tubercle or rounded elevation on the bony surface.

Endocrine gland (**en**-dah-krin) Type of gland without a duct, with the secretion being poured directly into the blood, which then carries the secretion to the region being used.

Epicondyle (ep-ee-**kon**-dyl) Small prominence that is located above or upon a condyle.

Epicranial aponeurosis (ep-ee-**kray**-nee-all ap-o-new-**row**-sis) Scalpal tendon from which the frontal belly of the epicranial muscle arises.

Epicranial muscle (ep-ee-**kray**-nee-al) Muscle of facial expression in the scalp region that has a frontal and an occipital belly.

Ethmoid bone (**eth**-moid) Single midline cranial bone of the skull.

Ethmoid sinuses (**eth**-moid **sy**-nuses) Paired paranasal sinuses located in the ethmoid bone, which are also called ethmoid air cells.

Exocrine gland (**ek**-sah-krin) Type of gland with an associated duct that serves as a passageway for the secretion to be emptied directly into the location where the secretion is to be used.

External (eks-**tern**-il) Outer side of the wall of a hollow structure.

External acoustic meatus (ah-**koos**-tik me-**ate**-us) Canal leading to the tympanic cavity.

External carotid artery (kah-**rot**-id) Artery that arises from the common carotid artery and supplies the extracranial tissues of the head and neck, including the oral cavity.

External jugular lymph nodes (**jug**-you-lar) Superficial cervical nodes located along the external jugular vein.

External jugular vein (**jug**-you-lar) Vein that forms from the posterior division of the retromandibular vein.

External nasal nerve (**nay**-zil) Nerve from portions of the nose skin that converges with other branches to form the nasociliary nerve.

External oblique line (**ob**-leek) Crest on the lateral side of the mandible, where the ramus joins the body.

Extrinsic tongue muscles (eks-**trin**-sik) Tongue muscles with different origins outside the tongue.

Eyelids Movable upper and lower tissues that cover and protect each eyeball.

F

Facial (**fay**-shal) Structures closest to the facial surface.

Facial artery Anterior arterial branch from the external carotid artery with a complicated path as it gives off the ascending palatine, submental, inferior and superior labial, and angular arteries.

Facial bones Skull bones that create the face and include the lacrimal bone, nasal bone, vomer, inferior nasal concha, zygomatic bone, maxilla, and mandible.

Facial lymph nodes Superficial nodes located along the facial vein that include the malar, nasolabial, buccal, and mandibular nodes.

Facial nerve Seventh cranial nerve(VII) that serves the muscles of facial expression, posterior suprahyoid muscles, lacrimal gland, sublingual and submandibular salivary glands, tongue portion, and portion of skin through its greater petrosal, chorda tympani, and posterior auricular nerves and muscular branches.

GLOSSARY OF KEY WORDS AND ANATOMICAL TERMS *Continued*

Facial paralysis (pay-**ral**-i-sis) Loss of action of the facial muscles.

Facial vein Vein that drains into the internal jugular vein after draining the facial areas.

Fascia, fasciae (**fash**-e-ah, **fash**-e-ay) Layers of fibrous connective tissue that underlie the skin and also surround the muscles, bones, vessels, nerves, organs, and other structures of the body.

Fascial spaces (**fash**-e-al) Potential spaces between the layers of fascia in the body.

Fauces (**faw**-seez) Faucial isthmus or junction between the oral region and oropharynx.

Filiform lingual papillae (**fil**-i-form) Papillae that give the tongue its velvety texture.

Fissure (**fish**-er) Opening in bone that is narrow and cleftlike.

Fistula, fistulae (**fis**-chool-ah, **fis**-chool-ay) Passageway in the skin, mucosa, or even bone allowing drainage of an abscess at the surface.

Foliate lingual papillae (**fo**-le-ate) Ridges of papillae on lateral tongue surface.

Foramen, foramina (for-**ay**-men, for-**am**-i-nah) Short windowlike opening in bone.

Foramen cecum (for-**ay**-men **se**-kum) Depression on the dorsal surface of the tongue where the sulcus terminalis points backward toward the pharynx.

Foramen lacerum (lah-**ser**-um) Foramen between the sphenoid, occipital, and temporal bones that is filled with cartilage.

Foramen magnum (**mag**-num) Foramen in the occipital bone that carries the spinal cord, vertebral arteries, and eleventh cranial nerve.

Foramen ovale (**ova**-lee) Foramen in the sphenoid bone for the mandibular division of the trigeminal or fifth cranial nerve.

Foramen rotundum (row-**tun**-dum) Foramen in the sphenoid bone that carries the trigeminal or fifth cranial nerve.

Foramen spinosum (**spine**-o-sum) Foramen in the sphenoid bone for the middle meningeal artery.

Fossa, fossae (**fos**-ah, **fos**-ay) Depression on a bony surface.

Frontal bone (**frunt**-il) Single cranial bone that forms the forehead and a portion of the orbits.

Frontal eminence (**frunt**-il **em**-i-nins) Prominence of the forehead.

Frontal nerve (**frunt**-il) Nerve from the merger of the supraorbital and supratrochlear nerves that continues into the ophthalmic nerve when joined by the lacrimal and nasociliary nerves.

Frontal plane (**frunt**-il) Plane created by an imaginary line that divides the body at any level into anterior and posterior parts.

Frontal process of the maxilla (**frunt**-il) Process that forms a portion of the orbital rim.

Frontal process of the zygomatic bone (**frunt**-il) Process that forms a portion of the orbital wall.

Frontal region (**frunt**-il) Region of the head that includes the forehead and supraorbital area.

Frontal section (**frunt**-il) Section of the body through any frontal plane.

Frontal sinuses (**frunt**-il **sy**-nuses) Paired paranasal sinuses located internally in the frontal bone.

Frontonasal duct (frunt-il-**na**-zil dukt) Drainage canal of each frontal sinus to the nasal cavity.

Fungiform lingual papillae (**fung**-i-form) Papillae with a mushroom-shaped appearance.

G

Ganglion, ganglia (**gang**-gle-in, **gan**-gle-ah) Accumulation of neuron cell bodies outside the central nervous system.

Genial tubercles (ji-**ni**-il) Midline bony projections or the mental spines on the inner aspect of the mandible.

Genioglossus muscle (ji-nee-o-**gloss**-us) Extrinsic tongue muscle that arises from the genial tubercles.

Geniohyoid muscle (ji-nee-o-**hi**-oid) Anterior suprahyoid muscle that is deep to the mylohyoid muscle.

GLOSSARY OF KEY WORDS AND ANATOMICAL TERMS *Continued*

Gingiva, gingivae (jin-**ji**-vah, jin-**ji**-vay) Mucosa surrounding the maxillary and mandibular teeth.

Glabella (glah-**bell**-ah) Smooth elevated area on the frontal bone between the supraorbital ridges.

Gland (gland) Structure that produces a chemical secretion necessary for normal body functioning.

Glossopharyngeal nerve (**gloss**-oh-fah-**rin**-je-al) Ninth cranial nerve that serves the parotid salivary gland, a pharyngeal muscle, and a tongue portion.

Goiter (**goit**-er) Enlarged thyroid gland due to a disease process.

Greater cornu Pair of projections from the sides of the body of the hyoid bone.

Greater palatine artery (**pal**-ah-tine) Arterial branch from the maxillary artery that travels to the palate.

Greater palatine block (**pal**-ah-tine) Local anesthetic block that achieves anesthesia for the lingual tissues of the maxillary posterior teeth and posterior palatal tissues.

Greater palatine foramen (**pal**-ah-tine) Foramen in the palatine bone that carries the greater palatine nerve and blood vessels.

Greater palatine nerve (**pal**-ah-tine) Nerve that serves the posterior hard palate and posterior lingual gingiva and then joins the maxillary nerve.

Greater petrosal nerve (peh-**troh**-sil) Branch of the facial nerve that serves the lacrimal gland, nasal cavity, and minor salivary glands of the hard and soft palate.

Greater wing of the sphenoid bone (**sfe**-noid) Posterolateral process of the body of the sphenoid bone.

H

Hamulus (**ha**-mu-lis) Process of the medial pterygoid plate of the sphenoid bone.

Hard palate (**pal**-it) Anterior portion of the palate formed by the palatine processes of the maxilla and the horizontal plates of the palatine bones.

Head (hed) Rounded surface projecting from a bone by a neck.

Helix (**heel**-iks) The superior and posterior free margin of the auricle.

Hematoma (hee-mah-**toe**-mah) Bruise that results when a blood vessel is injured and a small amount of blood escapes into the surrounding tissue and clots.

Hemorrhage (**hem**-ah-rij) Large amounts of blood that escape into the surrounding tissue without clotting when a blood vessel is seriously injured.

Hilus (**hi**-lus) Depression on one side of the lymph node where lymph flows out by way of an efferent lymphatic vessel.

Horizontal plane (hor-i-**zon**-tal) Plane created by an imaginary line that divides the body at any level into superior and inferior parts.

Horizontal plates of the palatine bones Plates that form the posterior portion of the hard palate.

Hyoglossus muscle (hi-o-**gloss**-us) Extrinsic tongue muscle that originates from the hyoid bone.

Hyoid bone (**hi**-oid) Bone suspended in the neck that allows the attachment of many muscles.

Hyoid muscles (**hi**-oid) Muscles that attach to the hyoid bone and can be classified by whether they are superior or inferior to the hyoid bone.

Hypoglossal canal (hi-poh-**gloss**-al) Canal in the occipital bone that carries the twelfth cranial nerve.

Hypoglossal nerve (hi-poh-**gloss**-al) Twelfth cranial nerve (XII) that serves the muscles of the tongue.

I

Incisive artery (in-**sy**-ziv) Arterial branch from the inferior alveolar artery that divides into dental and alveolar branches to supply the pulp tissue and periodontium of the mandibular anterior teeth.

Incisive block (in-**sy**-ziv) Local anesthetic block that achieves anesthesia of the pulp and facial tissues of the mandibular anterior and premolar teeth.

Incisive foramen (in-**sy**-ziv) Foramen in the maxilla that carries branches of the right and left nasopal-

GLOSSARY OF KEY WORDS AND ANATOMICAL TERMS *Continued*

atine nerves and blood vessels and is marked by the incisive papilla.

Incisive nerve (in-**sy**-ziv) Nerve that is formed from dental and interdental branches of the mandibular anterior teeth and merges with the mental nerve to form the inferior alveolar nerve.

Incisive papilla (in-**sy**-ziv pah-**pil**-ah) Bulge of tissue on the hard palate over the incisive foramen.

Incisura (in-si-**su**-rah) Indentation or notch at the edge of the bone.

Inferior (in-**fere**-ee-or) Area that faces away from the head and toward the feet of the body.

Inferior alveolar artery (al-**ve**-o-lar) Arterial branch from the maxillary artery that supplies the mandibular posterior teeth and branches into the mental and incisive arteries.

Inferior alveolar (IA) block (al-**ve**-o-lar) Local anesthetic block that achieves anesthesia of the pulp and lingual tissues of the mandibular teeth, as well as facial tissues of the anterior and premolar mandibular teeth.

Inferior alveolar (IA) nerve (al-**ve**-o-lar) Nerve formed from the merger of the incisive and mental nerves that serves the tissues of the chin, lower lip, and labial mucosa of the mandibular anterior teeth and later joins the posterior trunk of the mandibular division of the trigeminal nerve.

Inferior alveolar vein (al-**ve**-o-lar) Vein formed by the merger of the dental, alveolar, and mental branches that drains the pulp tissue and periodontium of the mandibular teeth, as well as the tissues of the chin.

Inferior articular processes (ar-**tik**-you-lar) Processes of the first and second cervical vertebrae that allow articulation with the vertebrae below.

Inferior labial artery Arterial branch from the facial artery that supplies the lower lip tissues.

Inferior labial vein Vein that drains the lower lip and then drains into the facial vein.

Inferior nasal conchae (**nay**-zil **kong**-kay) Paired facial bones that project inwardly from the maxilla to form walls of the nasal cavity.

Inferior orbital fissure (**or**-bit-al) Fissure between the greater wing of the sphenoid bone and maxilla that carries the infraorbital and zygomatic nerves, as well as the infraorbital artery and inferior ophthalmic vein.

Infrahyoid muscles (in-frah-**hi**-oid) Hyoid muscles that are inferior to the hyoid bone.

Infraorbital artery (in-frah-**or**-bit-al) Arterial branch from the maxillary artery that gives off the anterior superior alveolar artery and branches to the orbit.

Infraorbital (IO) block (in-frah-**or**-bit-al) Local anesthetic block that achieves anesthesia in the tissues supplied by the middle and anterior superior alveolar nerves, including the pulp and facial tissues of the maxillary anterior and premolar teeth.

Infraorbital canal (in-frah-**or**-bit-al) Canal off the infraorbital sulcus that terminates on the surface of the maxilla as the infraorbital foramen.

Infraorbital foramen (in-frah-**or**-bit-al) Foramen of the maxilla that transmits the infraorbital nerve and blood vessels.

Infraorbital (IO) nerve (in-frah-**or**-bit-al) Nerve that is involved in forming the maxillary nerve and is formed from branches of the upper lip, cheek portion, lower eyelid, and side of the nose.

Infraorbital region (in-frah-**or**-bit-al) Region of the head that is located below the orbital region and lateral to the nasal region.

Infraorbital rim (in-frah-**or**-bit-al) Inferior rim of the orbit.

Infraorbital sulcus (in-frah-**or**-bit-al) Groove in the floor of the orbital surface.

Infratemporal crest (in-frah-**tem**-poh-ral) Crest that divides each greater wing of the sphenoid bone into temporal and infratemporal surfaces.

Infratemporal fossa (in-frah-**tem**-poh-ral) Fossa inferior to the temporal fossa and infratemporal crest on the greater wing of the sphenoid bone.

Infratemporal space (in-frah-**tem**-poh-ral) Space that occupies the infratemporal fossa.

GLOSSARY OF KEY WORDS AND ANATOMICAL TERMS *Continued*

Infratrochlear nerve (in-frah-**trok**-lere) Nerve from the medial eyelid and side of the nose that converges with other branches to form the nasociliary nerve.

Innervation (in-er-**vay**-shin) Supply of nerves to tissues or organs.

Insertion (in-**sir**-shun) End of the muscle that is attached to the more movable structure.

Interdental gingiva (in-ter-**den**-tal) Attached gingiva between the teeth.

Intermediate tendon (in-ter-**me**-dee-it **ten**-don) Tendon between two muscle bellies, such as the anterior and posterior bellies of the digastric muscle.

Internal (in-**tern**-il) Inner side of the wall of a hollow structure.

Internal acoustic meatus (ah-**koos**-tik) Bony meatus in the temporal bone that carries the seventh and eighth cranial nerves.

Internal carotid artery (kah-**rot**-id) Artery off the common carotid artery that gives rise to the ophthalmic artery and also supplies intracranial structures.

Internal jugular vein (**jug**-you-lar) Vein that travels in the carotid sheath from the jugular foramen and drains the tissues of the head and neck.

Internal nasal nerves (**nay**-zil) Nerves from the nasal cavity that converge with other branches to form the nasociliary nerve.

Intrinsic tongue muscles (in-**trin**-sik) Muscles located entirely inside the tongue.

Investing fascia (in-**vest**-ing) Most external layer of the deep cervical fascia.

Ipsilateral (ip-see-**lat**-er-il) Structures on the same side of the body.

Iris (eye-ris) Central area of coloration of the eyeball.

J

Joint (joint) Site of a junction or union between two or more bones.

Joint capsule of the temporomandibular joint Fibrous capsule that encloses the temporomandibular joint.

Jugular foramen (**jug**-you-lar) Foramen between the occipital and temporal bones that carries the internal jugular vein and ninth, tenth, and eleventh cranial nerves.

Jugular notch of the occipital bone (**jug**-you-lar) Occipital or medial portion of the jugular foramen.

Jugular notch of the temporal bone (**jug**-you-lar) Temporal or lateral portion of the jugular foramen.

Jugular trunk (**jug**-you-lar trungk) Lymphatic vessel that drains one side of the head and neck and then empties into that side's lymphatic duct.

Jugulodigastric lymph node (jug-you-lo-di-**gas**-trik) Superior deep cervical node located below the posterior belly of the digastric muscle.

Jugulo-omohyoid lymph node (jug-you-lo-o-mo-**hi**-oid) Inferior deep cervical node located at the crossing of the omohyoid muscle and internal jugular vein.

L

Labial (**lay**-be-al) Structures closest to the lips.

Labial commissure (**kom**-i-shoor) Corner of the mouth where the upper and lower lips meet.

Labial frenum (**free**-num) Fold of tissue or frenulum located at the midline between the labial mucosa and alveolar mucosa of the maxilla and mandible.

Labial mucosa Lining of the inner portions of the lips.

Labiomental groove (lay-bee-o-**ment**-il) A groove that separates the lower lip from the chin.

Lacrimal bones (**lak**-ri-mal) Paired facial bones that help form the medial wall of the orbit.

Lacrimal fluid (**lak**-ri-mal) Tears or watery fluid excreted by the lacrimal gland.

Lacrimal fossa (**lak**-ri-mal) Fossa of the frontal bone that contains the lacrimal gland.

GLOSSARY OF KEY WORDS AND ANATOMICAL TERMS *Continued*

Lacrimal gland (**lak**-ri-mal) Gland in the lacrimal fossa of the frontal bone that produces lacrimal fluid or tears.

Lacrimal nerve (**lak**-ri-mal) Nerve that serves the lateral part of the eyelid and other eye tissues and joins the frontal and nasociliary nerves to form the ophthalmic nerve.

Lambdoidal suture (lam-**doid**-al) Suture between the occipital bone and both parietal bones.

Laryngopharynx (lah-ring-gah-**far**-inks) Inferior portion of pharynx, close to the laryngeal opening.

Lateral (**lat**-er-il) Area that is farther away from the median plane of the body or structure.

Lateral canthus, canthi (**kan**-this, **kan**-thy) Outer corner of the eye or outer canthus where the upper and lower eyelids meet.

Lateral deviation of the mandible (de-vee-**ay**-shun) Shifting of the lower jaw to one side.

Lateral masses (masz) Lateral portions of the first cervical vertebra where it articulates with the occipital bone above and the axis below.

Lateral pterygoid muscle (**teh**-ri-goid) Muscle of mastication that lies in the infratemporal fossa.

Lateral pterygoid nerve (**teh**-ri-goid) Muscular branch from the anterior trunk of the mandibular division of the trigeminal nerve that serves the lateral pterygoid muscle.

Lateral pterygoid plate (**teh**-ri-goid) Portion of the pterygoid process.

Lateral surface of the tongue Side of the tongue.

Lesser cornu Pair of projections of the hyoid bone.

Lesser palatine artery (**pal**-ah-tine) Arterial branch from the maxillary artery that travels to the soft palate.

Lesser palatine foramen (**pal**-ah-tine) Foramen in the palatine bone that transmits the lesser palatine nerve and blood vessels.

Lesser palatine nerve (**pal**-ah-tine) Nerve that serves the soft palate and palatine tonsilar tissues along with the posterior nasal cavity and then joins the maxillary nerve.

Lesser petrosal nerve (peh-**troh**-sil) Parasympathetic fibers from the ninth cranial nerve that exit the skull through the foramen ovale of the sphenoid bone.

Lesser wing of the sphenoid bone (**sfe**-noid) Anterior process of the body of the sphenoid bone.

Levator anguli oris muscle (le-**vate**-er **an**-gu-lie **or**-is) Muscle of facial expression in the mouth region that elevates the angle of the mouth.

Levator labii superioris alaeque nasi muscle (le-**vate**-er **lay**-be-eye soo-per-ee-**or**-is **a**-lah-cue **naz**-eye) Muscle of facial expression in the mouth region that elevates the upper lip and ala of the nose.

Levator labii superioris muscle (le-**vate**-er **lay**-be-eye soo-per-ee-**or**-is) Muscle of facial expression in the mouth region that elevates the upper lip.

Levator veli palatini muscle (le-**vate**-er **vee**-lie pal-ah-**teen**-ee) Muscle of the soft palate that raises the soft palate to close off the nasopharynx.

Ligament (**lig**-ah-mint) Band of fibrous tissue connecting bones.

Line (line) Straight small ridge of bone.

Lingual (**ling**-gwal) Structures closest to the tongue.

Lingual artery Anterior arterial branch from the external carotid artery that supplies tissues superior to the hyoid bone, as well as the tongue and floor of the mouth.

Lingual frenum (**free**-num) Midline fold of tissue or frenulum between the ventral surface of the tongue and floor of the mouth.

Lingual nerve Nerve that serves the tongue, floor of the mouth, and lingual gingiva of the mandibular teeth and joins the posterior trunk of the mandibular division of the trigeminal nerve.

Lingual papillae (pah-**pil**-ay) Small elevated structures covering the dorsal surface of the body of the tongue.

Lingual tonsil (**ton**-sil) Indistinct layer of lymphoid tissue located on the dorsal surface of the tongue's base.

Lingual veins Veins that include the deep lingual, dorsal lingual, and sublingual veins.

GLOSSARY OF KEY WORDS AND ANATOMICAL TERMS *Continued*

Lingula (**ling**-gul-ah) Bony spine overhanging the mandibular foramen.

Lobule (**lob**-yule) Inferior fleshy protuberance from the helix of the auricle.

Local infiltration (**lo**-kal in-fil-**tray**-shun) Type of injection that anesthetizes a small area, one or two teeth and associated structures when the anesthetic solution is deposited near terminal nerve endings.

Ludwig's angina (**lood**-vigz an-**ji**-nah) Serious infection of the submandibular space, with a risk of spread to the neck and chest.

Lymph (limf) Tissue fluid that drains from the surrounding region and into the lymphatic vessels.

Lymphadenopathy (lim-fad-in-**op**-ah-thee) Process in which there is an increase in the size and a change in the consistency of lymphoid tissue.

Lymphatic ducts (lim-**fat**-ik dukts) Larger lymphatic vessels that drain smaller vessels and then empty into the venous system.

Lymphatics (lim-**fat**-iks) Part of the immune system.

Lymphatic vessels (lim-**fat**-ik **ves**-els) System of channels that drain tissue fluid from the surrounding regions.

Lymph nodes (limf nodes) Organized bean-shaped lymphoid tissue that filters the lymph by way of lymphocytes to fight disease and is grouped into clusters along the connecting lymphatic vessels.

M

Major salivary glands Large paired glands with associated named ducts that include the parotid, submandibular, and sublingual glands.

Malar lymph nodes (**may**-lar) Superficial nodes of the face located in the infraorbital region.

Mandible (**man**-di-bl) Single facial bone that articulates bilaterally with the temporal bones at the temporomandibular joints.

Mandibular canal (man-**dib**-you-lar) Canal in the mandible where the inferior alveolar nerve and blood vessels travel.

Mandibular foramen (man-**dib**-you-lar) Foramen of the mandible that allows the inferior alveolar nerve and blood vessels to exit or enter the mandibular canal.

Mandibular lymph nodes (man-**dib**-you-lar) Superficial nodes of the face located over the surface of the mandible.

Mandibular nerve (man-**dib**-you-lar) Mandibular division of the trigeminal nerve that is formed by the merger of posterior and anterior trunks and joins with the ophthalmic and maxillary nerves to form the trigeminal ganglion of the trigeminal nerve.

Mandibular notch (man-**dib**-you-lar) Notch located on the mandible between the condyle and coronoid process.

Mandibular teeth (man-**dib**-you-lar) Teeth of the mandible.

Marginal gingiva (**mar**-ji-nal) Nonattached gingiva at the gingival margin of each tooth.

Masseter muscle (**mass**-et-er **mus**-il) Most obvious and strongest muscle of mastication.

Masseteric artery (mass-et-**tehr**-ik) Arterial branch from the maxillary artery that supplies the masseter muscle.

Masseteric nerve (mass-et-**tehr**-ik) Muscular nerve branch from the anterior trunk of the mandibular division of the trigeminal nerve that serves the masseter muscle and temporomandibular joint.

Masseteric-parotid fascia (mass-et-**tehr**-ik-pah-**rot**-id) Deep fascia that is located inferior to the zygomatic arch and over the masseter muscle.

Masticator space (mass-ti-**kay**-tor) Space that includes the entire area of the mandible and muscles of mastication.

Mastoid air cells (**mass**-toid) Air spaces in the mastoid process of the temporal bone that communicate with the middle ear cavity.

Mastoid notch (**mass**-toid) Notch on the mastoid process of the temporal bone.

Mastoid process (**mass**-toid) Area on the petrous portion of the temporal bone that contains the air cells and on which the cervical muscles attach.

GLOSSARY OF KEY WORDS AND ANATOMICAL TERMS *Continued*

Maxilla, maxillae (mak-**sil**-ah, mak-**sil**-lay) Upper jaw that consists of two maxillary bones.

Maxillary artery (**mak**-sil-lare-ee) Terminal arterial branch from the external carotid artery.

Maxillary nerve (**mak**-sil-ar-ee) Second division of the sensory root of the trigeminal nerve that is formed by the convergence of many nerves, including the infraorbital nerve, and serves many maxillary tissues, such as the maxillary sinus, palate, nasopharynx, and overlying skin.

Maxillary process of the zygomatic bone (**mak**-sil-lare-ee) Process that forms a portion of the infraorbital rim and orbital wall.

Maxillary sinuses (**mak**-sil-lare-ee **sy**-nuses) Paranasal sinuses in each body of the maxilla.

Maxillary sinusitis (**si**-nu-**si**-tis) Infection of the maxillary sinus.

Maxillary teeth (**mak**-sil-lare-ee) Teeth of the maxilla.

Maxillary tuberosity (**mak**-sil-lare-ee) Elevation on the posterior aspect of the maxilla that is perforated by the posterior superior alveolar foramina.

Maxillary vein (**mak**-sil-lare-ee) Vein that after collecting from the pterygoid plexus merges with the superficial temporal vein to form the retromandibular vein.

Meatus (me-**ate**-us) Opening or canal in the bone.

Medial (**me**-dee-il) Area that is closer to the median plane of the body or structure.

Medial canthus, canthi (**kan**-this, **kan**-thy) Inner angle or canthus of the eye.

Medial pterygoid muscle (**teh**-ri-goid) Muscle of mastication that inserts on the medial surface of the mandible.

Medial pterygoid plate (**teh**-ri-goid) Portion of the pterygoid process.

Median (**me**-dee-an) Structure at the median plane.

Median lingual sulcus (**ling**-wal **sul**-kus) Midline depression on the dorsal surface of the tongue that corresponds to the deeper median septum.

Median palatine raphe (**pal**-ah-tine **ra**-fe) Midline fibrous band of the palate.

Median palatine suture (**pal**-ah-tine) Midline suture between the palatine processes of the maxillae and between the horizontal plates of the palatine bones.

Median pharyngeal raphe (fah-**rin**-je-al **ra**-fe) Midline fibrous band on the posterior wall of the pharynx.

Median plane (**me**-dee-an) Plane created by an imaginary line dividing the body into right and left halves.

Median septum (**sep**-tum) Midline fibrous structure that divides the tongue and corresponds to a midline depression, the median lingual sulcus, on the dorsal surface of the tongue.

Meningitis (men-in-**jite**-is) Inflammation of the meninges of the brain or spinal cord.

Mental artery (**ment**-il) Arterial branch from the inferior alveolar artery that exits the mental foramen and supplies the tissues of the chin.

Mental block (**ment**-il) Local anesthetic block that achieves anesthesia of the facial tissues of the mandibular premolars and anterior teeth.

Mental foramen (**ment**-il) Foramen between the apices of the mandibular first and second premolars that transmits the mental nerve and blood vessels.

Mental nerve (**ment**-il) Nerve that joins the incisive nerve to form the inferior alveolar nerve and serves the tissues of the chin and lower lip and labial mucosa of the mandibular anterior teeth.

Mental protuberance (**ment**-il pro-**too**-ber-ins) Mandibular bony prominence of the chin.

Mental region (**ment**-il) Region of the head where the major feature is the chin.

Mentalis muscle (men-**ta**-lis) Muscle of facial expression in the mouth region that raises the chin.

Metastasis (meh-**tas**-tah-sis) Spread of cancer from the original or primary site to another or secondary site.

GLOSSARY OF KEY WORDS AND ANATOMICAL TERMS *Continued*

Middle meningeal artery (meh-**nin**-je-al) Arterial branch from the maxillary artery that supplies the meninges of the brain by the way of the foramen spinosum.

Middle meningeal vein (meh-**nin**-je-al) Vein that drains blood from the meninges of the brain into the pterygoid plexus of veins.

Middle nasal conchae (**nay**-zil **kong**-kay) Lateral portions of the ethmoid bone in the nasal cavity.

Middle superior alveolar (MSA) block (al-**ve**-o-lar) Local anesthetic block that achieves anesthesia of the pulp and buccal tissues of the maxillary premolars and mesiobuccal root of the maxillary first molar.

Middle superior alveolar (MSA) nerve (al-**ve**-o-lar) Nerve that serves the maxillary premolar teeth and tissues, as well as the mesiobuccal root of the maxillary first molar, is formed from dental, interdental, and interradicular branches, and later joins the infraorbital nerve.

Middle temporal artery (**tem**-poh-ral) Arterial branch from the superficial temporal artery that supplies the temporalis muscle.

Midsagittal section (mid-**saj**-i-tl) Section of the body through the median plane.

Minor salivary glands Small glands scattered in the tissues of the buccal, labial, and lingual mucosa, soft and hard palate, and floor of the mouth, as well as associated with the circumvallate lingual papillae.

Motor root of the trigeminal nerve Root of the trigeminal nerve.

Mucobuccal fold (mu-**ko**-buk-al) Fold in the vestibule where the labial or buccal mucosa meets the alveolar mucosa.

Mucogingival junction (mu-ko-**jin**-ji-val). Border between the alveolar mucosa and attached gingiva.

Mucosa (mu-**ko**-sah) Mucous membrane, such as that lining the oral cavity.

Muscle (**mus**-il) Type of body tissue that shortens under neural control, causing soft tissue and bony structures to move.

Muscle of the uvula (**u**-vu-lah) Muscle of the soft palate that is within the uvula.

Muscles of facial expression Paired muscles that give the face expression and are located in the superficial fascia of the facial tissues.

Muscles of mastication (mass-ti-**kay**-shun) Pairs of muscles attached to and moving the mandible, including the temporalis, masseter, and medial and lateral pterygoid muscles.

Muscles of the pharynx (**far**-inks) Muscles that include the stylopharyngeus, pharyngeal constrictor, and soft palate muscles.

Muscles of the soft palate (**pal**-it) Muscles that include the palatoglossal, palatopharyngeus, levator veli palatini, and tensor veli palatini muscles and muscle of the uvula.

Muscles of the tongue Muscles of the tongue that can be further grouped according to whether they are intrinsic or extrinsic.

Mylohyoid artery (my-lo-**hi**-oid) Arterial branch from the inferior alveolar artery that supplies the floor of the mouth and mylohyoid muscle.

Mylohyoid groove (my-lo-**hi**-oid) Groove on the mandible where the mylohyoid nerve and blood vessels travel.

Mylohyoid line (my-lo-**hi**-oid) Line on the inner aspect of the mandible.

Mylohyoid muscle (my-lo-**hi**-oid) Anterior suprahyoid muscle that forms the floor of the mouth.

Mylohyoid nerve (my-lo-**hi**-oid) Nerve branch from the inferior alveolar nerve that serves the mylohyoid muscle and anterior belly of the digastric muscle.

N

Naris, nares (**nay**-ris, **nay**-rees) Nostril of the nose.

Nasal bones (**nay**-zil) Paired facial bones that form the bridge of the nose.

GLOSSARY OF KEY WORDS AND ANATOMICAL TERMS *Continued*

Nasal cavity (**nay**-zil **kav**-it-ee) Cavity of the nose.

Nasal conchae (**nay**-zil **kong**-kay) Projecting structures that extend inward from the lateral walls of the nasal cavity.

Nasal meatus (**nay**-zil) Groove beneath each nasal concha that contains openings for communication with the paranasal sinuses or nasolacrimal duct.

Nasal region (**nay**-zil) Region of the head where the main feature is the external nose.

Nasal septum (**nay**-zil **sep**-tum) Vertical partition of the nasal cavity.

Nasion (**nay**-ze-on) Midline junction between the nasal and frontal bones.

Nasociliary nerve (nay-zo-**sil**-eh-a-re) Nerve that joins the frontal and lacrimal nerves to form the ophthalmic nerve.

Nasolabial lymph nodes (nay-zo-**lay**-be-al) Superficial nodes of the face located near the nose.

Nasolabial sulcus (nay-zo-**lay**-be-al **sul**-kus) Groove running upward between the labial commissure and ala of the nose.

Nasolacrimal duct (nay-zo-**lak**-rim-al dukt) Duct formed at the junction of the lacrimal and maxillary bones that drains the lacrimal fluid or tears.

Nasopalatine block (nay-zo-**pal**-ah-tine) Local anesthetic block that achieves anesthesia of the anterior portion of the hard palate.

Nasopalatine nerve (nay-zo-**pal**-ah-tine) Nerve that serves the anterior hard palate and lingual gingiva of the maxillary anterior teeth and then joins the maxillary nerve.

Nasopharynx (nay-zo-**far**-inks) Portion of the pharynx that is superior to the level of soft palate.

Nerve (nurv) Bundle of neural processes outside the central nervous system, part of the peripheral nervous system.

Nerve block (nurv) Type of injection that anesthetizes a larger area than the local infiltration because the anesthetic solution is deposited near large nerve trunks.

Neuron (**noor**-on) Cellular component of the nervous system that is individually composed of a cell body and neural processes.

Normal flora (**flor**-ah) Resident microorganisms that usually do not cause infections.

Notch (noch) Indentation at the edge of the bone.

O

Occipital artery (ok-**sip**-it-tal) Posterior arterial branch from the external carotid artery that supplies the suprahyoid and sternocleidomastoid muscles and posterior scalp tissues.

Occipital bone (ok-**sip**-it-al) Single cranial bone in the most posterior portion of the skull.

Occipital condyles (ok-**sip**-it-al) Projections of the occipital bone that articulate with lateral masses of the first cervical vertebra.

Occipital lymph nodes (ok-**sip**-it-al) Superficial nodes located on the posterior base of the head.

Occipital region (ok-**sip**-it-al) Region of the head overlying the occipital bone and covered by the scalp.

Oculomotor nerve (ok-yule-oh-**mote**-er) Third cranial nerve (III) that serves some of the eye muscles.

Olfactory nerve (ol-**fak**-ter-ee) First cranial nerve (I) that transmits smell from the nose to the brain.

Omohyoid muscle (o-mo-**hi**-oid) Infrahyoid muscle with superior and inferior bellies.

Ophthalmic artery (of-**thal**-mic) Arterial branch that supplies the eye, orbit, and lacrimal gland.

Ophthalmic nerve (of-**thal**-mic) First division of the sensory root of the trigeminal nerve that arises from the frontal, lacrimal, and nasociliary nerves.

Ophthalmic veins (of-**thal**-mic) Veins that drain the tissues of the orbit.

Opportunistic infections (op-or-tu-**nis**-tik) Normal flora creating an infectious process because the body's defenses are compromised.

Optic canal (**op**-tik) Canal in the orbital apex between the roots of the lesser wing of the sphenoid bone.

GLOSSARY OF KEY WORDS AND ANATOMICAL TERMS *Continued*

Optic nerve (**op**-tik) Second cranial nerve (II) that transmits sight from the eye to the brain.

Oral cavity Inside of the mouth.

Oral region Region of the head that contains the lips, oral cavity, palate, tongue, and floor of the mouth and portions of the pharynx.

Orbicularis oculi muscle (or-bik-you-**laa**-ris **oc**-yule-eye) Muscle of facial expression that encircles the eye.

Orbicularis oris muscle (or-bik-you-**laa**-ris **or**-is) Muscle of facial expression that encircles the mouth.

Orbit (**or**-bit) Eye cavity that contains the eyeballs.

Orbital apex (**or**-bit-al) Deepest portion of the orbit composed of portions of the sphenoid and palatine bones.

Orbital plate of the ethmoid bone (**or**-bit-al) Plate that forms most of the medial orbital wall.

Orbital region (**or**-bit-al) Region of the head with the eyeball and all its supporting structures.

Orbital walls (**or**-bit-al) Walls of the orbit composed of portions of the frontal, ethmoid, lacrimal, maxillary, zygomatic, and sphenoid bones.

Origin (**or**-i-jin) End of the muscle that is attached to the least movable structure.

Oropharynx (or-o-**far**-inks) Portion of the pharynx that is between the soft palate and opening of the larynx.

Osteomyelitis (os-tee-o-my-il-**ite**-is) Inflammation of bone marrow.

Ostium, ostia (**os**-tee-um, **os**-tee-ah) Small opening in bone.

Otic ganglion (**ot**-ik) Ganglion associated with the lesser petrosal nerve and branches of the mandibular nerve.

P

Palatal (**pal**-ah-tal) Structures closest to the palate.

Palate (**pal**-it) Roof of the mouth.

Palatine bones (**pal**-ah-tine) Paired bones of the skull that consist of two plates, vertical and horizontal plates.

Palatine process of the maxilla (**pal**-ah-tine) Paired processes that articulate with each other and form the anterior portion of the hard palate.

Palatine rugae (**pal**-ah-tine **ru**-gay) Irregular ridges of tissues surrounding the incisive papilla on the hard palate.

Palatine tonsils (**pal**-ah-tine **ton**-sils) Tonsils located between the anterior and posterior tonsillar pillars.

Palatoglossal muscle (pal-ah-to-**gloss**-el) Muscle of the soft palate that forms the anterior tonsillar pillar.

Palatopharyngeus muscle (pal-ah-to-fah-**rin**-je-us) Muscle of the soft palate that forms the posterior tonsillar pillar.

Paranasal sinuses (pare-ah-**na**-zil **sy**-nuses) Paired air-filled cavities in bone that include the frontal, sphenoid, ethmoid, and maxillary sinuses.

Parapharyngeal space (pare-ah-fah-**rin**-je-al) Space located lateral to the pharynx.

Parasympathetic nervous system (pare-ah-sim-pah-**thet**-ik) Portion of the autonomic nervous system that is involved in "rest or digest."

Parathyroid glands (par-ah-**thy**-roid) Small endocrine glands located close to or even inside the thyroid gland.

Parathyroid hormone (par-ah-**thy**-roid) Hormone produced and secreted by the parathyroid glands directly into the blood to regulate calcium and phosphorus levels.

Paresthesia (par-es-**the**-ze-ah) Abnormal sensation from an area, such as burning or prickling.

Parietal bones (pah-**ri**-it-al) Paired cranial bones of the skull that articulate with each other and other skull bones.

Parietal region (pah-**ri**-it-al) Region of the head that overlies the parietal bones and is covered by the scalp.

Parotid duct (pah-**rot**-id) Duct associated with the parotid salivary gland that opens into the oral cavity at the parotid papilla.

GLOSSARY OF KEY WORDS AND ANATOMICAL TERMS *Continued*

Parotid fascia (pah-**rot**-id) Space created inside the investing fascia as it envelops the parotid salivary gland.

Parotid papilla (pah-**rot**-id pah-**pil**-ah) Small elevation of tissue that marks the opening of the parotid salivary gland and is located opposite the second maxillary molar on the inner cheek.

Parotid salivary gland (pah-**rot**-id) Major gland located over the mandibular ramus that is divided into superficial and deep lobes.

Parotid space (pah-**rot**-id) Space enveloping parotid gland.

Pathogens (**path**-ah-jens) Flora that are not normal body residents and can cause an infection.

Perforation (per-fo-**ray**-shun) Abnormal hole in a hollow organ, such as in the wall of a sinus.

Peripheral nervous system (per-**if**-er-al) Portion of the nervous system that consists of the spinal and cranial nerves.

Perpendicular plate (per-pen-**dik**-you-lar) Midline vertical plate of the ethmoid bone.

Petrotympanic fissure (pe-troh-tim-**pan**-ik) Fissure between the tympanic and petrosal portions of the temporal bone, just posterior to the articular fossa, through which the chorda tympani nerve emerges.

Petrous portion of the temporal bone (pe-**tros**) Inferior portion of the bone that contains the mastoid process and air cells.

Pharyngeal constrictor muscles (fah-**rin**-je-il kon-**strik**-tor) Three paired muscles that form the lateral and posterior walls of the pharynx.

Pharyngeal tonsil (fah-**rin**-je-il) Tonsil located on the posterior wall of the nasopharynx.

Pharynx (**far**-inks) Portion of both the respiratory and digestive tracts that is divided into the nasopharynx, oropharynx, and laryngopharynx.

Philtrum (**fil**-trum) Vertical groove in the midline of the upper lip.

Piriform aperture (**pir**-i-form) Anterior opening of the nasal cavity.

Plate (plate) Flat structure of bone.

Platysma muscle (plah-**tiz**-mah) Muscle of facial expression that runs from the neck to the mouth.

Plexus (**plek**-sis) Network of blood vessels, usually veins.

Plica fimbriata, plicae fimbriatae (**pli**-kah fim-bree-**ay**-tah, **pli**-kay fim-bree-**ay**-tay) Fold with fringelike projections on the ventral surface of the tongue.

Posterior (pos-**tere**-ee-or) Back of an area of the body.

Posterior arch Arch on the first cervical vertebra.

Posterior auricular artery (aw-**rik**-yule-lar) Posterior arterial branch from the external carotid artery that supplies the tissues around the ear.

Posterior auricular nerve (aw-**rik**-yule-lar) Branch of the facial nerve that serves the occipital belly of the epicranial muscle, stylohyoid muscle, and posterior belly of the digastric muscle.

Posterior cervical triangle (**ser**-vi-kal) Lateral region of the neck.

Posterior digastric nerve (di-**gas**-trik) Nerve that supplies the posterior belly of the digastric muscle.

Posterior nasal apertures (**nay**-zil) Posterior openings of the nasal cavity.

Posterior superior alveolar artery (al-**ve**-o-lar) Arterial branches from the maxillary artery that supply the pulp tissue and periodontium of the maxillary posterior teeth and maxillary sinus.

Posterior superior alveolar (PSA) block (al-**ve**-o-lar) Local anesthetic block that is used to achieve pulpal and buccal tissue anesthesia of the maxillary molars.

Posterior superior alveolar foramina (al-**ve**-o-lar) Foramina on the maxillary tuberosity that carry the posterior superior alveolar nerve and blood vessels.

Posterior superior alveolar (PSA) nerve (al-**ve**-o-lar) Nerve that directly joins the maxillary nerve after serving maxillary molars and tissues.

Posterior superior alveolar vein (al-**ve**-o-lar) Vein that is formed from the merger of dental and alveolar branches that drain the pulp tissue and periodontium of the maxillary teeth.

Posterior suprahyoid muscle group (soo-prah-**hi**-oid) Suprahyoid muscles posterior to the hyoid bone

GLOSSARY OF KEY WORDS AND ANATOMICAL TERMS *Continued*

that include the posterior belly of the digastric and stylohyoid muscles.

Posterior tonsillar pillar (**ton**-sil-ar **pil**-er) Vertical fold posterior to each palatine tonsil created by the palatopharyngeus muscle.

Postglenoid process (post-**gle**-noid) Process of the temporal bone.

Previsceral space (pre-**vis**-er-al) Space located between the visceral and investing fasciae.

Primary nodes Lymph node that drains lymph from a particular region.

Primary sinusitis (sy-nu-**si**-tis) Inflammation of the sinus.

Process (**pros**-es) General term for any prominence on the bony surface.

Protrusion of the mandible (pro-**troo**-shun) Bringing of the lower jaw forward.

Proximal (**prok**-si-mil) Area closer to the median plane of the body.

Pterygoid arteries (**teh**-re-goid) Arterial branches from the maxillary artery that supply the pterygoid muscles.

Pterygoid canal (**teh**-ri-goid) Small canal at the superior border of each posterior nasal aperture.

Pterygoid fascia (**teh**-ri-goid) Deep fascia located on the medial surface of the media pterygoid muscle.

Pterygoid fossa (**teh**-ri-goid) Fossa between the medial and lateral pterygoid plates of the sphenoid bone.

Pterygoid fovea (**teh**-ri-goid fo-**vee**-ah) Depression on the anterior surface of the condyle of the mandible.

Pterygoid plexus of veins (**teh**-ri-goid) Collection of veins around the pterygoid muscles and maxillary arteries that drain the deep face and alveolar veins into the maxillary vein.

Pterygoid process (**teh**-ri-goid) Portion of the sphenoid bone that forms the lateral borders of the posterior nasal apertures.

Pterygomandibular fold (teh-ri-go-man-**dib**-yule-lar) Fold of tissue in the oral cavity that covers the pterygomandibular raphe.

Pterygomandibular raphe (teh-ri-go-man-**dib**-yule-lar **ra**-fe) Fibrous structure that extends from the hamulus to the posterior end of the mylohyoid line.

Pterygomandibular space (teh-ri-go-man-**dib**-yule-lar) Space that is a portion of the infratemporal space.

Pterygopalatine fossa (teh-ri-go-**pal**-ah-tine) Fossa deep to the infratemporal fossa and between the pterygoid process and maxillary tuberosity.

Pterygopalatine ganglion (teh-ri-go-**pal**-ah-tine) Ganglion associated with the greater petrosal nerve and branches of the maxillary nerve.

Pupil (**pew**-pil) Area in the center of the iris that responds to changing light conditions.

Pustule (**pus**-tule) Small, elevated, circumscribed suppuration-containing lesion of either the skin or oral mucosa.

R

Ramus (**ray**-mus) Plate of the mandible that extends superiorly from the body of the mandible.

Regions of the head Regions that include the frontal, parietal, occipital, temporal, orbital, nasal, infraorbital, zygomatic, buccal, oral, and mental regions.

Regions of the neck Regions that include the anterior and posterior cervical triangles.

Retraction of the mandible (re-**trak**-shun) Bringing of the lower jaw backward.

Retroauricular lymph nodes (reh-tro-aw-**rik**-you-lar) Superficial nodes located posterior to the ear.

Retromandibular vein (reh-tro-man-**dib**-you-lar) Vein that is formed by the merger of the superficial temporal and maxillary veins and divides into anterior and posterior divisions below the parotid salivary gland.

Retromolar pad (re-tro-**moh**-lar) Dense pad of tissue distal to the last tooth of the mandible that covers the retromolar triangle.

Retromolar triangle (re-tro-**moh**-lar) Portion of the mandibular alveolar process just posterior to the most distal mandibular molar that is covered by the retromolar pad.

GLOSSARY OF KEY WORDS AND ANATOMICAL TERMS *Continued*

Retropharyngeal lymph nodes (ret-ro-far-**rin**-je-al) Deep nodes located near the deep parotid nodes and at the level of the first cervical vertebra.

Retropharyngeal space (ret-ro-far-**rin**-je-al) Fascial space located immediately posterior to pharynx.

Right lymphatic duct Duct formed from the convergence of the lymphatics of the right arm and thorax and the right jugular trunk that drains this side of the head and neck.

Risorius muscle (ri-**soh**-ree-us) Muscle of facial expression in the mouth region that is used when smiling widely.

Root of the nose Area of the nasal region between the eyes.

S

Sagittal plane (saj-i-tel) Any plane of the body created by an imaginary plane parallel to the median plane.

Sagittal suture (**saj**-i-tel) Suture between the paired parietal bones.

Saliva (sah-**li**-vah) Product produced by the salivary glands.

Salivary gland (**sal**-i-ver-ee) Gland that produces saliva that lubricates and cleanses the oral cavity and helps in digestion.

Scalp (skalp) Layers of soft tissue overlying the bones of the cranium.

Sclera (**skler**-ah) White area on the eyeball.

Secondary nodes Lymph node that drains lymph from a primary node.

Secondary sinusitis (sy-nu-**si**-tis) Inflammation of the sinus related to another source.

Sensory root of the trigeminal nerve Root of the trigeminal nerve that has ophthalmic, maxillary, and mandibular divisions.

Skull (skul) Structure composed of both the cranial bones or cranium and facial bones.

Soft palate (**pal**-it) Posterior nonbony portion of the palate.

Space of the body of the mandible (**man**-di-bl) Space formed by the periosteum covering the body of the mandible.

Sphenoid bone (**sfe**-noid) Single midline cranial bone with a body and several pairs of processes.

Sphenoid sinuses (**sfe**-noid **sy**-nuses) Paired sinuses located in the body of the sphenoid bone.

Sphenomandibular ligament (sfe-no-man-**dib**-you-lar) Ligament that connects the spine of the sphenoid bone with the lingula of the mandible.

Sphenopalatine artery (sfe-no-**pal**-ah-tine) Terminal arterial branch from the maxillary artery that supplies the nose, including a branch through the incisive foramen.

Spine (spine) Abrupt small prominence of bone.

Spine of the sphenoid bone Spine located at the posterior extremity of the sphenoid bone.

Squamosal suture (**skway**-mus-al) Suture between the temporal and parietal bones.

Squamous portion of the temporal bone (**skway**-mus) Portion that forms the braincase and portions of the zygomatic arch and temporomandibular joint.

Sternocleidomastoid muscle (SCM) (stir-no-klii-do-**mass**-toid **mus**-il) Paired cervical muscle that serves as a primary landmark of the neck.

Sternohyoid muscle (ster-no-**hi**-oid) Infrahyoid muscle that is located superficial to the thyroid gland and cartilage.

Sternothyroid muscle (ster-no-**thy**-roid) Infrahyoid muscle that inserts on the thyroid cartilage.

Stoma (**stow**-mah) Opening, such as a fistula.

Styloglossus muscle (sty-lo-**gloss**-us) Extrinsic tongue muscle that originates from the styloid process of the temporal bone.

Stylohyoid muscle (sty-lo-**hi**-oid) Posterior suprahyoid muscle that originates from the styloid process of the temporal bone.

Stylohyoid nerve (sty-lo-**hi**-oid) Branch of the facial nerve that supplies the stylohyoid muscle.

GLOSSARY OF KEY WORDS AND ANATOMICAL TERMS *Continued*

Styloid process (**sty**-loid) Bony projection of the temporal bone that serves as an attachment for muscles and ligaments.

Stylomandibular ligament (sty-lo-man-**dib**-you-lar) Ligament that connects the styloid process with the angle of the mandible.

Stylomastoid artery (sty-lo-**mass**-toid) Artery that is a branch from the posterior auricular artery and supplies the mastoid air cells.

Stylomastoid foramen (sty-lo-**mass**-toid) Foramen in the temporal bone that carries the facial or seventh cranial nerve.

Stylopharyngeus muscle (sty-lo-fah-**rin**-je-us) Paired longitudinal muscle of the pharynx arising from the styloid process.

Subclavian artery (sub-**klay**-vee-an) Artery that arises from the aorta on the left and the brachiocephalic artery on the right and gives off branches to supply both intracranial and extracranial structures, as well as the arm.

Subclavian vein (sub-**klay**-vee-an) Vein from the arm that drains the external jugular vein and then joins with the internal jugular vein to form the brachiocephalic vein.

Sublingual artery (sub-**ling**-gwal) Arterial branch from the lingual artery that supplies the sublingual salivary gland, floor of the mouth, and mylohyoid muscle.

Sublingual caruncle (sub-**ling**-gwal **kar**-unk-el) Papilla near the midline of the floor of the mouth where the sublingual and submandibular ducts open into the oral cavity.

Sublingual duct (sub-**ling**-gwal) Duct associated with the sublingual salivary gland that opens at the sublingual caruncle.

Sublingual fold (sub-**ling**-gwal) Fold of tissue on the floor of the mouth where other smaller ducts of the sublingual salivary gland open into the oral cavity.

Sublingual fossa (sub-**ling**-gwal) Fossa on the medial surface of the mandible, above the mylohyoid line, that contains the sublingual salivary gland.

Sublingual salivary gland (sub-**ling**-gwal) Major gland located in the sublingual fossa.

Sublingual space (sub-**ling**-gwal) Space located below the oral mucosa, thus making this tissue its roof.

Subluxation (sub-luk-**ay**-shun) Acute episode of temporomandibular joint disorder where both joints become dislocated, often due to excessive mandibular protrusion and depression.

Submandibular fossa (sub-man-**dib**-you-lar) Fossa on the medial surface of the mandible, below the mylohyoid line, that contains the submandibular salivary gland.

Submandibular ganglion (sub-man-**dib**-you-lar) Ganglion superior to the deep lobe of the submandibular salivary gland that communicates with the chorda tympani and lingual nerves.

Submandibular lymph nodes (sub-man-**dib**-you-lar) Superficial cervical nodes located at the inferior border of the ramus of the mandible.

Submandibular salivary gland (sub-man-**dib**-you-lar) Major gland that is located in the submandibular fossa.

Submandibular space (sub-man-**dib**-you-lar) Space located lateral and posterior to the submental space on each side of the jaws.

Submandibular triangle (sub-man-**dib**-you-lar) Portion of the anterior cervical triangle formed by the mandible and anterior and posterior bellies of the digastric muscle.

Submasseteric space (sub-mas-et-**tehr**-ik) Space located between the masseter muscle and external surface of the vertical ramus.

Submental artery (sub-**men**-tal) Arterial branch from the facial artery that supplies the submandibular lymph nodes, submandibular salivary glands, and mylohyoid and digastric muscles.

Submental lymph nodes (sub-**men**-tal) Superficial cervical nodes located inferior to the chin.

Submental space (sub-**men**-tal) Space located midline between the symphysis and hyoid bone.

Submental triangle (sub-**men**-tal) Unpaired midline portion of the anterior cervical triangle created by the right and left anterior bellies of the digastric muscle and the hyoid bone.

GLOSSARY OF KEY WORDS AND ANATOMICAL TERMS *Continued*

Submental vein (sub-**men**-tal) Vein that drains the tissues of the chin and then drains into the facial vein.

Sulcus, sulci (**sul**-kus, **sul**-ky) Shallow depression or groove, such as that on the bony surface or between the tooth and inner surface of the marginal gingiva.

Sulcus terminalis (**sul**-kus **ter**-mi-nal-is) V-shaped groove on the dorsal surface of the tongue.

Superficial (soo-per-**fish**-al) Structures located toward the surface of the body.

Superficial parotid lymph nodes (pah-**rot**-id) Nodes located just superficial to the parotid salivary gland.

Superficial temporal artery (**tem**-poh-ral) Terminal arterial branch from the external carotid artery that arises in the parotid salivary gland and gives off the transverse facial and middle temporal arteries, as well as frontal and parietal branches.

Superficial temporal vein (**tem**-poh-ral) Vein that drains the side of the scalp and goes on to form the retromandibular vein along with the maxillary vein.

Superior (soo-**pere**-ee-or) Area that faces toward the head of the body, away from the feet.

Superior articular processes (ar-**tik**-you-lar) Processes from the vertebrae that allow articulation with the vertebra above.

Superior labial artery Arterial branch from the facial artery that supplies the upper lip tissues.

Superior labial vein Vein that drains the upper lip and then drains into the facial vein.

Superior nasal conchae (**nay**-zil **kong**-kay) Lateral portions of the ethmoid bone in the nasal cavity.

Superior orbital fissure (**or**-bit-al) Fissure between the greater and lesser wings of the sphenoid bone that transmits structures from the cranial cavity to the orbit.

Superior thyroid artery (**thy**-roid) Anterior arterial branch from the external carotid artery that supplies the tissues inferior to the hyoid bone, including the thyroid gland.

Superior vena cava (**vee**-na **kay**-va) Vein formed from the union of the brachiocephalic veins that empties into the heart.

Supraclavicular lymph nodes (soo-prah-klah-**vik**-you-ler) Deep cervical nodes located along the clavicle.

Suprahyoid muscles (soo-prah-**hi**-oid) Hyoid muscles located superior to the hyoid bone that can be further divided by their anterior or posterior relationship to the hyoid bone.

Supraorbital nerve (soo-prah-**or**-bit-al) Nerve from the forehead and anterior scalp that merges with the supratrochlear nerve to form the frontal nerve.

Supraorbital notch (soo-prah-**or**-bit-al) Notch on the frontal bone located on the supraorbital ridge.

Supraorbital ridge (soo-prah-**or**-bit-al) Ridge on the frontal bone located over the orbit.

Supraorbital vein (soo-prah-**or**-bit-al) Vein that joins the supratrochlear vein to form the facial vein in the frontal region.

Supratrochlear nerve (soo-prah-**trok**-lere) Nerve from the nose bridge and medial parts of the upper eyelid and forehead that merges with the supraorbital nerve to form the frontal nerve.

Supratrochlear vein (soo-prah-**trok**-lere) Vein that joins the supraorbital vein to form the facial vein in the frontal region.

Suture (**su**-cher) Generally immovable articulation in which bones are joined by fibrous tissue.

Sympathetic nervous system (sim-pah-**thet**-ik) Portion of the autonomic nervous system that is involved in "fight or flight."

Symphysis (**sim**-fi-sis) Midline articulation where bones are joined by fibrocartilage, such as the midline ridge on the mandible that fuses together in early childhood.

Synapse (**sin**-aps) Junction between two neurons or between a neuron and an effector organ where neural impulses are transmitted by electrical or chemical means.

Synovial cavities of the temporomandibular joint (sy-**no**-vee-al **kav**-i-tees) Upper and lower

GLOSSARY OF KEY WORDS AND ANATOMICAL TERMS *Continued*

spaces created by the division of the joint by the disc.

Synovial fluid of the temporomandibular joint (sy-**no**-vee-al) Fluid secreted by the membranes lining the synovial cavities.

T

T-cell lymphocytes (**lim**-fo-sites) White blood cells of the immune system that mature in the gland in response to stimulation by thymus hormones.

Temporal bones (**tem**-poh-ral) Paired cranial bones that form the lateral walls and articulate with the mandible at the temporomandibular joint.

Temporal fascia (**tem**-poh-ral) Deep fascia covering the temporalis muscle down to the zygomatic arch.

Temporal fossa (**tem**-poh-ral) Fossa on the lateral surface of the skull that contains the body of the temporalis muscle.

Temporalis muscle (tem-poh-**ral**-is) Muscle of mastication that fills the temporal fossa.

Temporal lines (**tem**-poh-ral) Two separate parallel ridges, superior and inferior, on the lateral surface of the skull.

Temporal process of the zygomatic bone (**tem**-poh-ral) Process that forms part of the zygomatic arch.

Temporal region (**tem**-poh-ral) Region of the head where the external ear is a prominent feature.

Temporal space (**tem**-poh-ral) Space formed by the temporal fascia covering the temporalis muscle.

Temporomandibular disorder (TMD) (tem-poh-ro-man-**dib**-you-lar) Disorder involving one or both temporomandibular joints.

Temporomandibular joint (TMJ) (tem-poh-ro-man-**dib**-you-lar) Articulation between the temporal bone and mandible that allows for movement of the mandible.

Temporomandibular joint ligament (tem-poh-ro-man-**dib**-you-lar) Ligament associated with the temporomandibular joint.

Temporozygomatic suture (tem-por-oh-zi-go-**mat**-ik) Suture between the temporal and zygomatic bones.

Tensor veli palatini muscle (**ten**-ser **vee**-lie pal-ah-**teen**-ee) Muscle of the soft palate that stiffens it.

Thoracic duct (tho-**ras**-ik) Lymphatic duct draining the lower half of the body and left side of the thorax and draining the left side of the head and neck through the left jugular trunk.

Thrombus, thrombi (**throm**-bus, **throm**-by) Clot that forms on the inner blood vessel wall.

Thymus gland (**thy**-mus) Endocrine gland located inferior to the thyroid gland and deep to the sternum.

Thyrohyoid muscle (thy-ro-**hi**-oid) Infrahyoid muscle that appears as a continuation of the sternothyroid muscle.

Thyroid gland (**thy**-roid) Endocrine gland having two lobes and located inferior to the thyroid cartilage.

Thyroxine (thy-**rok**-sin) Hormone produced and secreted by the thyroid gland directly into the blood.

Tonsillar tissue (**ton**-sil-lar) Masses of lymphoid tissue located in the oral cavity and pharynx to protect the body against disease processes.

Tonsils (**ton**-sils) Tonsillar tissue that includes the palatine, lingual, pharyngeal, and tubal tonsils.

Tragus (**tra**-gus) Flap of tissue that is part of the auricle and anterior to the external acoustic meatus.

Transverse facial artery Arterial branch from the superficial temporal artery that supplies the parotid salivary gland.

Transverse foramen Foramen on the transverse processes of each cervical vertebra that carries the vertebral artery.

Transverse palatine suture (**pal**-ah-tine) Suture between the palatine processes of the maxillae and horizontal plates of the palatine bones.

Transverse process Lateral projections of the cervical vertebrae.

Transverse section (**trans**-vers) Section of the body through any horizontal plane.

GLOSSARY OF KEY WORDS AND ANATOMICAL TERMS *Continued*

Trapezius muscle (trah-**pee**-zee-us) Cervical muscle that covers the lateral and posterior surfaces of the neck.

Trigeminal ganglion (try-**jem**-i-nal **gang**-gle-on) Sensory ganglion located intracranially on the petrous portion of the temporal bone.

Trigeminal nerve (try-**jem**-i-nal) Fifth cranial nerve (V) that serves the muscles of mastication and cranial muscles through its motor root and serves the teeth, tongue, and oral cavity and most of the facial skin through its sensory root.

Trigeminal neuralgia (try-**jem**-i-nal noor-**al**-je-ah) Type of lesion of the trigeminal nerve involving facial pain.

Trochlear nerve (**trok**-lere) Fourth cranial nerve (IV) that serves an eye muscle.

Tubal tonsil (**tube**-al) Tonsil located in the nasopharynx near the auditory tube.

Tubercle (**too**-ber-kl) Eminence or small rounded elevation on the bony surface.

Tubercle of the upper lip (**too**-ber-kl) Thicker area in the upper lip where the philtrum terminates.

Tuberosity (too-beh-**ros**-i-tee) Large, often rough, prominence on the surface of bone.

Tympanic portion of the temporal bone (tim-**pan**-ik) Portion that forms most of the external acoustic meatus.

U

Uvula of the palate (**u**-vu-lah) Midline muscular structure that hangs from the posterior margin of the soft palate.

V

Vagus nerve (**vay**-gus) Tenth cranial nerve (X) that serves the muscles of the soft palate, pharynx, and larynx, a portion of the ear skin, and many organs of the thorax and abdomen.

Vein (vane) Type of blood vessel that travels to the heart carrying blood.

Venous sinus (**vee**-nus **sy**-nus) Blood-filled space between two layers of tissue.

Ventral (**ven**-tral) Front of an area of the body.

Ventral surface of the tongue Underside of the tongue.

Venule (**ven**-yule) Smaller vein that drains the capillaries of the tissue area and then joins larger veins.

Vermilion border (ver-**mil**-yon) Outline of the entire lip from the surrounding skin.

Vermilion zone (ver-**mil**-yon) Darker appearance of the lips.

Vertebral fascia (**ver**-teh-brahl) Deep cervical fascia that covers the vertebrae, spinal column, and associated muscles.

Vertebral foramen (**ver**-teh-brahl) Central foramen in the vertebrae for the spinal cord and associated tissues.

Vertical plates of the palatine bones Plates that form part of the lateral wall of the nasal cavity and orbital apex.

Vestibular space of the mandible (**man**-di-bl) Space of the lower jaw.

Vestibular space of the maxilla (mak-**sil**-ah) Space of the upper jaw.

Vestibules (**ves**-ti-bules) Upper and lower spaces between the cheeks, lips, and gingival tissues in the oral region.

Vestibulocochlear nerve (ves-tib-you-lo-**kok**-lear) Eighth cranial nerve (VIII) that serves to convey signals from the inner ear to the brain.

Visceral fascia (**vis**-er-al) Deep cervical fascia that is a single midline tube running down the neck.

Vomer (**vo**-mer) Single facial bone that forms the posterior portion of the nasal septum.

Z

Zygomatic arch (zy-go-**mat**-ik) Arch formed by the union of the temporal process of the zygomatic bone and zygomatic process of the temporal bone.

GLOSSARY OF KEY WORDS AND ANATOMICAL TERMS *Continued*

Zygomatic bones (zy-go-**mat**-ik) Paired facial bones that form the cheek bones.

Zygomatic nerve (zy-go-**mat**-ik) Nerve that is formed from the merger of the zygomaticofacial and zygomaticotemporal nerves and joins the maxillary nerve.

Zygomatic process of the frontal bone (zy-go-**mat**-ik) Process lateral to the orbit.

Zygomatic process of the maxilla (zy-go-**mat**-ik) Process that forms a portion of the infraorbital rim.

Zygomatic region (zy-go-**mat**-ik) Region of the head that overlies the cheek bone.

Zygomaticofacial nerve (zy-go-**mat**-i-ko-**fay**-shal) Nerve that serves the skin of the cheek and joins with the zygomaticotemporal nerve to form the zygomatic nerve.

Zygomaticotemporal nerve (zy-go-**mat**-i-ko-**tem**-poh-ral) Nerve that serves the skin of the temporal region and joins with the zygomaticofacial nerve to form the zygomatic nerve.

Zygomaticus major muscle (zy-go-**mat**-i-kus **may**-jer) Muscle of facial expression in the mouth region that is used when smiling.

Zygomaticus minor muscle (zy-go-**mat**-i-kus **my**-ner) Muscle of facial expression in the mouth region that elevates the upper lip.

Index

Note: Page numbers in *italics* refer to illustrations; page numbers followed by t refer to tables.

B

C

D

E

F

N

R

S